Neuroscience Research Progress

Cell Biology Research Progress

nova
science publishers
www.novapublishers.com

Neurosurgery Research Progress

Reticular Concept of Nervous System Physiology
Oleg Sotnikov, PhD (Editors)
2022. ISBN: 978-1-68507-996-3 (Hardcover)
2022. ISBN: 979-8-88697-350-1 (eBook)

Neuroscience and Technology Transform the Educational Ecosystem
Franco F. Orsucci (Editor)
Nicoletta Sala (Editor)
2022. ISBN: 979-8-88697-157-6 (Hardcover)
2022. ISBN: 979-8-88697-248-1 (eBook)

More information about this series can be found at
https://novapublishers.com/product-category/series/neurosurgery-research-progress/

Cell Biology Research Progress

Transcriptomics and Their Importance
Sarah Naiyer, PhD (Editor)
Khalid Hussain Bhat, PhD (Editor)
2022. ISBN: 979-8-88697-044-9 (Hardcover)
2022. ISBN: 979-8-88697-203-0 (eBook)

Endothelial Progenitor Cells in Health and Disease
Amankeldi A. Salybekov, MD, PhD (Editor)
Shuzo Kobayashi, MD, PhD (Editor)
Takayuki Asahara, MD, PhD (Editor)
2022. ISBN: 978-1-68507-794-5 (Hardcover)
2022. ISBN: 978-1-68507-948-2 (eBook)

More information about this series can be found at
https://novapublishers.com/product-category/series/cell-biology-research-progress/

Ignacio González-Burgos
Editor

Dendritic Spines

An Update

nova
science publishers

www.novapublishers.com

Library of Congress Cataloging-in-Publication Data

ISBN: 979-8-89113-080-7

Published by Nova Science Publishers, Inc. † New York

Contents

Foreword 1

Dendritic spines have intrigued neuroscientists since their discovery by Santiago Ramón y Cajal in 1888. These specialized neuronal structures are involved in the reception of inputs from other neurons and play essential roles in the mechanisms of excitatory synaptic transmission. The book "Dendritic Spines. An Update" provides an actualized narrative on the state of the art on dendritic spine research, which is addressed from complementary perspectives in the different chapters.

The first chapter presents an historical overview on the interpretation of the structure, development, and function of dendritic spines, starting by their discovery in the precise historical moment when a passionate scientific debate opposed the new ideas on the organization of the brain proposed by Cajal with the old and prevalent reticular hypothesis supported, among others, by Camillo Golgi. The next 3 consecutive chapters provide a central basic core of the book and deal with the molecular and electrophysiological mechanisms of dendritic spine function and plasticity. Spines present a head, where the synaptic connection is located, and a neck that connects the head to the dendrite. Both structures, head and neck, are plastic and changes in their shape modify the electric properties of dendritic spines, playing an essential role in the plasticity of neuronal circuits. Chapter 2 includes a considerable amount of information and a detailed description of the molecular mechanisms involved in the maintenance of the morphological and functional organization of dendritic spines. Chapter 3 presents an analysis of the electric properties conferred by the spine shape, in particular the electrical resistance provided by the neck, and discusses the physiological implications of spine electric properties for synaptic function and plasticity. Chapters 2 and 3 are complementary to chapter 4, which provides an updated review of the molecular signaling pathways involved in the control of spine plasticity. These plastic properties make dendritic spines good candidates for participating in learning and memory formation and this possibility has been addresses by numerous experimental studies. An excellent discussion on this interesting topic is presented in chapter 5, which is devoted to the implications of spine synaptic plasticity on cognitive processing. Directly related with this issue is the next chapter (chapter 6), focusing on the possible role of spine synapses on the effects of estradiol on memory consolidation, a question of particular relevance for women's health.

The participation of glial cells in synaptic transmission has received considerable attention in recent years. The editor has made an excellent choice by including a discussion on this topic in chapter 7, which describes the different mechanisms involved in the participation of astrocytes and microglia in the regulation of synaptic plasticity. Finally, chapter 8 deals with the alterations in dendritic spines under pathological conditions, emphasizing the consequences of malnutrition during gestation and lactation for dendritic spine morphology and plasticity.

Overall, this book presents an attractive and informative narrative, which is complemented by excellent illustrations. Clearly, it is a highly recommendable reading not only for neuroscientist but also for a general audience interested on how the brain works.

Luis Miguel García-Segura
Honorary professor at the Autonoma University of Madrid
and former research professor at the Cajal Institute, CSIC, Madrid, Spain

Foreword 2

As its title implies, the book "Dendritic Spines. An Update", edited by Ignacio González-Burgos, is a high quality work composed of excellent descriptions of the developmental, anatomical, and electrophysiological characteristics of dendritic spines; these are followed by an account of studies on synaptic mechanisms of memory formation and their relationships to pathological conditions, such as those derived from malnutrition. A timely discussion on the involvement of membrane receptors in the rapid effects of estrogens on the facilitation of memory consolidation is also presented, together with a report on the important role of astrocytes and microglia in synaptic plasticity. Surely, this book will be of interest to those interested in neuroplasticity, particularly in cognition, both normal and pathological.

Chapter 1 provides an excellent account of the history of the discovery of dendritic spines followed by an outline of modern findings regarding the development of these cytoplasmic protrusions, their functional activity, their structural shapes, and diverse functional characteristics.

A comprehensive overview is presented in Chapter 2 about the molecular mechanisms underlying dendritic spine plasticity, and how different proteins shape the actin cytoskeleton of dendritic spines, as well as the interactions among these molecules, that drive the formation and elimination, size and shape, stability, and motility of spines.

The third chapter is devoted to the electrical properties of dendritic spines. The author reviews the intrinsic electrical properties of dendritic spines and the increasingly sophisticated approaches used to estimate the electrical coupling between spine and dendrite. It is concluded that biophysically, dendritic spines are relatively simple structures whose most salient feature is the electrical resistance provided by their spine necks.

In Chapter 4, the author offers a detailed description of signaling pathways in dendritic spines related to synaptic plasticity, with emphasis on LTP and LTD, as well as on the signaling pathways activated upon induction of the

NMDA receptor-dependent synaptic plasticity within dendritic spines and resulting in both functional and structural alterations at synapses.

After defining neuroplasticity and discussing the main theoretical approaches to studying this phenomenon, the authors of Chapter 5 describe the different types of synaptic plasticity and synaptic transmission. At a higher level of complexity, they turn to the topics of dendritic spine plasticity and synaptic mechanisms of memory formation and their relationships to pathological conditions.

In Chapter 6 the authors provide convincing evidence that estradiol enhances memory, in both female and male subjects, by increasing the density of dendritic spines; and the signaling cascades leading to spinogenesis and memory enhancements are detailed.

The authors of Chapter 7 present an up-to-date review of the important role of astrocytes and microglia in synaptic plasticity. They also review the molecular basis of the modulation of neural activity by these cells. An outline of neuropsychiatric, neurodevelopmental, and aging disorders associated with astrocytic and microglial dysfunction is also offered.

Chapter 8 reviews the effects of malnutrition on dendritic spine formation and morphology in critical stages of life and discusses how malnutrition-induced changes in dendritic spines contribute to the appearance of neurodevelopmental disorders as well as neuropsychiatric illnesses (schizophrenia and Alzheimer's disease).

Roberto A. Prado-Alcalá
Senior Researcher, Department of Behavioral and Cognitive Neurobiology
Institute of Neurobiology, National Autonomous University of Mexico
Querétaro, México

Preface

In the brains of vertebrate organisms, dendritic spines are the door that separates the exterior and interior worlds. Since their discovery, scientific research has succeeded in characterizing some of the mechanisms, events, and processes through which these spines contribute to the neural processing of information acquisition, impression and concept formation, memories, and behavior that, ultimately, determine the interaction –successful or unsuccessful– of an individual with his environment. The structural features of dendritic spines underlie variations in the processing of afferent synaptic information to neurons as they translate the information generated by changes in the potential of the membrane that separates them from the intersynaptic space. The bioelectrical phenomena that support synaptic transmission mediated by these spines condition the plastic changes they undergo due to conformational modifications in the protein complexes that structure them.

The information that underlies manifestations of behavior patterns related to cognitive and instrumental processes is transmitted variably through the neural circuits in which the brain is organized. The expression of such plastic events is aided by the neuromodulatory participation of various molecules and cellular entities whose activities condition the excitatory and inhibitory effects of synaptic inputs to postsynaptic neurons mediated by dendritic spines. The plastic synaptic events in which these spines participate play a crucial role in the brain's ability to efficiently organize afferent information. This is particularly important in the context of the expression of behaviors, which may be adaptive or maladaptive. For example, cognitive processes depend strongly on the plastic dynamism of dendritic spines which, under certain conditions, can express aberrant plastic modifications that result in an erratic organization of information that eventually leads to the expression of maladaptive behavior patterns, such as pathological conduct.

"When protoplasmic expansions are studied using the Golgi Method, certain morphological details can be observed that should be made known, for over time they may come to have physiological significance. One is the presence of certain short appendages or collateral spines that emerge at a right angle to the contour of the dendritic expansions and end in a rounded or ellipsoid thickening."

Santiago Ramón y Cajal

In: Histologie du Système Nerveux de L'homme et des Vertébrés, 1909

Chapter 1

Spinogenesis and the Morphology of Dendritic Spines

Ignacio González-Burgos[1,*]
and Dulce A. Velázquez-Zamora[2]

[1]Laboratorio "Burgos" de Investigaciones Psicobiológicas, Zapopan, Jalisco, Mexico
[2]Research Coordination, Tlajomulco University Center,
Tlajomulco de Zúñiga, Jalisco, Mexico

Introduction

Dendritic spines are the excitatory contact sites *per excellence* in all neurons of the central nervous system (CNS). From the earliest descriptions, by Santiago Ramón y Cajal in the late 19th century, it has been postulated that they could be related to communication between neurons. Over the years, not only has this assertion been confirmed, but many of the characteristics related to their functional activity have been elucidated, various structural "forms" based on their geometry have been described, and diverse functional characteristics have been attributed to them. As their name suggests, dendritic spines are cytoplasmic protrusions that emerge from the dendritic trunk. Their distribution along the trunk is highly variable, depending on the neuronal line that carries them, and the synaptic relations established with the afferent pathways that innervate them. Other types of postsynaptic protrusions exist that also perform the function of receiving synaptic inputs, but they are atypical variations of spines due to their uniquely different geometric structure compared to the spines themselves.

* Corresponding Author's Email: igonbur@hotmail.com.

In: Dendritic Spines
Editor: Ignacio González-Burgos
ISBN: 979-8-89113-080-7

Various neurobiological events have been described in relation to the origin, development, and neoformation of dendritic spines. However, it is clear that, while there is evidence of differential processes, they all refer to the plastic capacity of nerve tissue, and attest to the importance of these postsynaptic structures in the processing of the afferent information to neurons that allows the cerebral organization of behavior and its normal variants, as well as conducts that eventually become pathological.

Brain Structure and Functioning: Theoretical and Methodological Debates

The history of the discovery of dendritic spines and their functional importance can be traced back to the early 19th century, a time that can be considered the golden age of modern and contemporary neuroscience. In the early years of that century, Franz Joseph Gall, his assistant Johann Spurzheim, and, later, Charles Caldwell, proposed the Theory of Phrenology, which postulated that brain functions were compartmentalized in circumscribed areas of the skull and cerebral cortex, especially in people they referred to as called "extreme" members of society (e.g., writers, poets, statesmen), and among the mentally unbalanced, criminals, and other classifications of people. However due, in part, to its clinical bias, the concept of these "functions" lacked scientific support, and phrenology was questioned severely until it was practically discarded in the late 1820s.

Although Gall set out from the solid idea of attributing functions to the brain, the method he adopted was inadequate. The idea of the "localization" of brain functions, however, was not dismissed completely. In fact, it was taken up by Marie-Jean-Pierre Fluorens, a strong detractor of phrenology who, using a methodological approach based on comparative anatomy and studies of neurostimulation and brain damage, proposed –almost unintentionally it turned out– the beginnings of the *Localizationist Theory* of brain functions. This theoretical current was strongly supported by the work of Paul Broca in 1861 in relation to the location of the faculty of spoken language in the frontal cortex, among other clinical studies carried out later involving the motor cortex by Eduard Hitzig and Gustav Fritsch in 1870 and Sir David Ferrier in 1881.

The development of techniques for recording brain electrical activity spurred great advances in our understanding of the relations between brain structure and function. Richard Caton, David Ferrier's fellow student, was likely the first to record spontaneous electrical activity in the brain. In 1875, he postulated that cerebral electrical changes occur as a function of specific peripheral stimuli (Finger, 1994). The second half of the 19th century represented a watershed in terms of attributing specific functions to circumscribed regions, first, in the cerebral cortex, later in subcortical regions of the brain. Neuroscience benefitted greatly from the development of those new theoretical-methodological approaches.

Cytoarchitectural Studies

The use of achromatic objectives –invented in 1826– significantly improved the ability to observe microscopic structures. Thus, for example, in 1837, Jan Purkinje recognized the somas and part of the extensions of the cerebellar neurons that now bear his name (Shepherd, 1991). It is important to note that Gabriel Gustav Valentin, Purkinje's student, published the first microscopic image of a nerve cell, in which the nucleus and nucleolus were clearly observable (López-Muñoz et al., 2006). Although in 1880 the Zeiss house launched a 2,000x-magnification microscope that equaled the maximum resolution of the human eye (Lopera-Chaves, 2011), the techniques then in vogue –Ehrlich staining, Held, hematoxylin, carmine, and gold chloride, among others– allowed only very thin, two-dimensional sections to be observed, significantly limiting the possibility of making theoretical interpretations regarding the organization of nerve tissue. In that setting, a heated debate over the continuity or discontinuity of nerve cell extensions raged between two groups of antagonistic scientists.

Reticular Theory

The prevailing idea at that time was *Reticular Theory*, which postulated that nerve cell processes gave rise to a network that spread throughout the nerve tissue. Despite the existence of Cell Theory, Virchow himself spoke of a fundamental mass that connected and united all nerve centers (Virchow, 1854). Though unintentionally, that proposal represented a useful argument for the incipient Reticular Theory, which enjoyed the support of such important figures as Bethe, Held, and Von Apathy, who erroneously believed they had demonstrated an apparently continuous network in the ganglia of invertebrates. Otto Friederich Karl Deiters is another figure who stands out. In his observations, published posthumously in 1865, he stated that only the longest emerging prolongation of nerve cells –which he called "axis cylinders"– remained undivided, unlike the other "protoplasmic processes" that Wilhelm His called "dendrites" in 1889 (Finger, 1994). However, Josef von Gerlach was the scientist who most strongly defended reticularism, calling the organization of nerve tissue "diffuse interstitial networks" (Ramón y Cajal, 1952) and stating that the filaments that emerged from this mesh came together to form nerve fibers that reached the white matter, before arriving at the nerve cells of the spinal cord, all in a continuous manner (López-Muñoz et al., 2006).

Camilo Golgi was one of the staunchest defenders of reticularism. In the 1880s, histology of the nervous system was dominated by the Italian school with Golgi leading a host of researchers that included Sala, Fusari, Mondino, Veratti, and Martinotti, among others. Partly by design and partly through serendipity, by 1873 Golgi had managed to develop a tissue staining technique –silver chromate– that made it possible to observe cells and their black-stained extensions in three dimensions (his famous *reazione nera*) (Golgi, 1873). The technique consisted in immersing samples of nerve tissue in potassium dichromate and then in a silver nitrate solution to precipitate silver chromate, which was deposited in the nerve tissue. While perfecting his so-called "slow technique," Golgi also developed –among other successes– the "fast" version of his method by substituting pure potassium dichromate for a mixture that contained this substance and osmium tetroxide. This technique was known then (and still today) as the "rapid Golgi method." Between 1883 and 1886, Golgi published a series of papers based on studies using his impregnation method, in which he stated that protoplasmic expansions ended freely, but that some axis cylinders lost their individuality before reaching the white matter by fusing in a terminal arborization that split the sensitive fibers (*rete nervosa diffusa*).

Neuronal Theory

An equally large group of researchers defended a distinct proposal; namely, that nerve tissue was composed of independent cells, a posture in line with Virchow's Cell Theory. *Neuronal Theory* proposed that, although the brain was an extremely complex organ made up of cell bodies and extensions that emanated from them, it consisted of independent cells. In this theory, the representative cell of the CNS was the "neuron," a term, incidentally, coined by Wilhelm von Waldeyer in 1891, at a time when Neuronal Theory was mounting a strong challenge to the Reticular approach. The new theory was based on the pioneering work of researchers like His, Forel, Kölliker, Duval, and Place. The case of Waldeyer deserves special mention, because the credit for developing Neuronal Theory has been attributed erroneously to his work, since he never conducted experimental studies or made exhaustive empirical observations, but simply compiled a large body of experimental evidence that showed the strength of that theory.

The growing evidence in favor of Neuronal Theory was, at the microscopic level, an extension of the Localizationism current that reigned in the 19th century. At the dawn of the 20th, several scientific figures, including Oscar Vogt, Cécile Vogt, and Korbinian Brodmann, were studying the fine anatomy of the cerebral cortex. In their reports they described the regional pattern of many cortical areas, noting such aspects as cortical thickness, horizontal lamination, the relative thickness of cortical sheets, cell arrangement, the presence and proportion of different cell types, structural variations of cell types, the degree of overlap between cell strata, the degree of change in the size of the same strain in the same layer and between layers, cell density, and the affinity of the cells for the dyes used to observe them.

The Golgi/Cajal Case

Golgi's influence as a defender of Reticular Theory was overwhelming, despite the inconsistencies of that theory and the contradictions that arose among its adherents. Those weaknesses were put to the test by his detractors, including Santiago Ramón y Cajal (Figure 1). Cajal was a Spanish neurohistologist who had recently arrived on the scientific scene. In 1884, he was appointed to the Chair of Anatomy at the University of Valencia, where the psychiatrist and neurologist Luis Simarro showed him sections of nerve tissue treated with Golgi's silver chromate technique. By 1888, Cajal had laid the foundations of Neuronal Theory, based on histological findings obtained by applying the Golgi method (with some modifications). His studies demonstrated convincingly that the relation among nerve cells was based not on continuity, but on contiguity.

A transcendental event in this story occurred at the Congress of the German Anatomical Society in 1889, where Cajal shared his findings with the main "neuronist" scientific figures of the time, including Waldeyer, Retzius, His, and, especially, Kölliker. It turned out that the evidence he presented coincided with the contributions of August Forel, who in 1891 claimed merit for postulating the new Neuronal Theory. The overwhelming number of studies that Cajal published between 1890 and 1891 convincingly demonstrated the scientific objectivity of Neuronal Theory. The combined weight of those findings made it clear that an anatomical compartmentalization of neurons was necessary. The terms "protoplasmic extensions" and "axis cylinders," proposed previously by Deiters, were renamed "dendrites" by His in 1889, and "axons" by Kölliker in 1896, respectively.

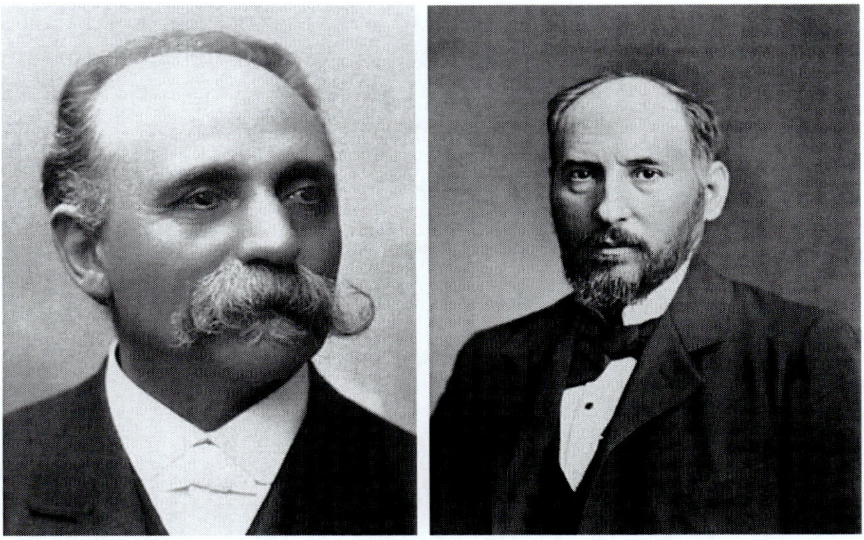

Figure 1. Photographs of Camilo Golgi (left) and Santiago Ramón y Cajal (right).

Ramón y Cajal's extensive neurohistological studies were collected in his monumental work *Texture of the Nervous System of Man and Vertebrates*, originally published in Spanish between 1899 (first part) and 1904 (second part), and in full in French in 1909, with complementary findings by the author. The scientific contributions of Ramón y Cajal in favor of Neuronal Theory earned him the Nobel Prize in Medicine in 1906. In a paradox of fate, the award was shared with Camilo Golgi, the recalcitrant defender of Reticular Theory. That act recognized the value of Golgi's work, not so much for his theoretical stance as for his methodological contribution that made Cajal's findings possible. Despite this, in the speech he delivered to the Swedish Academy, Golgi persisted in his reticularist postulates and in his opposition to

Neuronal Theory, concluding that the nerve network he defended explained "the anatomical and functional continuity between nerve cells. And this is the reason why I have not been able to accept the idea of the independence of each nerve cell."

Though some pockets of resistance persisted, by that time Neuronal Theory was almost universally accepted by the scientific community; in fact, it turned out to be fully congruent with Virchow's Cell Theory. According to Cajal, Golgi's work was tremendously fruitful in terms of developing a method, but failed totally in its morphological and physiological interpretation of findings. Cajal did, however, benevolently recognize Golgi's role in his classic compendium published in 1933, *Neuronism or Reticularism?*, where he wrote that, "It is a sad truth that almost no one can escape entirely from the tradition and spirit of his age." The objectivity, strength, and scientific solidity of the evidence in favor of Neuronal Theory was revealed in that text: "And let us not fear future technical inventions, because if the facts have been well observed, they will endure, even if interpretations change."

Without doubt. Cajal's arduous and meticulous observations revealed the need to extend the limits of classic Localizationism by making it laxer in terms of the regional constituency proposed. The now solidly-established fact that neurons were well-defined cellular structures led Cajal to suggest that the nervous system was organized into regions, strata, nuclei, and areas that, basically, contain the neuronal somas which, though relatively circumscribed, interconnect through both dendritic and axonal processes; that is, through what Sherrington called "synapses" (Sherrington, 1897).

General Systems Theory

The numerous electrophysiological studies that characterized the development of neuroscience based on Neuronal Theory during, above all, the first half of the 20th century, served to strengthen this theoretical position. Together with other nascent disciplines, neuroanatomical and electrophysiological studies led Alexander Luria (1970, 1972) to propose his *General Systems Theory*, elaborated from an evolutionary theoretical position and based methodologically on neuropsychological criteria. Luria proposed that the cerebral organization of the fundamental neuropsychological processes (attention, memory, language, emotion, motivation, sensory perception, learning, thought) are structured through the implementation of three systems. He theorized that the ascending reticular activating pathway in the brain stem (system 1) triggers the formation of behavioral conditioning in the primary sensory association cortices involved in sensory perception (system 2) which, in turn, may underlie the organization of individual behaviors that are organized, planned, and executed in the frontal cortex of the brain (system 3). It is important to emphasize that this theory has been of great importance for the understanding of psychic activity and, consequently, for the clinical and psychiatric care of individuals who suffer behavioral disorders. By that time, Localizationism as such had come to an end.

Dendritic Spines

The incalculable value of the Golgi Method (and his own subsequent modifications) enabled Cajal to characterize the histological structure of virtually the entire mammalian brain. Over the course of his observations, Cajal identified, for the first time in history, certain structures that seemed to constitute an integral element of the surface of dendrites –particularly in Purkinje cells– based on studies and descriptions of the cerebellum of birds (Ramón y Cajal, 1888). Other authors had already observed some dendritic protrusions, but had interpreted them as artifices of the staining techniques used. In fact, Cajal himself recognized this dilemma in one of his aforementioned publications, where he wrote:

> "…the surface [of the dendrites of Purkinje cells] appears to be covered with thorns or short spines. At first, we thought that these eminences were the result of a tumultuous precipitation of the silver, but the constancy of their existence and presence, even in preparations in which the reaction appears to be very delicate in the remaining elements, incline us to believe this to be a normal condition."

The multiple repetitions of histological experiments based on the diverse staining techniques available at the time, together with the modifications of the Golgi Method developed by Cajal (Garcia-Lopez et al., 2010; Ramón y Cajal and De Castro 1972), led researchers to think, with ever greater conviction, that those "artificial" structures were really elements of dendritic processes. The consistency, objectivity, and solidity of the evidence led Cajal to publish his results in the first issue of the scientific journal that he founded and financed out of his own pocket. More than mere descriptions of the cerebellum of birds, that report presented a proposal that was to become one of the most promising pillars of contemporary neuroscience; namely, the objective existence of dendritic spines (Figure 2).

Though in 1888 Cajal lacked valid arguments on which to base his proposals, he envisioned a possible relation between the neuronal structure and the functions in which the spines might be involved. It is important to note at this point that, although these structures are known as "dendritic spines," in reality they are also observed occasionally in the neuronal soma (Figure 3) and at the base of the axon hillock of some neurons.

Figure 2. Photomicrograph of the dendritic arbor of a Purkinje cell, showing the large amount of spines that protrude from all the dendritic branches. Modification of the Golgi method (González-Burgos et al., 1992). Scale bar: 50 μm.

Through the final thirty years of the 20th century, neuroscience was largely dominated by Cajal's findings (in fact, he is still one of the most oft-cited authors in history). From his first illustrated descriptions, the innumerable illustrations on the histology and cytology of the nervous system (see DeFelipe, 2010, 2018) attested to the objective existence of dendritic spines. Cajal's scientific conviction concerning these neural structures was clearly reflected in a letter he wrote to one of his most valued disciples, Rafael Lorente de Nó, two days before his death on October 17, 1934 (see Figure 4).

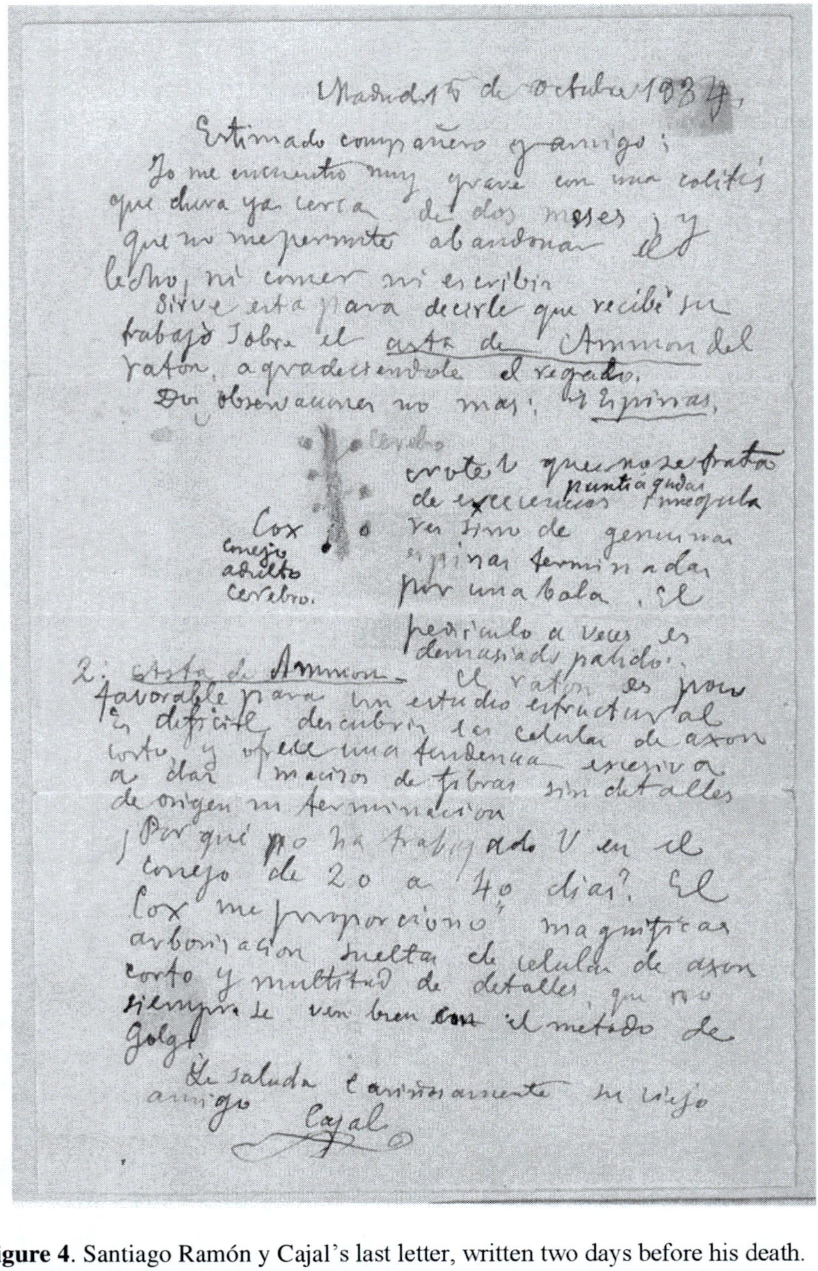
Madrid 15 de octubre 1934.

Estimado compañero y amigo:

Yo me encuentro muy grave con una colitis que dura ya cerca de dos meses y que no me permite abandonar el lecho, ni comer ni escribir.

Sirve esta para decirle que recibí su trabajo sobre el asta de Ammon del ratón, agradeciéndole el regalo.

Dos observaciones no mas: 1ª Espinas.

cerebro

Creo que no se trata de excrecencias [puntiagudas] irregulares sino de genuinas espinas terminadas por una bola. El pediculo a veces es demasiado palido.

Cox
conejo
adulto
cerebro.

2ª asta de Ammon. El raton es poco favorable para un estudio estructural es dificil descubrir las celulas de axon corto, y ofrece una tendencia excesiva a dar macizos de fibras sin detalles de origen ni terminacion

¿Por qué no ha trabajado V. en el conejo de 20 a 40 dias? El Cox me proporcionó magnificas arborizacion suelta de celulas de axon corto y multitud de detalles que no siempre se ven bien con el método de Golgi.

Le saluda cariñosamente su viejo amigo

Cajal

Figure 4. Santiago Ramón y Cajal's last letter, written two days before his death. The letter was sent, and delivered by hand, to his esteemed disciple Rafael Lorente de Nó. In it, he mentions some scientific concepts, including an allusion to the structure of dendritic spines.

The existence of dendritic spines as cytoplasmic extensions of neurons was confirmed unequivocally in the 1950s with the advent of electron microscopy (DeFelipe, 2015). In addition to studies of the nervous system by Palade, Palay, and DeRobertis, among others, the work published by Gray (1959a) stood out, for it demonstrated definitively that dendritic spines were postsynaptic structures. Subsequent ultrastructural studies, combined with the use of the Golgi Method, later showed that the spines of impregnated neurons did in fact establish effective and functional synaptic contacts (Fairén et al., 1977).

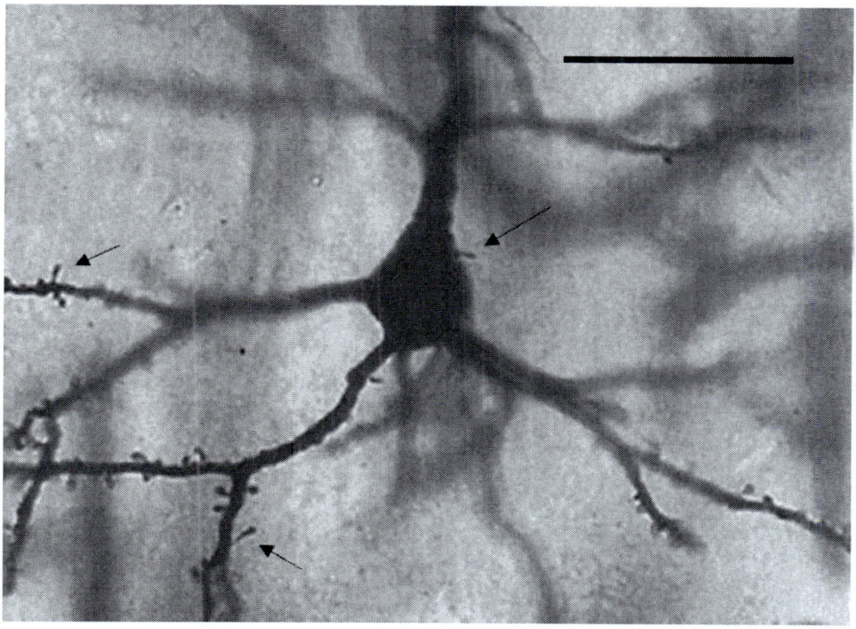

Figure 3. Metallic impregnation of a pyramidal neuron of the prefrontal cortex. Along with the dendritic spines seen in the basal dendrites (small arrows), a spine clearly protrudes from the cell soma (large arrow). Modification of the Golgi method (González-Burgos et al., 1992). Scale bar: 50 µm.

Development of Dendritic Spines

Generic studies in rodents have shown that dendritic spines originate from the first weeks of postnatal life, reaching maturity in the juvenile stage (Papa and Segal, 1996), though research shows that the processes underlying spinogenesis and the associated process of synaptogenesis depend on the neuronal type in question and the area of the nervous system where it is located (Yuste and Bonhoeffer, 2004). Attempt to categorize those processes have generated and proposed various models of spine development based on extensive experimental evidence: 1) the Sotelo Model; 2) the Miller/Peters Model, and 3) the Filopodial Model (Portera-Cailliau and Yuste, 2001).

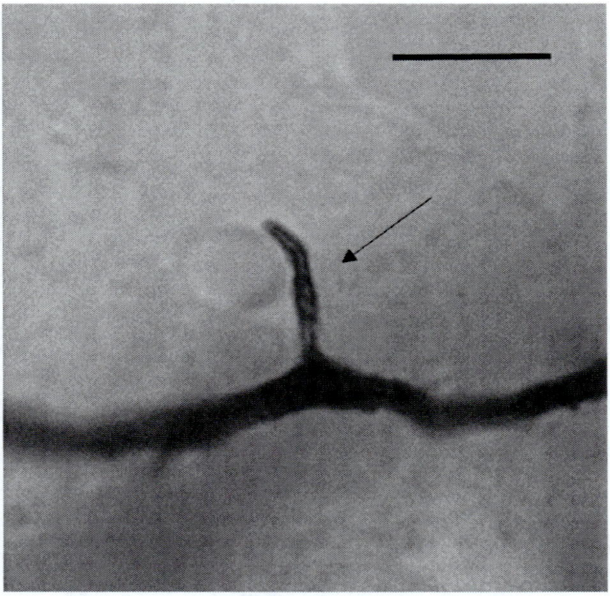

Figure 5. Photomicrograph of a filopodium (arrow) emerging from a dendrite. Modification of the Golgi method (González-Burgos et al., 1992). Scale bar: 10 μm.

The Sotelo Model is based on evidence which shows that the development of spines in the dendritic trunk of neurons occurs independently of afferent-mediated presynaptic stimulation. This model was based primarily on observations of spinogenesis in cerebellar Purkinje cells.

Support for the Miller/Peters Model comes from several brain regions, though it was initially described in the cerebral cortex. This model is based on the fact that when axonal fibers come into contact with the dendritic trunk of the postsynaptic neurons, they induce the generation of a synaptic contact that gradually evolves from the initial creation of a stubby spine (considered biologically immature) toward the formation of mature spines, of either the thin or mushroom-shaped types (see below).

The Filopodial Model is based on the existence of cytoplasmic appendages called "filopodia" that cover the undifferentiated dendrites (neurites) of developing neurons (Portera-Cailliau and Yuste, 2001) (Figure 5). It should be noted that they are also observed in neurons of the adult brain, though in a significantly lower proportion than in immature neurons. Filopodia are filamentous cytoplasmic sprouts measuring 3-40 μm in length. They were described by Cajal (Ramón y Cajal, 1892), but it was not until the late 1960s that they garnered significant attention among researchers (Morest, 1969). Filopodia and dendritic spines are distinct structures. Cytologically, filopodia lack the bulbous end that characterizes the mature spine, while ultrastructurally they are more electron-dense than spines due to the presence of a large amount of biological material associated with the cytoskeleton, made up specifically of actin filaments (Fiala et al., 1998). The filopodia are highly dynamic structures, even more so than spines. They have the ability to extend and retract in a matter of minutes, so the property of being precursors –direct or indirect– of the formation of dendritic spines has been attributed to them (Ziv and Smith, 1996).

The Filopodia Model holds that spinogenesis can occur through two somewhat different mechanisms. 1) Spine formation can proceed from filopodia as a direct precursor; that is, a spine differentiates from a filopodia by virtue of its direct contact, either terminal or *en passant*, with a synaptic bouton until it completes its transformation into a mature spine (Sorra and Harris, 2000; Ziv and Smith, 1996). 2) Spinogenesis begins through contact with a terminal axonal button or, *en passant*, with a given site of the filopodia, which induces the retraction of the latter. Upon retraction, this pseudosynaptic contact transits to the base of the filopodium, where a dendritic spine begins to form with which the terminal button –still in contact– establishes a functional synapse (Fiala et al., 1998). Although the filopodia disappear as the neurons mature (Velázquez-Zamora et al., 2011), there is no proportionality between the number of filopodia that retract and the number of spines that form (Fiala et al., 1998). These findings clearly show that spinogenesis is not a phenomenon that depends exclusively on the activity of filopodia (Portera-Cailliau and Yuste, 2001) (Figure 6).

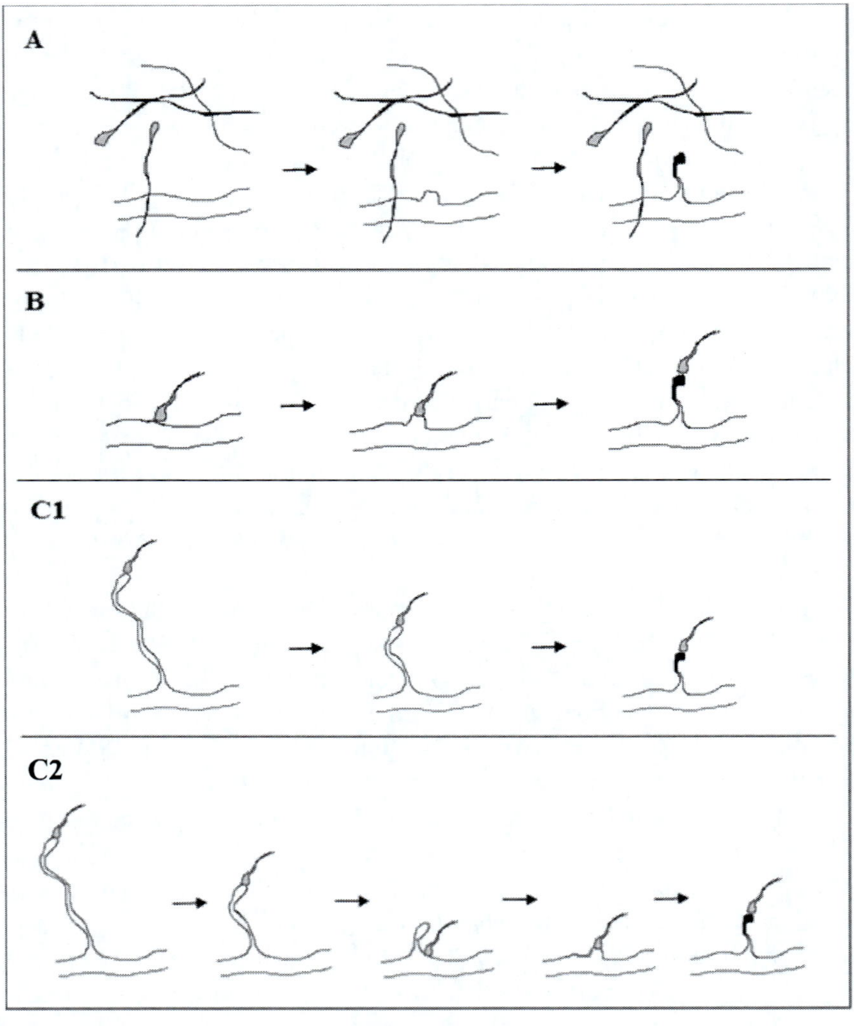

Figure 6. Schematic representation of the different ways that spinogenesis can occur. A: From the dendritic arbor, independently from presynaptic stimulation or filopodial influence. B: Spinogenesis from the contact between a presynaptic bouton with the dendritic arbor. C1: Spinogenesis from a direct contact between a presynaptic bouton and the tip of a filopodium. C2: Sinogenesis from the contact between a presynaptic terminal and a filopodium, which guides the axonal bouton to the parent dendrite from which a dendritic spine finally surge.

Geometric Structure of Dendritic Spines

Dendritic spines are cytoplasmic projections that, clearly, emerge from the dendrites of neurons. They constitute 80% of excitatory synaptic contact sites. They measure 0.1-2μm in length (Harris and Stevens, 1989) and, though with slight angular variations, are arranged perpendicular to the longitudinal axis of the corresponding parent dendrites.

Morphologically, the geometric structure of most spines is characterized by the presence of two regions: the head and neck (Figure 7) (Harris et al., 1989). This anatomical property confers distinct functional properties to these spines (González-Burgos, 2009, 2012, 2015a, 2022).

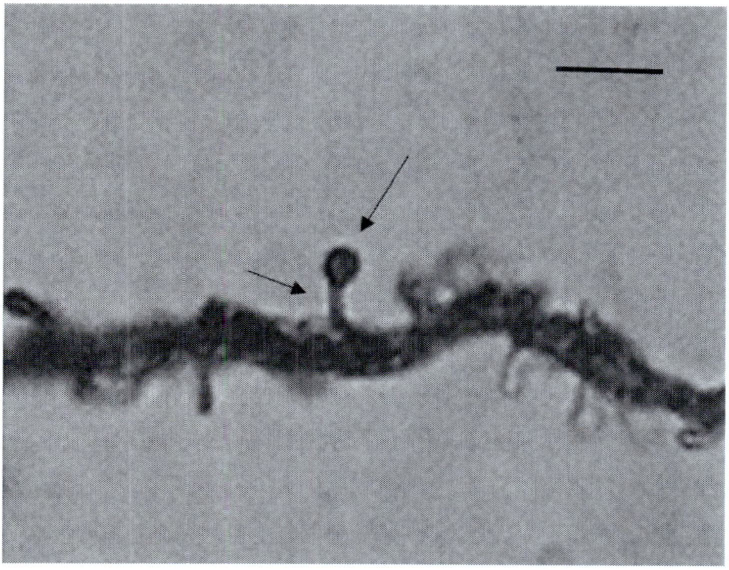

Figure 7. Dendritic spine emerging from its parent dendrite. The neck (small arrow) and head (large arrow) of the spine are clearly distinguishable. Modification of the Golgi method (González-Burgos et al., 1992). Scale bar: 2 μm.

Ultrastructurally, an electron density that corresponds to the postsynaptic density is observed at the apex of the head of the spines, which is composed of a complex network of actin filaments (cytoskeleton). The upper region of the head, and the area below the cell membrane, contain terminals where intracellular transmembrane proteins that act as receptors are inserted in an organized manner, together with proteins associated with the actin cytoskeleton and various biomolecules related to the signaling pathways of presynaptic stimuli that correspond to the various neurotransmission systems (Rao et al., 1998; González-Burgos, 2022).

The heads of dendritic spines also present structural sub-specializations related to the precise sites where the postsynaptic receptors are located in the intersynaptic space, corresponding to the biochemical systems involved in the transduction of synaptic signals into the postsynaptic density. Thus, there are: 1) macular or simple synapses, compact and irregular in shape that occupy the more-or-less central region of the spine's head and constitute the functional area of the synaptic complex; 2) perforated synapses, that have a larger diameter and, as a distinctive feature, a synaptically "silent" zone in the center of the macula, which is surrounded by a functional ring of the synapse; 3) segmented synapses, whose irregular structure has multiple, isolated, synaptically-functional regions (Figure 8).

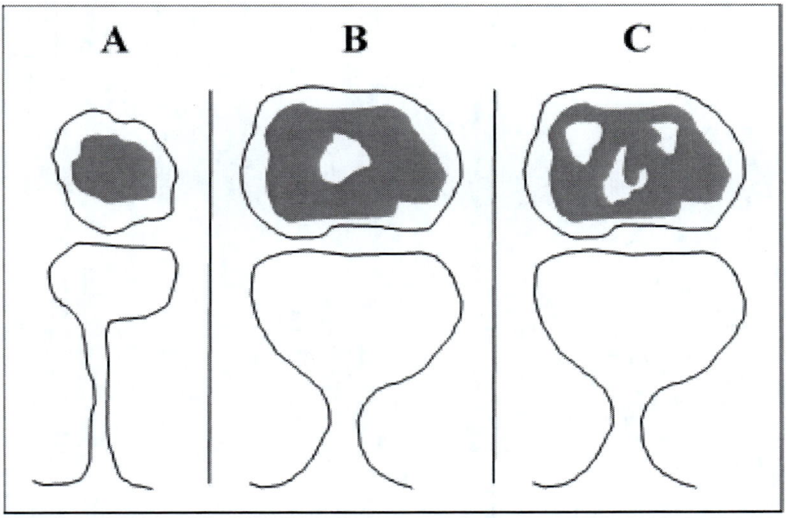

Figure 8. Types of synapses on the heads of dendritic spines, according to their surface structure. Figure A represents the macular synapse that typically occurs on small spines. Figure B shows a perforated synapse. Figure C shows a segmented synapse. Perforated and segmented synapses occur mainly on large spines. Gray regions represent functional synaptic regions.

Macular densities are seen more frequently on small-headed spines, while perforated and segmented densities occur more often on large-headed ones (Spaceck and Harris, 1997). Within the category of large-headed spines, some are elongated, flat cisterns with a smooth endoplasmic reticulum internalized from the parent dendrite, called the spine apparatus (Gray, 1959b). Recent evidence suggests that, in addition to mediating calcium signaling in the spinal apparatus, and modulating calcium and lipid homeostasis, the endoplasmic reticulum of neurons might also be involved in regulating vesicular traffic and the events of proteostasis that underlie the autonomous functions of numerous subcellular compartments. However, based on its neuron-wide geometric complexity, the endoplasmic reticulum also integrates the activity of autonomous compartments throughout the neuron and coordinates their interactions with the soma; thus it also underlies the expression of plasticity phenomena (Chanadai and Kavalali, 2022). The presence of free polyribosomes, endosomes, and multivesicular bodies is another distinguishing feature of these large spines, which have been associated with local protein degradation (Sorra and Harris, 2000). The absence of mitochondria also stands out, so the energy necessary for the transduction of synaptic signals could come (i) from the activity of mitochondria in the parent dendrite; (ii) from the diffusion of ATP; or (iii) through glycolytic production of ATP that could occur at the synapse (Sorra and Harris, 2000; Wu et al., 1997).

The necks of spines are of variable length, though their diameter varies only within a relatively small range. Once the synaptic impulse is transmitted, the neck participates in regulating the passage of current mediated by calcium ions. This action has led to the proposal that these spines may perform a modulating function of the synaptic currents related to the excitatory afferent information that is transmitted to neurons (Koch et al., 1992; Koch and Zador, 1993). Thus, the neck of the spines represents a kind of "pass valve" for the current that heads toward the parent dendrite and, eventually, the soma. The physiological and functional significance of the neck of the dendritic spines is explained in detail in later chapters of this book.

After the discovery of the relation of dendritic spines to the transmission of excitatory impulses (Gray, 1959a) mediated by glutamate (Johnson, 1972), subsequent studies suggested that, in addition to the density of spines and the proximity between them (Harris and Kater, 1994), the transduction of excitatory messages could be influenced by their geometric shape (Peters and Kaiserman-Abramof, 1970). Based on that work, and to facilitate the interpretation of the functional activity of the spines, it was determined that they could be classified by the characteristics of their head and neck. Based on ultrastructural studies, they were initially classified as thin, mushroom, and stubby types (Harris et al., 1989; Peters and Kaiserman-Abramof, 1970).

Later, the category of branched spines was added (Chicurel and Harris, 1992; Harris et al., 1992), while more recent, extensive experimental research has identified two additional types: wide and double spines (Beas-Zárate et al., 2002; del Valle et al., 2012; Flores-Soto et al., 2021; García-Chávez et al., 2008; González-Burgos, 2009, 2012, 2022; González-Burgos and Feria-Velasco, 2008; González-Burgos et al., 2001, 2004, 2005, 2006, 2007, 2009, 2010, 2011, 2012, 2015a, 2015b, 2017; González-Ramírez et al., 2014; González-Tapia et al., 2015a, 2015b, 2020a, 2020b; Martínez-Torres et al., 2018, 2021; Moralí et al., 2012; Niño et al., 2021; Pérez-Vega et al., 2000; Portilla-de Buen et al., 2017; Reyes-Corona et al., 2017; Tarelo-Acuña et al., 2000; Vázquez-Hernández et al., 2017, 2021, 2022; Velázquez-Zamora et al., 2011a, 2011b, 2012a, 2012b, 2016) (Figure 9).

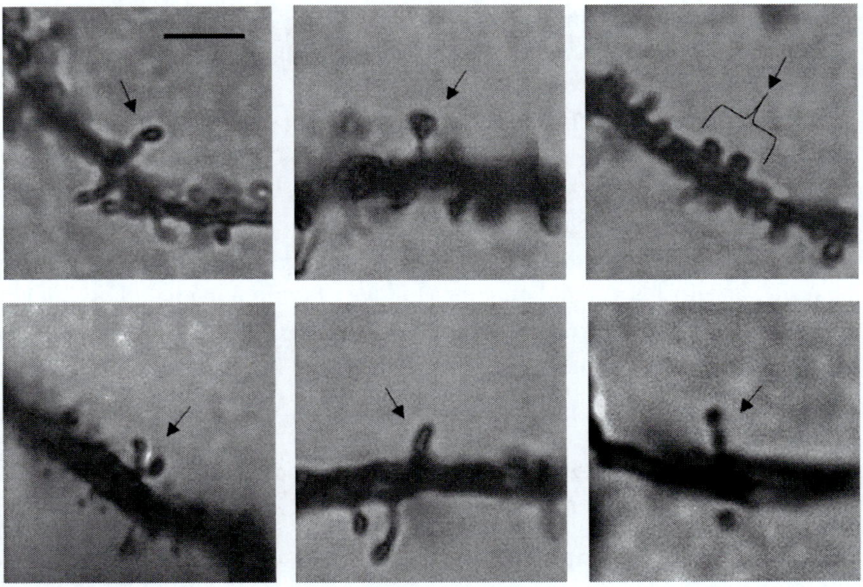

Figure 9. Composition of the six types of dendritic spines (arrows) reported in the literature. Upper left: thin spine. Upper center: mushroom spine. Upper right: stubby spine (two). Lower left: branched spine. Lower center: wide spine. Lower right: double spine. Modification of the Golgi method (González-Burgos et al., 1992). Scale bar: 2 μm.

Spines that do not have a well-defined neck (the stubby and wide types) are commonly referred to as sessile spines, while those with a well-differentiated neck (thin, mushroom-shaped, branched, and double types) are often called pedunculated spines (Fiala and Harris, 1999). Thin spines are characterized by a neck that is longer than the length of the head and a diameter slightly narrower than that of the head. Mushroom spines have the most contrasting structural components, as the diameter of the neck is very small in relation to that of the head, and short in comparison to the length of the head. Stubby spines show no distinction between neck and head. Their total length is practically equal to, or may be less than, their diameter. The features of the necks of branched spines are similar to those of the mushroom type, but the head of this kind of spine is double. The two heads are of similar proportions, but are separated by an invagination that allows them to make synaptic contact with two different axonal buttons. Like branched spines, the double type has two heads. The difference is that the heads of double spines are aligned longitudinally and separated by a small-diameter neck similar to the one that separates the lowest head from the parent dendrite. Wide spines, finally, show geometric characteristics similar to the stubby type in that they do not have a clearly separable head and neck; however, their length is greater than their diameter (González-Burgos, 2009, 2012, 2022). It should be noted that, although the proportions are preserved, the length of the different types of spines can vary so. for example, it is common to see thin spines that are smaller than some spines of the mushroom type, and *vice versa*.

The proportional density of each type of spine varies according to the neuronal lineage in question. However, evidence to date suggests that thin spines are the most abundant ones in projection neurons, such as the pyramidal cells of the hippocampal CA1 field (Harris et al., 1992) and layer II-III of the parietal cortex (Peters and Kaiserman-Abramof, 1970), or those of the Purkinje cells in the cerebellum (Lee et al., 2005). They are followed by the mushroom, stubby, wide, branched, and, finally, double types (González-Burgos, 2012, 2022; González-Tapia et al., 2015a).

Other Dendritic Protrusions

In addition to these "spines," other cytoplasmic protrusions are also capable of mediating the transmission of bioelectrical signals in synapses. Among others, there are *claw endings*, which are observed in the terminal region of the dendrites of the cerebellar granule cells. They receive afferent synaptic stimulation represented by both the axon endings of the extracerebellar mossy fibers and Golgi cell terminals, thus constituting the so-called "cerebellar glomerulus." Besides, the so-called *thorny excrescences* are located in the proximal dendritic portion of the pyramidal neurons of the CA3 field of the hippocampus. They receive afferents from the granule cells of the dentate gyrus of the hippocampus, as part of the hippocampal trisynaptic circuit. *Racemose appendages* are present on type I and II neurons of the inferior medullary olive. They establish synaptic contacts with multiple structures to, ultimately, send information to the cerebellar cortex. *Coralline appendages* can be found in the neurons of the dentate nucleus of the cerebellum. They represent the output of the information integrated in the cerebellar cortex (González-Burgos, 2022) (Figure 10). There are also *paintbrush bunch terminals*, found in the brush neurons in the granule layer of the cerebellum, establish multiple synaptic contacts with Golgi cells, dendrites of granule cells, and cerebellar glomeruli from the cerebellar granular layer (Mugnaini and Floris, 1994; Mugnaini et al., 1994).

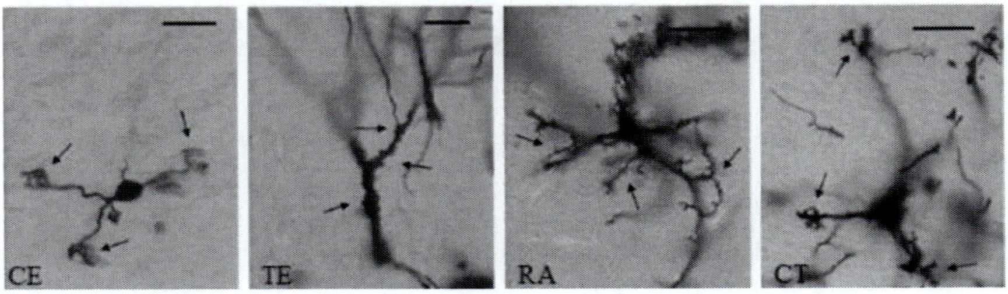

Figure 10. Photomicrographs of: claw endings (arrows) of a cerebellar granule cell (CE, scale bar: 20 μm); thorny excrescences (arrows) of a pyramidal neuron of field CA3 of the hippocampus (TE, scale bar: 50 μm); racemose appendages (arrows) of a neuron in the inferior olivary nucleus (RA, scale bar: 30 μm), and; coralline terminals (arrows) of a neuron in the dentate nucleus of the cerebellum (CE, scale bar: 30 μm). Modification of the Golgi method (González-Burgos et al., 1992).

Conclusion and Future Perspectives

From the confirmation of the Neuronal Theory, throughout the 20th century and up to the present, the visionary conclusions of Don Santiago Ramón y Cajal about the role played by the morphophysiology of dendritic spines in interneuronal communication have been confirmed. Furthermore, one of the most significant advances in this sense is represented by the diversity of geometric shapes that their plastic capacity allows them to be assigned and that is related to the differential processing of afferent excitatory information of both cognitive and non-cognitive content, in addition to their implication in the regulation of neuronal excitability.

To date, a large number of scientific reports on plastic morphological modifications of dendritic spines in punctual and circumscribed regions of neurons have been reported. One of the most important perspectives waiting to be addressed and understood is how the excitatory (and inhibitory) information coming from the different afferent systems to neurons is integrally processed, which in fact results in coherent postsynaptic responses according to the demands that the external and internal environment imposes on the individual from their interactive relationship.

References

Beas-Zárate, C., Pérez-Vega, M. I., González-Burgos, I., 2002. Neonatal exposure to monosodium L-glutamate induces loss of neurons and cytoarchitectural alterations in hippocampal CA1 pyramidal neurons of adult rats. *Brain Res.* 952, 275-281.

Chanaday, N. L., Kavalali, E. T., 2022. Role of the endoplasmic reticulum in synaptic transmission. *Curr. Op. Neurobiol.* 73, 102538.

Chicurel, M. E., Harris, K. M., 1992. Three-dimensional analysis of the structure and composition of CA3 branched dendritic spines and their synaptic relationships with mossy fiber boutons in the rat hippocampus. *J. Comp. Neurol.* 325, 169-182.

DeFelipe, J., 2010. *Cajal's Butterflies of the Soul.* Oxford University Press, USA.

DeFelipe, J., 2015. The dendritic spines tory: an intriguing process of discovery. Front. Neuroanat. 9(art. 14), 1-13.

DeFelipe, J., 2018. *Cajal's Neuronal Forest. Science and Art.* Oxford University Press, USA.

del Valle, J., Bayod, S., Camins, A., Beas-Zárate, C., Velázquez-Zamora, D. A., González-Burgos, I., Pallàs, M., 2012. Dendritic spine abnormalities in hippocampal CA1 pyramidal neurons underlying memory deficits in the SAMP8 mouse model of Alzheimer's disease. *J. Alzheimers Dis.* 32, 233-240.

Fairén, A., Peters, A., Saldanha, J., 1977. A new procedure for examining Golgi impregnated neurons by light and electron microscopy. *J. Neurocytol.* 6, 311-337.

Fiala, J. C., Feinberg, M., Popov, V., Harris, K. M., 1998. Synaptogenesis via dendritic filopodia in developing hippocampal área CA1. *J. Neurosci.* 18, 8900-8911.

Fiala, J. C., Harris, K. M., 1999. *Dendrite structure.* In: Stuart, G., Spruston, N., Häusser, M. (Eds.), Dendrites. Oxford University Press, New York, pp. 1-34.

Finger, S., 1994. *Origins of Neuroscience. A History of Explorations into Brain Function.* Oxford University Press, New York.

Flores-Soto, M., Romero-Guerrero, Ch., Vázquez-Hernández, N., Tejeda-Martínez, A., Martín-Amaya-Barajas, F. L., Orozco-Suárez, S., González-Burgos, I., Pentylenetetrazol-induced seizures in adult rats are associated with plastic changes to the dendritic spines on hippocampal CA1 pyramidal neurons. *Behav. Brain Res.* 406, 113198.

García-Chávez, D., González-Burgos, I., Letechipía-Vallejo, G., López-Loeza, E., Moralí, G., Cervantes, M., 2008. Long-term evaluation of cytoarchitectonic characteristics of prefrontal cortex pyramidal neurons, following global cerebral ischemia and neuroprotective melatonin treatment, in rats. *Neurosci. Lett.* 448, 148-152.

García-López, P., García-Marín, V., Freire, M., 2010. Dendritic spines and development: Towards a unifying model of spinogenesis-A present day review of Cajal's histological slides and drawings. *Neural Plast.* 769207, 29 pages.

Golgi, C., 1873. Sulla struttura della sostanza grigia del cervello [On the structure of the gray matter of the brain]. *Gaz. Med. Ital.* 6, 244-246.

González-Burgos, I., 2009. *Dendritic spines plasticity and learning / memory processes: Theory, evidence and perspectives.* In: Baylog, L.R. (Ed.), Dendritic Spines. Biochemistry, Modelling and Properties. Nova Science Publishers, New York, pp. 163-186.

González-Burgos, I., 2012. From synaptic transmission to cognition: an intermediary role for dendritic spines. *Brain Cogn.* 80, 177-183.

González-Burgos, I., 2022. *Psychobiological Principles of the Memory Process.* Nova Science Publishers, Inc. New York. 2022.

González-Burgos, I., Alejandre-Gómez, M., Cervantes, M., 2005. Spine-type densities of hippocampal CA1 neurons vary in proestrus and estrus rats. *Neurosci. Lett.* 379, 52-54.

González-Burgos, I., Alejandre-Gómez, M., Olvera-Cortés, M. E., Pérez-Vega, M. I., Evans, S., Feria-Velasco, A., 2006. Prenatal-through-postnatal exposure to moderate levels of ethanol leads to damage on the hippocampal CA1 field of juvenile rats: a stereology and Golgi study. *Neurosci. Res.* 56, 400-408.

González-Burgos, I., Feria-Velasco, A., 2008. Serotonin/dopamine interaction in memory formation. *Prog. Brain Res.* 172, 603-622.

González-Burgos, I., García-Martínez, S., Velázquez-Zamora, D. A., Ponce-Rolón, R., 2010. Cytoarchitectural impairments in the medium spiny neurons of the Nucleus Accumbens core of hyperactive juvenile rats. *Int. J. Dev. Neurosci.* 28, 475-480.

González-Burgos, I., González-Tapia, D., Feria-Velasco, A., 2015a. Plasticidad neuronal asociada al aprendizaje y la memoria [Neuronal plasticity associated with learning and memory]. In: González-Burgos, I. (Ed.), Psicobiología de la Memoria. Un Enfoque Interdisciplinario [Psychobiology of Memory. An Interdisciplinary approach]. Bios-Médica, Guadalajara, pp. 159-190.

González-Burgos, I., González-Tapia, D., Velázquez Zamora, D. A., Feria-Velasco, A., Beas-Zárate, C., 2011. Guided motor training induces dendritic spine plastic changes in adult rat cerebellar purkinje cells. *Neurosci. Let.* 491, 216-220.

González-Burgos, I., Letechipía-Vallejo, G., López-Loeza, E., Moralí, G., Cervantes, M., 2007. Long-term study of dendritic spines from hippocampal CA1 pyramidal cells, after neuroprotective melatonin treatment following global cerebral ischemia in rats. *Neurosci. Lett.* 423, 162-166.

González-Burgos, I., López-Vázquez, M. A., Beas-Zárate, C., 2004. Density, but not shape, of hippocampal dendritic spines varies after a seizure-inducing acute dose of monosodium glutamate in rats. *Neurosci. Lett.* 363, 22-24.

González-Burgos, I., Pérez-Vega, M. I., Beas-Zárate, C., 2001. Neonatal exposure to monosodium glutamate induces cell death and dendritic hypotrophy in rat prefrontocortical pyramidal neurons. *Neurosci. Lett.* 297, 69-72.

González-Burgos, I., Rivera-Cervantes, M. C., Velázquez-Zamora, D.A., Feria-Velasco, A., Garcia-Segura, L.M., 2012. Selective estrogen receptor modulators regulate dendritic spine plasticity in the hippocampus of male rats. *Neural Plast.* 2012, 309494.

González-Burgos., I., Tapia-Arizmendi, G., Feria-Velasco, A., 1992. Golgi method without osmium teroxide for the study of the central nervous system. *Biotech. Histochem.* 67, 288-296.

González-Burgos, I., Velázquez-Zamora, D. A., Beas-Zárate, C., 2009. Damage and plasticity in adult rat hippocampal trisynaptic circuit neurons after neonatal exposure to glutamate excitotoxicity. *Int. J. Dev. Neurosci.* 27, 741-745.

González-Burgos, I., Velázquez-Zamora, D. A., González-Tapia, D., Cervantes, M., 2015b. A Golgi study of the plasticity of dendritic spines in the hypothalamic ventromedial nucleus during the estrous cycle of female rats. *Neuroscience* 9, 298-274.

González-Burgos, I., Velázquez-Zamora, D. A., González-Tapia, D., Vázquez-Hernández, N., Martínez-Torres, N. I., 2017. Plasticity of dendritic spines. Not only for cognitive processes. In: Heinbockel, Th. (Ed.), Synaptic Plasticity. *InTech*, Croacia, pp. 153-166.

González-Ramírez, M. M., Velázquez-Zamora, D. A., Olvera-Cortés, M. E., González-Burgos, I., 2014. Changes in the plastic properties of hippocampal dendritic spines underlie the attenuation of place learning in healthy aged rats. *Neurobiol. Learn. Mem.* 109, 94-103.

González-Tapia, D., González-Ramírez, M. M., Vázquez-Hernández, N., González-Burgos, I., 2020b. Motor learning induces plastic changes in Purkinje cell dendritic spines in the rat cerebellum. *Neurologia* 35, 451-457.

González-Tapia, D., González-Tapia, D. C., Vázquez-Hernández, N., Martínez-Torres, N. I., Flores-Soto, M., González-Burgos, I., 2020a. Modifications to cytoskeleton-associated proteins in dendritic spines underlie the adaptive plasticity involved in long term reference memory. *Neurobiol. Learn. Mem.* 172, 107247.

González-Tapia, D., Velázquez-Zamora, D. A., González-Burgos, I., 2015a. *The molecular biology of dendritic spine plasticity in memory processing*. In: González-Burgos, I. (Ed.), Synaptic Fundamentals of Memory Performance. Nova Science Publishers, New York, pp. 59-80.

González-Tapia, D., Velázquez-Zamora, D. A., Olvera-Cortés, M. E., González-Burgos, I., 2015b. The motor learning induces plastic changes in dendritic spines of Purkinje cells from the neocerebellar cortex of the rat. Restor. *Neurol. Neurosci.* 33, 639-645.

Gray, E. G., 1959a. Electron microscopy of synaptic contacts on dendritic spines of the cerebral cortex. *Nature* 163, 1592-1593.

Gray, E. G., 1959b. Axo-somatic and axo-dendritic synapses of the cerebral cortex: an electron microscope study. *J. Anat.* 93(Pt 4), 420-433.

Harris, K. M., Jensen, F. E., Tsao, B. H., 1989. *Ultrastructure development, and plasticity of dendritic spine synapses in área CA1 of the rat hippocampus: extending our visión with serial electron microscopy and three-dimensional analysis*, In: Chan-Palay, V., Kohler, Ch. (Eds.), The Hippocampus – new vistas. Alan R. Liss, USA, pp. 33-52.

Harris, K. M., Jensen, F. E., Tsao, B. H., 1992. Three-dimensional structure of dendritic spines and synapses in rat hippocampus (CA1) at postnatal day 15 and adult ages: implications for the maturation of synaptic physiology and long-term potentiation. *J. Neurosci.* 12, 2685-2705.

Harris, K. M., Kater, S. B., 1994. Dendritic spines: cellular specializations imparting both stability and flexibility to synaptic function. *Annu. Rev. Neurosci.* 17, 341-371.

Harris, K. M., Stevens, J. K., 1989. Dendritic spines of CA 1 pyramidal cells in the rat hippocampus: serial electron microscopy with reference to their biophysical characteristics. *J. Neurosci.* 9, 2982-2997.

Johnson, J. L., 1972. Glutamic acid as a synaptic transmitter in the nervous system. A review. *Brain Res.* 37, 1-19.

Koch, C., Zador, A., 1993. The function of dendritic spines: devices subserving biochemical rather than electrical compartmentalization. *J. Neurosci.* 13, 413-422.

Koch, C., Zador, A., Brown, T. H., 1992. Dendritic spines: convergence of theory and experiment. *Science* 256, 973-974.

Lee, K. J., Kim, H., Rhyu, I. J., 2005. The roles of dendritic spine shapes in Purkinje cells. *Cerebellum* 4, 97-104.

Lopera-Chaves, M. C., 2011. La Teoría Neuronal: La tecnología como soporte de un debate científico. Un capítulo en la historia de la ciencia [The Neuronal Theory: Technology as support for a scientific debate. A chapter in the history of science]. *Cien. Tecnol. Soc.* 4, 131-139.

López-Muñoz, F., Álamo, C., García-García, P., Boya, J., 2006. *Psiq. Biol.* 13, 167-182.

Luria, A. R., 1970. The functional organization of the brain. *Sci. Am.* 222, 66-72.

Luria, A. R., 1982. Introducción Evolucionista a la Psicología [Evolutionary Introduction to Psychology]. Fontanella, Spain.

Martínez-Torres, N. I., González-Tapia, D., Vázquez-Hernández, N., González-Burgos, I., 2018. Atomoxetine prevents working memory loss in hyperactive rats, mediating plastic changes in prefrontal cortex pyramidal neurons. *Pharmacol. Biochem. Behav.* 175, 116-122.

Martínez-Torres, N. I., Vázquez-Hernández, N., Martín-Amaya-Barajas, F. L., Flores-Soto, M., González-Burgos, I., 2021. Ibotenic acid induced lesions impair the modulation of dendritic spine plasticity in the prefrontal cortex and amygdala, a phenomenon that underlies working memory and social behavior. *Eur. J. Pharmacol.* 896, 173883.

Moralí, G., Montes, P., González-Burgos, I., Velázquez-Zamora, D. A., Cervantes, C., 2012. Cytoarchitectural characteristics of hippocampal CA1 pyramidal neurons of rats, four months after global cerebral ischemia and progesterone treatment. *Restor. Neurol. Neurosci.* 30, 1-8.

Morest, D. K., 1969. The differentiation of cerebral dendrites: A study of the post-migratory neuroblast in the medial nucleus of the trapezoid body. *Z. Anat. Entwicklungsgesch* 128, 271-289.

Mugnaini, E., Floris, A., 1994. The unipolar brush cell: a neglected neuron of the mammalian cerebellar cortex. *J. Comp. Neurol.* 339, 174-180.

Mugnaini, E., Floris, A., Wright-Goss, M., 1994. Extraordinary synapses of the unipolar brush cell: an electron microscopic study in the rat cerebellum. *Synapse* 16, 284-311.

Niño, S. A., Vázquez-Hernández, N., Arevalo-Villalobos, J., Chi-Ahumada, E., Martín-Amaya-Barajas, F. L., Díaz-Cintra, S., Martel-Gallegos, G., González-Burgos, I., Jiménez-Capdeville, M.E., 2021. Cortical synaptic reorganization under chronic arsenic exposure. *Neurotox. Res.* 39, 1970-1980.

Papa, M., Segal. M., 1996. Morphological plasticity in dendritic spines of cultured hippocampal neurons. *Neuroscience* 71, 1005-1011.

Pérez-Vega, M. I., Feria-Velasco, A., González-Burgos, I., 2000. Prefrontocortical serotonin depletion results in plastic changes of prefrontocortical pyramidal neurons, underlying a greater efficiency of short-term memory. *Brain Res. Bull.* 53, 291-300.

Peters, A., Kaiserman-Abramof, I. R., 1970. The small pyramidal neuron of the rat cerebral cortex. The perikaryon, dendrites and spines. *Am. J. Anat.* 127, 321-355.

Portera-Cailliau, C., Yuste, R., 2001. Sobre la función de los filopodios dendríticos [On the function of dendritic filopodia]. *Rev. Neurol.* 33, 1158-1166.

Portilla-de Buen, E., Salgado-Ceballos, H., González-Tapia, D., Leal-Cortés, C., Mondragón-Lozano, R., Sánchez-Torres, S., Álvarez-Mejía, L., Fabela-Sánchez, O., Martínez-Torres, N. I., González-Ramírez, M. M., Vázquez-Hernández, N., González-Burgos, I., 2017. Spinogenesis and plastic changes in the dendritic spines of spinal cord motoneurons after traumatic injury in rats. *Arch. Med. Res.* 48, 609-615.

Ramón y Cajal, S., 1888. Estructura de los centros nerviosos de las aves [Structure of the nerve centers of birds]. *Rev. Trim. Histol. Norm. Patol.* 1:1-10.

Ramón y Cajal, S., 1892. El nuevo concepto de la histología de los centros nerviosos [The new concept of histology of nerve centers]. Revista de Ciencias Médicas. Barcelona 18, 361–376, 457–476, 505–520, 529—541.

Ramón y Cajal, S., 1952. ¿Neuronismo o Reticularismo? Las pruebas objetivas de la unidad anatómica de las células nerviosas [Neuronism or Reticularism? Objective evidence of the anatomical unit of nerve cells], 2nd ed. Instituto Ramón y Cajal [Ramón y Cajal Institute] CSIC. Madrid.

Ramón y Cajal, S., De Castro, F., 1972. Elementos de Técnica Micrográfica del Sistema Nervioso [Elements of Micrographic Technique of the Nervous System], 2nd ed. Salvat, Barcelona.

Rao, A., Kim, E., Sheng, M., Craig, A. M., 1998. Heterogeneity in the molecular composition of excitatory postsynaptic sites during development of hippocampal neurons in culture. *J. Neurosci.* 18, 1217-1229.

Reyes-Corona, D., Vázquez-Hernández, N., Escobedo, L., Orozco-Barrios, C. E., Ayala-Davila, J., Moreno, M. G., Amaro-Lara, M. E., Flores-Martinez, Y. M., Espadas-Alvarez, A. J., Fernandez-Parrilla, M. A., Gonzalez-Barrios, J. A., Gutierrez-Castillo, M. E., González-Burgos, I., Martinez-Fong, D., 2017. Neurturin overexpression in dopaminergic neurons induces presynaptic and postsynaptic structural changes in rats with chronic 6-hydroxydopamine lesion. *PLoS One* 12, e0188239.

Shepherd, G., 1991. *Foundations of the Neuron Doctrine.* Oxford University Press, New York.

Sorra, K. E., Harris, K. M., 2000. Overview on the structure, composition, function, development, and plasticity of hippocampal dendritic spines. *Hippocampus* 10, 501-511.

Spacek, J., Harris, K. M., 1997. Three-dimensional organization of smooth endoplasmic reticulum in hippocampal CA1 dendrites and dendritic spines of the immature and mature rat. *J. Neurosci*. 17, 190-203.

Tarelo-Acuña, L., Olvera-Cortés, M. E., González-Burgos, I., 2000. Prenatal and postnatal exposure to ethanol induces changes in the shape of the dendritic spines from hippocampal CA1 pyramidal neurons of the rat. *Neurosci. Lett*. 286, 13-16.

Vázquez-Hernández, N., González-Tapia, D. C., Martínez-Torres, N. I., González-Tapia, D., González-Burgos, I., 2017. Different patterns of motor activity induce differential plastic changes in pyramidal neurons in the motor cortex of rats: A Golgi study. *Neurosci. Lett*. 657, 27-31.

Vázquez-Hernández, N., Martínez-Torres, N. I., González-Burgos, I., 2021. Plastic changes to dendritic spines in the cerebellar and prefrontal cortices underlie the decline in motor coordination and working memory during successful aging. *Behav. Brain Res.* 400, 113014.

Vázquez-Hernández, N., Martínez-Torres, N. I., Tejeda-Martínez, A., Flores-Soto, M., Salgado-Ceballos, H., González-Burgos, I., 2022. Rehabilitation on a treadmill induces plastic changes in the dendritic spines of spinal motoneurons associated with improved execution after a pharmacological injury to the motor cortex in rats. *J. Chem. Neuroanat*. 125, 102159.

Velázquez-Zamora, D. A., Garcia-Segura, L. M., González-Burgos, I., 2012a. Effects of selective estrogen receptor modulators on allocentric working memory performance and on dendritic spines in medial prefrontal cortex pyramidal neurons of ovariectomized rats. *Horm. Behav*. 61, 512-517.

Velázquez-Zamora, D. A., González-Ramírez, M. M., Beas-Zárate, C., González-Burgos, I., 2011b. Egocentric working memory impairment and dendritic spine plastic changes in prefrontal neurons after NMDA receptor blockade in rats. *Brain Res*. 1402, 101-108.

Velázquez-Zamora, D. A., González-Tapia, D., González-Ramírez, M. M., Flores-Soto, M., Vázquez-Valls, E., Cervantes, M., González-Burgos, I., 2012b. Plastic changes in dendritic spines of hippocampal CA1 pyramidal neurons from ovariectomized rats after estradiol treatment. *Brain Res*. 1470, 1-10.

Velázquez-Zamora, D. A., Martínez-Degollado, M., González-Burgos, I., 2011a. Morphological development of dendritic spines on rat cerebellar Purkinje cells. *Int. J. Dev. Neurosci.* 29, 515-520.

Velázquez-Zamora, D. A., Martínez-Torres, N. I., Cervantes, M., González-Burgos, I., 2016. Administration of Anastrozole to ovariectomized rats impairs working memory in association with plastic changes to dendritic spines on prefrontal third-layer pyramidal neurons. *J. Steroids & Horm. Sci*. 7, 1000173 (7 pages).

Virchow, R., 1854. Ueber eine im Gehirn und Rückenmark des Menschen aufgefundene Substanz mit der chemsichen Reaction der Cellulose [On a substance found in the human brain and spinal cord with the chemical reaction of cellulose]. *Virchow's Archiv*. 6, 135-138.

Wu, K., Aoki, C., Elste, A., Rogalski-Wilk, A. A., Siekevitz, P., 1997. The synthesis of ATP by glycolytic enzymes in the postsynaptic density and the effect of endogenously generated nitric oxide. *Proc. Natl. Acad. Sci. USA* 94, 13273-13278.
Yuste, R., Bonhoeffer, T., 2004. Genesis of dendritic spines: insights from ultrastructural and imaging studies. *Nat. Rev. Neurosci.* 5, 24-34.
Ziv, N., Smith, S., 1996. Evidence for a role of dendritic filopodia in synaptogenesis and spine formation. *Neuron* 17, 91-102.

Chapter 2

Molecular Mechanisms and Cytoskeleton-Associated Protein Dynamics Underlying Dendritic Spine Plasticity

David González-Tapia[1,*]
and Mario Flores-Soto[2]

[1] División de Salud, Centro Universitario de Tlajomulco, Universidad de Guadalajara. Tlajomulco de Zúñiga, Jalisco. México

[2]División de Neurociencias, Centro de Investigación Biomédica de Occidente, IMSS. Guadalajara, Jalisco, México

Introduction

As the years go by, scientists in the field of neuroscience research continue to unmask, in ever greater detail, the biological bases of synaptic plasticity. The implementation of techniques of molecular biology has contributed significantly to identifying the components that give the synaptic complex its form and function. The use of techniques like electron microscopy, immunodetection, gene silencing, and metallic impregnation has made it possible to recognize not only the elements that make up neuronal synapses but also the molecules inside postsynaptic terminals and, especially, dendritic spines.

This chapter begins by explaining certain key mechanisms that regulate protein expression inside neurons and describing the importance of the so-called transcription factors that mold the architecture of neurons in general, and of dendritic spines, specifically. The activity of these molecules is now known to be determined not only by a wide range of intrinsic phenomena but also by environmental stimuli. Given this, we attempt to orient the reader in

In: Dendritic Spines
Editor: Ignacio González-Burgos
ISBN: 979-8-89113-080-7

the tiny space of the synaptic complex while providing an account of the numerous molecules that exist there. According to the interactions that occur among these proteins, we explain how some maturational processes take place, given that the proportion of receptors located on their surface can vary. Concepts addressed include identifying the key molecules involved in the study of dendritic spine-mediated synaptic plasticity, especially the importance of postsynaptic proteins as we delve inside dendritic spines to analyze the complex molecular machinery that woven around the actin cytoskeleton. These structural components ultimately determine the size, shape, and/or movement of all cells, especially those of the nerve tissue of dendritic spines, and participate in regulating neuronal excitability at the synaptic level. In addition to providing experimental background, we describe the interactions among and between these actin cytoskeleton-associated proteins and propose research perspectives that may enhance our understanding of the complex biochemical flexibility expressed in dendritic spines as major morphological regulators of synaptic plasticity. Overall, we exhibit the complex degree of biochemical compartmentalization inside dendritic spines and the elaborate molecular mechanisms that maintain them. Finally, we can only marvel at what seems to be one of the smallest biological units in nature, yet one tremendously important for understanding the physiology of the nervous system.

Epigenetic Mechanisms in Dendritic Spine Plasticity

Recent studies show that signaling pathways and certain molecular mechanisms, like epigenetics and the participation of transcription factors, play essential roles in regulating the synaptic plasticity of dendritic spines. It is important to note that the density and morphology of these spines vary as a function of their location in the brain, local microenvironments, and pathological states. This knowledge is extremely important for our understanding, given the way in which dendritic spines respond to injury, and their specific responses to agents –pharmacological among others– that can modulate their density, morphology, and plasticity.

Waddington coined the term epigenetics to refer to the set of processes involved in the regulation of gene expression that (i) do not undergo change in the DNA nucleotide sequence, and (ii) have the quality of being heritable (Waddington, 1952). The two main mechanisms of gene regulation are: 1) DNA modification (methylation); and 2) histone modification (e.g., acetylation), because they trigger changes in chromatin condensation, which regulates the entry of the transcriptional machinery and, hence, gene expression (Day and Sweatt, 2012; Malecová et al., 2010; Saxena et al., 2011).

On the topic of histone modification, it is important to mention that the amino-terminal ends of histones, which protrude from the chromatin structure, participate in a process of signal integration that gives rise to a specific pattern of post-translational modifications, and forms a 'histone code,' which directs the activity of numerous factors and cofactors of the transcriptional machinery. To date, four histone modifications involved in epigenetic marking have been described: 1) acetylation; 2) methylation; 3) ubiquitination; and 4) phosphorylation (Levenson and Sweatt, 2009).

Histone acetylation and deacetylation have been associated with gene transcription regulation through the loosening of heterochromatin. Regarding histone deacetylase 2 (HDAC2), which is expressed predominantly in neurons (MacDonald and Roskams, 2008), these phenomena play an important role in the dynamic regulation of synaptic plasticity, associated gene expression, and memory formation. For example, specific overexpression of HDAC2 –but not HDAC1– at the neuronal level decreased hippocampal synaptic plasticity (i.e., the number of dendritic spines and synapses) and impaired memory formation, while an HDAC2 deficiency increased the number of dendritic spines and synapses in the hippocampus, and improved memory formation in mice (Bramham et al., 2010) (Figure 1).

Epigenetic mechanisms have also been implicated in regulating the expression of the brain-derived neurotrophic factor gene (BDNF), a key neurotrophin for the development of dendritic spines and, consequently, synaptic plasticity. BDNF signaling induces expression of proteins associated with the activity-regulated cytoskeleton (Arc) (Bramham et al., 2010). BDNF and Arc have been shown to play causal roles in both anxiety and alcoholism (Pandey, 2008). Studies of alcohol preference using murine models have shown low levels of BDNF and Arc, accompanied by a decrease in the density of dendritic spines in both the central and medial nucleus of the amygdala (MeA) (Pandey, 2008). Those findings were accompanied by an increased expression of HDAC2 and low levels of H3-K9 histone acetylation, which may be responsible for the deficits in the expression of BDNF and Arc in the amygdaloid structures of rats that show a preference for alcohol. These data demonstrate the role of epigenetic mechanisms mediated by HDAC2 in modulating BDNF expression in models of anxiety and alcoholism, so HDAC2 inhibition can be a useful pharmacotherapeutic tool for treatment.

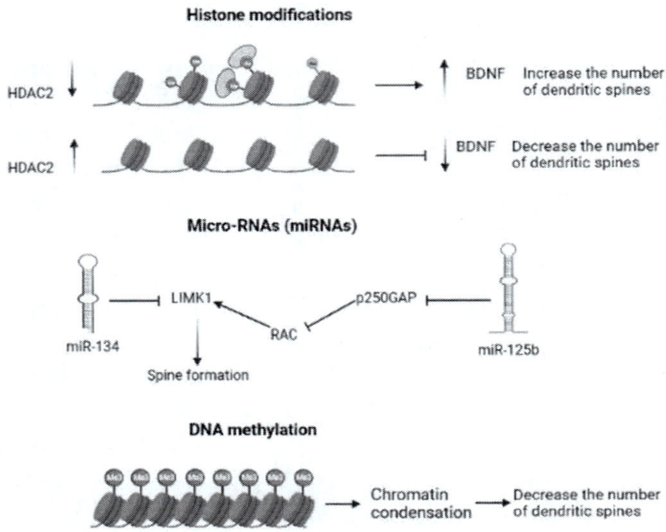

Figure 1. Epigenetic mechanisms involved in dendritic spine plasticity. Epigenetic regulation of gene expression can be mediated by histone modifications, by noncoding RNAs or by DNA methylation.

Another factor is environmental enrichment, a phenomenon that can affect visual cortical plasticity at the epigenetic level. Mice reared in an enriched environment show greater dendritic spine density, dependent on the Arc and Egr-1 genes (early growth response protein 1), and the myocyte enhancer 2 (MEF2C) transcription factor. That study demonstrated increased binding of MEF2C to the promoter regions of the Arc and Egr1 genes. In addition, pups reared in an enriched environment showed a reduction in HDAC5 and in its binding to promoters of the Mef2c, Arc, and Egr-1 genes, results that suggest a possible molecular mechanism underlying the enriched environment through MEF2C and HDAC5, which drive Arc and Egr-1. It may be that these mechanisms lead to the observed increase in dendritic spine density, along with complex stimulation driven by an early enriched environment (Puang et al., 2020).

DNA methylation is another type of epigenetic modification that participates in the formation of dendritic spines. It is important to understand that this process is usually associated with decreased gene expression, since histone methylation affects this depending on the residue where methylation occurs and the type of mark that influences whether or not gene silencing occurs. DNA presents regions of 1,000-1,500 base pairs rich in CpG dinucleotides (CpG islands), which are recognized by DNA-methyltransferase enzymes. During DNA replication, the cytokines of the newly-synthesized strand are methylated, so the memory of the methylated state is maintained in the daughter DNA molecule. This modification usually becomes stable and is inherited as a clonal pattern of methylation. Over-methylation of these regions maintains the condensed structure of chromatin and prevents transcription of the genes involved (Day and Sweatt, 2012; Levenson and Sweatt, 2005). In this regard, one study showed that rodents exposed repeatedly to cocaine increase the density of dendritic spines in the nucleus accumbens, though this effect was mitigated by the lack of lysine dimethyl transferase G9a, an enzyme that transfers methyl groups to cysteine residues in DNA (Maze et al., 2010).

Transcriptional and epigenetic regulation are interconnected processes responsible for turning gene expression on and off. This fine modulation of gene expression determines the phenotype of different cell types, as well as their morphology, functionality, and ability to respond to various conditions. Efforts to understand epigenetic mechanisms have been extended to synaptic plasticity and dendritic spines, leading to the identification of transcription factors and genes involved in these processes. One recent study identified the MeCP2 gene as important in regulating the expression of thousands of genes, including BDNF (Zhou et al., 2006). Among other functions, this neurotrophin is known to promote neuronal and synaptic maturation (Carvalho et al., 2008), increase dendritic spine density, and improve synaptic plasticity, learning, and memory (Luine and Frankfurt, 2013). MeCP2 binds to the promoter region of the BDNF gene and modulates its expression in an activity-dependent manner (Chahrour and Zoghbi, 2007; Zhou et al., 2006). Lower levels of BDNF mRNA and the BDNF protein, as well as altered BDNF trafficking and activity-dependent release, have been highlighted as pathophysiological mechanisms of the progression of RTT disease, where a decrease in MeCP2 expression results in a downregulation of BDNF gene expression and, consequently, of its BDNF protein (Xu et al., 2014). In fact, BDNF overexpression rescues several cellular and behavioral deficits in mice that lack the MeCP2 gene (Chahrour and Zoghbi, 2007).

These studies indicate the crucial role that BDNF plays in neurological deficits in MeCP2 knockout mice. Several studies have shown that BDNF is involved in synaptic plasticity and is critical for the formation and maturation of dendritic spines (Vigers et al., 2012). For example, exogenously-applied BDNF increases dendritic spine density in pyramidal neurons in the CA1 region (Ji et al., 2005). BDNF also alters the proportions of the various morphological types of dendritic spines in the hippocampus (Chapleau et al., 2008), while overexpression in cultured hippocampal neurons rescued dendritic atrophy caused by shRNA-mediated Mecp2 knockdown (Larimore et al., 2009). These effects of BDNF on dendritic spines are mediated by the tropomyosin-related kinase B (TrkB) receptor and subsequent extracellular signal-regulated kinase (ERK) activation of phosphatidylinositol 3-kinase (PI3K) and phospholipase C-γ (PLC-γ), which leads to the opening of transient receptor potential canonical (TRPC) channels that contain the TRPC3 subunit (Chapleau et al., 2009; Li et al., 2012; Luine and Frankfurt, 2013).

Another transcription factor that actively participates in the regulation of dendritic spines is the aforementioned Egr-1, a member of the EGR family, also known as NEFI-A, Zif268, Krox-24, and TIS85. This transcription factor regulates various cellular processes, including immune function, cell proliferation/growth, female reproduction, apoptosis, vascular function, learning, and memory (Knapska and Kaczmarek 2004), as well as transcription through its Cys2His2 zinc finger motifs that recognize, and bind to, the Egr-9 response elements (EREs) of its target genes. Chromatin immunoprecipitation studies show binding sites of the Egr-1 transcription factor in the promoter region for drebrin. Overexpression of Egr-1 in vitro in primary hippocampal neurons, or in vivo in hippocampal brain homogenates, shows reduced levels of the expression of drebrin mRNA and protein. Drebrin, significantly, is an actin-binding protein that is expressed abundantly in neurons and localizes to dendritic spines to regulate several changes in their morphology and density (Sekino et al., 2007). It is necessary for the morphogenesis and maturation of dendritic spines, for it recruits the postsynaptic scaffolding protein PSD-95 –discussed later– and the N-methyl-D-aspartate type glutamate receptor (NMDAR), two crucial components in the functioning of dendritic spines (Takahashi et al., 2006).

Downregulation of drebrin expression in vitro using antisense nucleotides or shRNA suppresses F-actin accumulation inside dendritic spines, thus preventing spine formation and, consequently, decreasing their density (Yamazaki et al., 2014). In vitro studies have shown that low drebrin expression reduces glutamatergic and GABAergic neurotransmission in hippocampal neurons. Similarly, mice deficient in the drebrin gene show deficits in memory function during contextual fear conditioning with a decrease in the number of mushroom-shaped, thin, and wide dendritic spines (Jung et al., 2015). Moreover, decreased drebrin expression has been reported in the brains of postmortem patients with cognitive disorders like Down syndrome (Shim and Lubec, 2002), in people with mild cognitive impairment

(Counts et al. 2006; Harigaya et al., 1996; Hatanpaa et al., 1999), in postmortem patients with Alzheimer's disease (AD) (Counts et al. 2012; Shim and Lubec, 2002), and in animal models of AD (Calon et al., 2004; Lacor et al., 2007; Mahadomrongkul et al., 2005; Zhao et al., 2006). Despite their functional importance, however, the regulatory factors of drebrin expression remain largely uncharacterized.

Immunocytochemical and Golgi staining analyses have revealed a reduction in the drebrin protein accompanied by a decrease in dendritic spine density and low expression of synaptic markers in Egr-1 overexpressed in hippocampal neurons in vitro. However, studies of mice deficient in the Egr-1 gene have detected an increase in drebrin expression that correlates with greater dendritic spine density. These data provide evidence that Egr-1 is a regulator of drebrin expression, which is related to changes in the density and morphology of dendritic spines. Thus, the downregulation of drebrin by Egr-1 may prevent PSD-95 accumulation in spines, whereas Egr-1 upregulation may result in reduced co-localization of drebrin and PSD-95. These data indicate that Egr-1 upregulation can directly downregulate PSD-95 expression and indirectly prevent PSD-95 accumulation in dendritic spines by downregulating drebrin. Loss of PSD-95 accumulation, in turn, can disrupt the recruitment of AMPA and NMDA glutamate receptors to the postsynaptic membrane (Ma et al., 2015). Both drebrin downregulation and PSD-95 may contribute to a decrease in dendritic spine density (Migaud et al., 1998). Other data regarding these proteins will be explained in detail below.

Micro-RNAs (miRNAs) can also regulate dendritic spine plasticity. In fact, they are considered essential elements for normal brain development (Buchsbaum, et al., 2003) and are known to play an important role in the formation of dendritic spines. Studies by Edbauer and cols. (2010) demonstrated that miR-125b and miR-132 are associated with the fragile X mental retardation protein (FMRP). That study showed that the NR2A subunit of the NMDA receptor is the target of these two miRNAs, though its effect on spinal morphology seems to be contradictory. While overexpression of miR-125b induced long, narrow spines, and correlated with reduced synaptic transmission, overexpression of miR-132 led to enlarged spine heads. MiR-132 has been associated with another mechanism: through translational suppression of p250GAP, it induces spine formation (Figure 1). A similar effect is seen with the deletion of p250GAP. Inhibition of miR-132 results in smaller dendritic spines and decreased, miniature excitatory postsynaptic currents (mEPSC) (Impey et al., 2010). Another cytoskeletal regulatory molecule that is regulated at the miRNA level is LIMK1. Studies have determined that miR-134 suppresses LIMK1, but that exposure of cultured cells to BDNF alleviates this suppression and contributes to dendritic spine stabilization (Schratt et al., 2006). In addition, LIMK1 inhibition has been shown to rescue Aβ-induced loss of hippocampal dendritic spines and morphological aberrations. These data suggest that LIMK1 inhibition may modulate the synaptic plasticity of dendritic spines in the presence of Aβ and, hence, benefit cognitively normal patients who are at high risk of developing dementia.

Although the field of the epigenetic control of dendritic spine plasticity has only begun to emerge, the importance of control at the RNA level has already been demonstrated. Undoubtedly, more epigenetic mechanisms involved in the regulation of dendritic spines will soon be elucidated.

Synaptic Proteins and Actin Cytoskeleton Dynamics

The synaptic complex is represented as a physiological unit for communication among neurons made up of three main regions: the presynaptic buttons, synaptic cleft, and postsynapse. The latter region comprises all the sites where afferent neurons receive information from other cells. Proteins in the presynaptic buttons, normally represented by axon terminals, play an important role in the vesiculation and release of neurotransmitters into the synaptic cleft (Südhof, 2004), while anchoring proteins coupled to the cell membrane support and stabilize the synapse (Sheng and Hoogenraad, 2007). It has been suggested that the proteins inside dendritic spines participate predominantly in sculpting their morphology, given their interactions with the actin cytoskeleton. Not necessarily indirectly, but certainly at a quite miniaturized level, these proteins control neuronal excitability by regulating the interconversion processes among the different types of spines identified, according to the nature of synaptic stimuli they receive.

There is evidence that interactions occur between some adhesion molecules and some actin cytoskeleton-associated proteins inside dendritic spines. An example of this is the association of N-catenin with β-catenin or δ-catenin, respectively. δ-catenin establishes communication directly or indirectly with the actin filaments inside dendritic spines. When this is indirect, the participation of α-catenin becomes crucial since it forms a fundamental link between synaptic adhesion molecules and actin polymers in the postsynaptic density. Both spinogenesis and dendritic spine stability are reported in relation to this (Calabrese et al., 2006). In contrast, these spines reduce motility when α-catenin is overexpressed because this seems to accelerate the process of spine maturation (Abe et al., 2004).

Since dendritic spines are motile structures, and it has been suggested that their movement facilitates their connectivity with axons (Yuste, 2010), exploring their interior and understanding the mechanisms that underlie their transformation and movement requires, first, "knocking on the door" of the membrane. The cell membrane of these spines has mainly ionotropic and metabotropic glutamatergic receptors. It is through the activation of these receptors, or their subunits, that the proteins inside are activated differentially, causing the start-up of signaling cascades that lead to the polymerization, depolymerization, or stability of actin filaments (F-actin). Thus, to comprehend how the latter processes take place, we now describe the pathways that generate protein phosphorylation-dephosphorylation in the postsynaptic density and, later, in deeper regions of dendritic spines.

The postsynaptic density (PSD) comprises the inner portion, located immediately below the cell membrane at the head of dendritic spines. This is a rich space that contains a wide variety of proteins functionally associated with the activation of glutamatergic ionotropic receptors, such as NMDA and AMPA, as well as membrane-bound metabotropic receptors (mGLUR) (Rao et al., 1998). It is important to mention that all the molecular interactions that occur between receptors and PSD proteins play essential roles in determining whether the signaling cascades lead to the assembly or depolymerization of actin filaments. Depending on the degree of phosphorylation and dephosphorylation of these proteins, the cytoskeleton will acquire a structure that molds dendritic spines into specific shapes. This process entails a certain degree of biochemical competition between certain proteins to establish "molecular clusters" that produce geometrical transformations of dendritic spines or, in some cases, may favor spinogenesis (Takahashi et al., 2003). This group of PSD proteins includes PSD-95, Homer, CaMKII, Shank, and GKAP, among others (Figure 2).

Beyond actin filament dynamics, it is clear that the proteins in the postsynaptic density also participate significantly in regulating synaptic plasticity. In fact, since the 1980s, CaMKII has been considered the major constituent of PSD (Kelly et al., 1984; Kennedy et al., 1983). Regulated by the protein spinophilin (Alen, 1997), phosphatase PP1 and dephosphorylate CaMKII provoke its disassembly from the PSD (Yoshimura et al., 1999). We also know that PP1 phosphatase and PKA kinase regulate the activity of the NMDA receptor channel (Westphal et al., 1999). Upon stimulation of NMDA receptors, several signaling cascades are triggered in the upper part of the PSD that generally lead to activation of the aforementioned PSD-95 protein. This protein belongs to a family of membrane-associated guanylate kinases, and has specific domains for bonding to certain adhesion molecules, including neuroligins, K+ channels, and subunits of Kainate receptors. In general, PSD-95 constitutes a key protein between NMDA receptors and cytoskeleton-related scaffolding proteins (Sheng and Sala, 2001; Yoshii and Constantine-Paton, 2007).

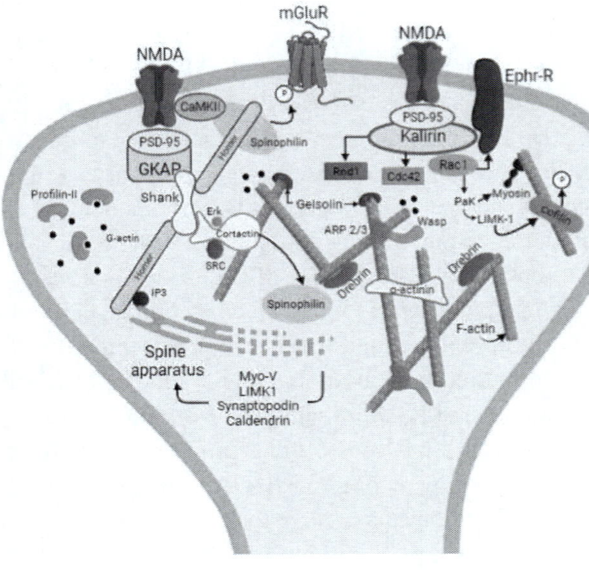

Figure 2. Diagram of actin cytoskeleton-associated proteins and related molecules orchestrating dendritic spine morphology. NMDA receptor activation triggers signaling pathways involving postsynaptic density proteins (CaMKII, PSD-95, GKAP, Kalirin, Shank, Homer, spinophilin, cortactin) to determine actin cytoskelton dynamics. Alternatively, ephrin (Ephr) and metabotropic glutamate (mGluR) receptors are regulated by spinophilin and Rac1, correspondingly. The formation of the spine apparatus involves the activity of proteins like Myosin-V, LIMK1, synaptopodyn and caldendrin. Profilin-II interacts with actin monomers (G-actin) whilst drebrin, ARP2/3, α-actinin, gelsolin, cofilin and myosin, configures the filamentous actin network (F-actin) to mold the cytoskeleton within dendritic spines.

The protein GKAP, meanwhile, binds through its PDZ domain to the GK domain of PSD-95, located upstream (Kim et al., 1997). Likewise, through its C-terminal region, GKAP establishes contact with the PDZ domain of the

Shank protein (Bellot et al., 2014; Naisbitt et al., 1999) in the lower part of the PSD (Valtschanoff and Weinberg, 2001). At that point, Shank binds to cortactin and, simultaneously, to the Homer protein (Naisbitt et al., 1999; Tu et al., 1999). Homer is associated primarily with activation of group 1 metabotropic glutamate receptors (mGLUR1) and IP3 receptors (IP3R) in the spinal apparatus through certain proline-rich regions (Tu et al.,1998). The Shank-Homer complex appears to stabilize the location of circumscribed glutamate receptors on the postsynaptic membrane (Tu et al., 1999), so, hypothetically, it is important in controlling neuronal excitability mediated by Ca++, though its direct relation to IP3R is not yet fully understood (Feng et al., 2002).

In addition to Homer, Shank binds to cortactin, a protein in the head of dendritic spines that becomes biologically active when phosphorylated by molecules like Src and Erk (Martin et al., 2006) (Figure 2). Cortactin has been related to the degree of polymerization of actin filaments during synaptogenesis (Pontrello and Ethell, 2009). One possible mechanism that may regulate branching of the actin cytoskeleton occurs through the activation of GTPases, where cortactin participates inside the pathway, as follows: cortactin → Rac1 → IRSp53 → WAVE (Ethell and Pascuale, 2005). Cortactin has also been shown to provide synaptic stability to mature spines, mainly through interactions with synaptic adhesion proteins like integrins, cadherins, and other cytoskeleton-associated proteins, including α-actinin, the Arp2/3 complex, and Dock4 (Durand et al., 2011; Helwani et al., 2004; Racz, 2004; Vuori and Rouslahti, 1995; Ueda et al., 2013; Weaver et al., 2001; Wyszynski et al., 1998). A study by our research group showed an increase of cortactin mRNA expression with overexpression of proteins like spinophilin and F-actin itself, in place cells of the hippocampus in a long-term spatial reference memory paradigm. These results correlated with a decreased density of dendritic spines, but a greater amount of thin spines. We hypothesized that overexpression of cortactin might support rapid synaptic transmission mechanisms in mature synapses, as could occur in some other proteins in the lower part of the PSD (González-Tapia et al., 2020). However, other studies show that cortactin knock-out reduces dendritic spine density, though those results were reported in vitro (Hering and Sheng, 2003). Apparently, these distinct results are due to a previously established mnemonic trace and the possible experience acquired during training. Clearly, additional research using distinct behavioral models or molecular markers needs to be conducted to clarify these issues.

After stimulation of NMDA receptors, along with the activation of GKAP, Shank, Homer, and cortactin, PSD-95 activity promotes phosphorylation of the kalirin-7 protein. This guanosine exchange factor (GEF) for proteins like Cdc42, Rnd1, and Rac1, is considered a key molecule in the regulation of signaling pathways associated with various phenomena, including the elongation of dendritic spines, synaptic adhesion efficacy, expression of AMPA receptors, phosphorylation of CaMKII derived from the stimulation of NMDA receptors, regulation of the activity of ephrine receptors (along with Rac1), the transition from filopodia to dendritic spines, increases in spine density in the cerebral cortex and even, under conditions of overexpression, spine sprouting in hippocampal interneurons that normally do not possess them, among other functions (Soria-Fregozo and Pérez-Vega, 2012; Penzes et al., 2001; Xie et al., 2007; Xie et al., 2008). As mentioned above, kalirin-7 activates Rac1 and coordinates Pak activation to control LIMK-I and myosin activity. Myosin-V participates in Ca++ dynamics inside dendritic spines and, along with such molecules as synaptopodin or caldendrin, fosters the formation of the spine apparatus (Konietzny et al., 2019; Konietzny et al., 2022) (Figure 2).

At this point, we are penetrating more deeply into the head of the spine. Here, the myosin-II isoform facilitates the cross-linking of actin filaments in the lower part of the spine head, close to the neck (Ryu et al., 2006; Zhang et al., 2005), though it also binds to proteins closer to the PSD, like drebrin, gelsolin, and actin filaments (Hayashi et al., 1996). Variations in the levels of myosin-II expression have been related to the number of mushroom-like and stubby spines. Other findings show that decreased myosin expression increases the number of filopodia (Ryu et al., 2006). According to our research in rats that performed a long-term memory task, along with low expression of some other cytoskeleton-associated proteins, decreased myosin expression correlated with a low number of mushroom spines but a high number of thin ones in CA1 pyramidal cells of the hippocampus (González-Tapia et al., 2020).

Like kalirin, drebrin is a protein that has been the focus of intense research due to the large number of plastic phenomena to which it has been linked and the proteins with which it is functionally related, including myosin, the Ras protein, profilin II, and gelsolin, (Biou et al., 2008; Mammoto et al., 1998), as will be discussed later. Recalling its association with transcription factor Egr-1, drebrin has been reported to promote spinogenesis, synaptogenesis, and dendritic spine elongation (Hayashi and Shirao, 1999; Takahashi et al., 2003; Takahashi et al., 2006). The latter could be related to drebrin's inhibitory activity on myosin under conditions of overexpression, which would reduce the contractility of actin filaments (Hayashi et al., 1996). Drebrin participates actively in remodeling the architecture, size, shape, and density of dendritic spines, at least in the hippocampus, as has been demonstrated in vitro (Ivanov et al., 2009). This protein allows the branching of actin filaments along with ARP2/3, a complex of proteins regulated by WASP and cortactin (Millard et al., 2004) (Figure 2). In addition, variations in the levels of drebrin expression are related to the location of NMDA receptors in dendritic spines. Activation of these receptors, in turn, determines the precise location of drebrin inside spines, through a rather complex process (Pontrelo and Ethell, 2009).

Immersed in the spine cytoplasm, actin monomers (G-actin) are kept available for assembly (F-actin) thanks to the activity of the profilin-II protein (Witke et al., 2001). However, there is conflicting evidence regarding whether profilin-II limits the availability of actin monomers, thus reducing spine elongation and spinogenesis (Perelroizen et al., 1994). In general, profilin-II is a nucleotide exchanger of ADP to ATP in regions that receive G-actin monomers for their subsequent polymerization (Pollard et al., 2000). This protein depends on Ca++ levels inside spines (Wolven et al., 2000) and gives the actin cytoskeleton stability during long-term potentiation (LTP) (Ackerman and Matus, 2003). More recent studies, however, suggest that profilin-II also impacts cytoskeletal dynamics during long-term depression (LTD) in a time activity-dependent manner (Corneilus et al., 2023). Additional evidence suggests that profilin-II relates not only to the actin network of spines but is also closely related to the cytoskeletal portion, composed of microtubules beyond the spines (Pinto-Costa et al., 2019). Along with profilin-II, gelsolin has been related to activation of the phosphatidyl-inositol pathway, which participates in the release of intracellular Ca++ through activation of metabotropic receptors. At the same time, it appears to promote actin polymerization due to activation of profilin (Yin and Janmey, 2003). This pathway also removes gelsolin from the barbed end of actin filaments, since it acts as a cap that prevents them from becoming longer and branched (Janmey and Lindberg, 2004).

The protein cofilin has been proposed to confer stability to the actin cytoskeleton of dendritic spines. When dephosphorylated –through a process that takes place during LTP– cofilin promotes structural and synaptic maturity (Adrianantoandro and Pollard, 2006; Chen et al., 2007; Frotscher et al., 2014; Pontrello and Ethell, 2009), though when overexpressed in hippocampal cultures it promotes the formation of immature filopodia and decreases spine density (Hotulainen et al., 2009). Some data show that in its inactive state cofilin promotes the stability of mature dendritic spines, but that when active it determines the formational arrangement of the actin cytoskeleton and favors spine elongation and the growth of filopodia (Shi et al., 2004).

Conclusion

With the initial descriptions by Cajal regarding neuron theory, it became clear that, beyond the collective activity of millions of individual cells that create functional systems inside the brain, understanding of the complexity and variety of animal and human behavior was still in a neonatal stage. Over the past century, exciting, even amazing, discoveries have enhanced our understanding of the physiology of the nervous system. Strikingly, the study of the molecular components related to dendritic spines and synaptic plasticity continues to raise interesting questions. This field has proven not only to enhance our curiosity but, thanks to the knowledge acquired, has shown that dendritic spines improve cerebral connectivity, possibly their most important role, in addition to functioning as a unique cellular unit of compartmentalization that, by default, entails a broad range of electrophysiological, biochemical, and even computational functions in the brain. However, more questions arise at every step along the way due to the unusual biological flexibility that is an inherent quality of dendritic spines.

The insights provided in this chapter seek to expound two major ideas: first, understanding the molecular mechanisms that underlie dendritic spine plasticity (whether intrinsic to neuron function or epigenetically guided) demands studies that match the degree of sophistication and fineness of the structures themselves. For this reason, all the knowledge acquired to date must be poured into experimental designs that will help answer our questions about the physiology of dendritic spines (e.g., omic sciences, computational models, biomechanics, pharmacological approaches, neural networks, and behavioral analysis, among others). Second, regarding the numerous proteins that shape the actin cytoskeleton of spines, the interactions among these molecules that drive the formation and elimination, size and shape, stability and motility of spines are so numerous that all paths of study seem to lead to dead ends. As key integrator inputs of synaptic transmission, dendritic spines are susceptible to adapting in accordance with the nature of the information that flows around specific neural networks yet, at the same time, individual processes take place per unit. This exquisite ability confers individual properties to these spines such that each one represents a computational element that modulates synaptic transmission and endows it with the efficiency it shows through brain structures and neural systems. Nature may be allowing us a quick glance at the molecular machinery of this elaborate process when we focus attention on the molecules that operate inside dendritic spines.

Several approaches need to be considered for future research on synaptic plasticity. Undoubtedly, the more minute the level of biological organization observed, the greater the challenge to develop novel technological tools and strategies and the need to define the course of modern neuroscience. As difficult as it is to understand the human mind, we have to take into consideration that as we delve more deeply into the molecular –even sub-atomic– realm of nature we will either come closer to, or draw farther away from, the goal of understanding human behavior. However, attempting to go deeper into the rabbit hole may well prove to be worth the risks involved.

References

Abe, K., Chisaka, O., Van Roy, F., Takeichi, M., 2004. Stability of dendritic spines and synaptic contacts is controlled by alpha N-catenin. *Nat. Neurosci.* 7, 357-363.

Ackermann, M., Matus, A., 2003. Activity-induced targeting of profilin and stabilization of dendritic spine morphology. *Nat. Neurosci.* 6, 1194– 1200.

Andrianantoandro, E., Pollard, T. D., 2006. Mechanism of actin filament turnover by severing and nucleation at different concentrations of ADF/cofilin. *Mol. Cell.* 24, 13-23.

Allen, P.B., Ouimet, C.C., Greengard, P., 1997. Spinophilin, a novel protein phosphatase 1 binding protein localized to dendritic spines. *Proc. Natl. Acad. Sci. USA* 94, 9956-9961.

Bellot, A., Guivernau, B., Tajes, M., Bosch-Morató, M., Valls-Comamala, V., Muñoz, F.J., 2014. The structure and function of actin cytoskeleton in mature glutamatergic dendritic spines. *Brain Res.* 1573, 1-16.

Biou, V., Brinkhaus, H., Malenka, R.C., Matus, A., 2008. Interactions between drebrin and Ras regulate dendritic spine plasticity. *Eur. J. Neurosci.* 27, 2847–2859.

Bramham, C.R., Alme, M.N., Bittins, M., Kuipers, S.D., Nair, R.R., Pai, B., Panja, D., Schubert, M., Soule, J., Tiron, A., Wibrand, K., 2010. The Arc of synaptic memory. *Exp. Brain Res.*, 200, 125–140.

Buchsbaum, R.J., Connolly, B.A., Feig, L.A., 2003. Regulation of p70 S6 kinase by complex formation between the Rac guanine nucleotide exchange factor (Rac-GEF) Tiam1 and the scaffold spinophilin. *J. Biol. Chem.*, 278, 18833–18841.

Calabrese, B., Wilson, M.S., Halpin S., 2006. Development and regulation of dendritic spine synapses. Physiology 21, 38-47.

Carvalho, A.L., Caldeira, M.V., Santos, S.D., Duarte, C.B., 2008. Role of the brain-derived neurotrophic factor at glutamatergic synapses. *Br. J. Pharmacol.,* 153 (Suppl 1), S310–S324.

Chahrour, M., Zoghbi, H. Y., 2007. The story of Rett syndrome: from clinic to neurobiology. *Neuron*, 56, 422–437.

Chapleau, C. A., Boggio, E. M., Calfa, G., Percy, A. K., Giustetto, M., Pozzo-Miller, L., 2012. Hippocampal CA1 pyramidal neurons of Mecp2 mutant mice show a dendritic spine phenotype only in the presymptomatic stage. *Neural Plast.* 2012, 976164.

Chapleau, C. A., Calfa, G. D., Lane, M. C., Albertson, A. J., Larimore, J. L., Kudo, S., Armstrong, D. L., Percy, A. K., Pozzo-Miller, L., 2009. Dendritic spine pathologies in hippocampal pyramidal neurons from Rett syndrome brain and after expression of Rett-associated MECP2 mutations. *Neurobiol. Dis.,* 35, 219–233.

Chen, L. Y., Rex, C. S., Casale, M. S., Gall, C. M., Lynch G., 2007. Changes in synaptic morphology accompany actin signaling during LTP. *J. Neurosci.* 27, 5363–5372.

Cornelius, J., Haak, S., Rothkegel, M., Korte, M., Michaelsen-Preusse, K., 2023. Phosphorylation of the actin-binding protein profilin2a at S137 modulates bidirectional structural plasticity at dendritic spines. *Front. Cell. Dev. Biol.* 11, 1107380.

Counts, S. E., He, B., Nadeem, M., Wuu, J., Scheff, S. W., Mufson, E. J., 2012. Hippocampal drebrin loss in mild cognitive impairment. *Neurodegener. Dis.* 10, 216–219.

Counts, S. E., Nadeem, M., Lad, S. P., Wuu, J., Mufson, E. J., 2006. Differential expression of synaptic proteins in the frontal and temporal cortex of elderly subjects with mild cognitive impairment. *J. Neuropathol. Exp. Neurol.,* 65, 592–601.

Day, J.J., Sweatt, J. D., 2012. Epigenetic treatments for cognitive impairments. *Neuropsychopharmacology* 37, 247–260.

Durand, C. M., Perroy, J., Loll, F., Perrais, D., Fagni, L., Bourgeron, T., Sans, N., 2011. SHANK3 mutations identified in autism lead to modification of dendritic spine morphology via an actin-dependent mechanism. *Mol. Psychiatry* 17, 71–84.

Edbauer, D., Neilson, J. R., Foster, K. A., Wang, C. F., Seeburg, D. P., Batterton, M. N., Tada, T., Dolan, B. M., Sharp, P. A., Sheng, M., 2010. Regulation of synaptic structure and function by FMRP-associated microRNAs miR-125b and miR-132. *Neuron* 65, 373–384.

Ethell, I. M., Pasquale, E. B., 2005. Molecular mechanisms in spine development and remodeling. *Prog. Neurobiol.* 75, 161-205.

Feng, W., Tu, J., Yang, T., Vernon, P.S., Allen, P.D., Worley, P.F., Pessah, I.N., 2002. Homer regulates gain of ryanodine receptor type 1 channel complex. *J. Biol. Chem.* 277, 722–744.

Frotscher, M., Studer, D., Graber, W., Chai, X., Nestel, S., Zhao, S., 2014. Fine structure of synapses on dendritic spines. *Front. Neuroanat.* 8, 94.

González-Tapia, D., González-Tapia, D.C., Vázquez-Hernández, N., Martínez-Torres, N.I., Flores-Soto, M., González-Burgos, I., 2020. Modifications to cytoskeleton-associated proteins in dendritic spines underlie the adaptive plasticity involved in long term reference memory. *Neurobiol. Learn. Mem.* 172, 107247.

González-Tapia, D., Velázquez-Zamora, D. A., González-Burgos, I., 2015. The molecular biology of dendritic spine plasticity in memory processing, in: *González-Burgos, I. (Ed.), Synaptic fundamentals of memory performance*. Nova Science Publishers, New York, pp. 59-79.

Hayashi, K., Ishikawa, R., Ye, L.H., He, X. L., Kohama, K., Shirao, T., 1996. Modulatory role of drebrin on the cytoskeleton within dendritic spines in the rat cerebral cortex. *J. Neurosci.* 16, 7161–7170.

Hayashi, K., Shirao, T., 1999. Change in the shape of dendritic spines caused by overexpression of drebrin in cultured cortical neurons. *J. Neurosci.* 19, 3918-3925.

Helwani, F.M., Kovacs, E.M., Paterson, A.D., Verma, S., Ali, R.G., Fanning, A.S., Yap, A.S., 2004. Cortactin is necessary for E-cadherin– mediated contact formation and actin reorganization. *J. Cell Biol.* 164, 899–910.

Hering, H., Sheng, M., 2003. Activity-dependent redistribution and essential role of cortactin in dendritic spine morphogenesis. *J. Neurosci.* 23, 11759-11769.

Hotulainen, P., Hoogenraad, C.C., 2010. Actin in dendritic spines: connecting dynamics to function. *J. Cell. Biol.* 189, 619-629.

Impey, S., Davare, M., Lesiak, A., Fortin, D., Ando, H., Varlamova, O., Obrietan, K., Soderling, T.R., Goodman, R. H., Wayman, G. A., 2010. An activity-induced microRNA controls dendritic spine formation by regulating Rac1-PAK signaling. *Mol. Cell. Neurosci.* 43, 146–156.

Ivanov, A., Esclapez, M., Ferhat, L., 2009. Role of drebrin A in dendritic spine plasticity and synaptic function. *Commun. Integr. Biol.* 2, 268-270.

Janmey, P.A., Lindberg, U., 2004. Cytoskeletal regulation: rich in lipids. *Nat. Rev. Mol. Cell. Biol.* 5, 658–666.

Ji, Y., Pang, P.T., Feng, L., Lu, B., 2005. Cyclic AMP controls BDNF-induced TrkB phosphorylation and dendritic spine formation in mature hippocampal neurons. *Nat. Neurosci.*, 8, 164–172.

Jung, G., Kim, E. J., Cicvaric, A., Sase, S., Gröger, M., Höger, H., Sialana, F.J., Berger, J., Monje, F.J., Lubec, G., 2015. Drebrin depletion alters neurotransmitter receptor levels in protein complexes, dendritic spine morphogenesis and memory-related synaptic plasticity in the mouse hippocampus. *J. Neurochem.* 134, 327–339.

Kelly, P. T., McGuinnes, T. L., Greengard, P., 1984. Evidence that the major postsynaptic density protein is a component of a Ca2+/calmodulin-dependent protein kinase. *Proc. Natl. Acad. Sci. USA* 81, 945-949.

Kennedy, M. B., Bennett, M. K., Erondu, N. E., 1983. Biomechanical and immunochemical evidence that the "major postsynaptic density protein" is a subunit of a calmodulin-dependent protein kinase. *Proc. Natl. Acad. Sci. USA* 80, 7357-7461.

Kim, E., Naisbit, S., Hsueh, Y. P., Rao, A., Rothschild, A., Craig, A. M., Sheng, M., 1997. GKAP, a novel synaptic protein that interacts with the guanylate kinase-like domain of the PSD-95/SAP90 family of channel clustering molecules. *J. Cell. Biol.* 136, 669-678.

Knapska, E., Kaczmarek, L., 2004. A gene for neuronal plasticity in the mammalian brain: Zif268/Egr-1/NGFI-A/Krox-24/TIS8/ZENK? Prog. *Neurobiol.* 74, 183–211.

Konietzny, A., González-Gallego, J., Bär, J., Perez-Alvarez, A., Drakew, A., Demmers, J. A. A., Dekkers, D. H. W., Hammer, J. A., Frotscher, M., Oertner, T. G., Wagner, W.,

Kneussel, M., Mikhaylova, M., 2019. Myosin V regulates synaptopodin clustering and localization in the dendrites of hippocampal neurons. *Cell Sci.* 132, 1–16.

Konietzny, A., Grendel, J., Kadek, A., Bucher, M., Han, Y., Hertrich, N., Dekkers, D.H.W. Demmers, J.A.A., Grünewald, K., Uetrecht, C., Mikhaylova, M., 2022. Caldendrin and myosin V regulate synaptic spine apparatus localization via ER stabilization in dendritic spines. *EMBO J.* 41, e106523.

Lacor, P. N., Buniel, M. C., Furlow, P. W., Clemente, A. S., Velasco, P. T., Wood, M., Viola, K. L., Klein, W. L., 2007. Abeta oligomer-induced aberrations in synapse composition, shape, and density provide a molecular basis for loss of connectivity in Alzheimer's disease. *J. Neurosci.* 27, 796–807.

Larimore, J. L., Chapleau, C. A., Kudo, S., Theibert, A., Percy, A.K., Pozzo-Miller, L., 2009. Bdnf overexpression in hippocampal neurons prevents dendritic atrophy caused by Rett-associated MECP2 mutations. *Neurobiol. Dis.*, 34, 199–211.

Levenson, J. M., Sweatt, J. D., 2005. Epigenetic mechanisms in memory formation. Nature reviews. *Neuroscience* 6, 108–118.

Li, W., Calfa, G., Larimore, J., Pozzo-Miller, L., 2012. Activity-dependent BDNF release and TRPC signaling is impaired in hippocampal neurons of Mecp2 mutant mice. *Proc. Natl. Acad. Sci. USA* 109, 17087–17092.

Luine, V., Frankfurt, M., 2013. Interactions between estradiol, BDNF and dendritic spines in promoting memory. *Neuroscience* 239, 34–45.

Ma, L., Li, Y., Wang, R., 2015. Drebrin and cognitive impairment. *Clin. Chim. Acta.*, 451(Pt B), 121–124.

MacDonald, J. L., Roskams, A. J., 2008. Histone deacetylases 1 and 2 are expressed at distinct stages of neuro-glial development. *Dev. Dynam.* 237, 2256–2267.

Malecová, B., Morris, K. V., 2010. Transcriptional gene silencing through epigenetic changes mediated by non-coding RNAs. *Curr. Opin. Mol. Ther.*, 12, 214–222.

Mammoto, A., Sasaki, T., Asakura, T., Hotta, I., Imamura, H., Takahashi, K., Matsuura, Y., Shirao, T. Takai, Y., 1998. Interactions of drebrin and gephyrin with profilin. Biochem. *Biophys. Res. Commun.* 243, 86–89.

Martin, K. H., Jeffery, E. D., Grigera, P. R., Shabanowitz, J., Hunt, D. F., Parsons, J. T., 2006. Cortactin phosphorylation sites mapped by mass spectrometry. *J. Cell. Sci.* 119, 2851-2853.

Maze, I., Covington, H. E., 3rd, Dietz, D. M., LaPlant, Q., Renthal, W., Russo, S. J., Mechanic, M., Mouzon, E., Neve, R. L., Haggarty, S. J., Ren, Y., Sampath, S.C., Hurd, Y. L., Greengard, P., Tarakhovsky, A., Schaefer, A., Nestler, E. J., 2010. Essential role of the histone methyltransferase G9a in cocaine-induced plasticity. *Science* 327, 213–216.

Migaud, M., Charlesworth, P., Dempster, M., Webster, L. C., Watabe, A. M., Makhinson, M., He, Y., Ramsay, M.F., Morris, R.G., Morrison, J. H., O'Dell, T. J., Grant, S. G., 1998. Enhanced long-term potentiation and impaired learning in mice with mutant postsynaptic density-95 protein. *Nature* 396, 433–439.

Millard, T. H., Sharp, S. J., Machesky, L. M., 2004. Signalling to actin assembly via the WASP (Wiskott-Aldrich syndrome protein)-family proteins and the Arp2/3 complex. *Biochem. J.* 380, 1-17.

Moonat, S., Sakharkar, A. J., Zhang, H., Pandey, S. C., 2011. The role of amygdaloid brain-derived neurotrophic factor, activity-regulated cytoskeleton-associated protein and dendritic spines in anxiety and alcoholism. *Addict. Biol.*, 16, 238–250.

Naisbitt, S., Kim, E., Tu, J.C , Xiao, B., Sala, C., Valtschanoff, J., WeinbergR. J., Worley, P.F., Sheng, M., 1999. Shank, a novel family of postsynaptic density proteins that binds to the NMDA receptor/PSD95/GKAP complex and cortactin. *Neuron* 3, 569-582.

Pandey, S.C., Zhang, H., Ugale, R., Prakash, A., Xu, T., Misra, K., 2008. Effector immediate-early gene arc in the amygdala plays a critical role in alcoholism. *J. Neurosci.* 28, 2589–2600.

Penzes, P., Johnson, R. C., Sattler, R., Zhang, X., Huganir, R.L., Kambampati, V., Mains, R.E. Eipper, B.A., 2001. The neural Rho-GEF Kalirin-7 interacts with PDZ domain-containing proteins and regulates dendritic morphogenesis. *Neuron* 29, 229-242.

Perelroizen, I., Marchand, J. B., Blanchoin, L., Didry, D., Carlier, M. F., 1994. Interaction of profilin with G-actin and poly(L-proline). *Biochemistry* 33, 8472–8478.

Pinto-Costa, R., Sousa, M. M., 2019. Profilin as a dual regulator of actin and microtubule dynamics. *Cytoskeleton* 77, 76-83.

Pollard, T. D., Blancjoin, L., Mullins, L. D., 2000. Molecular mechanisms controlling actin fillament dynamics in nonmuscle cells. *Annu. Rev. Biophys. Biomol. Struct.* 29, 545-576.

Pontrello, C. G., Ethell, I. M., 2009. Accelerators, brakes, and gears of actin dynamics in dendritic spines. *Open Neurosci. J.* 3, 67-86.

Puang, S. J., Elanggovan, B., Ching, T., Sng, J. C. G., 2020. MEF2C and HDAC5 regulate Egr1 and Arc genes to increase dendritic spine density and complexity in early enriched environment. *Neuronal Signal.* 4, NS20190147.

Racz, B., 2004. The subcellular organization of cortactin in hippocampus. *J. Neurosci.* 24, 10310–10317.

Rao, A., Kim, E., Sheng, M., Craig, A. M., 1998. Heterogeneity in the molecular composition of excitatory postsynapatic sites during development of hippocampal neurons in culture. *J. Neurosci.* 18, 1217-1229.

Ryu, J., Liu, L., Wong, T.P., Wu, D.C., Burette, A., Weinberg, R., Wang, Y.T., Sheng, M., 2006. A critical role for myosin IIb in dendritic spine morphology and synaptic function. *Neuron* 49, 175–182.

Saxena, A., Carninci, P., 2011. Long non-coding RNA modifies chromatin: epigenetic silencing by long non-coding RNAs. *BioEssays* 33, 830–839.

Schratt, G. M., Tuebing, F., Nigh, E. A., Kane, C. G., Sabatini, M. E., Kiebler, M., Greenberg, M. E., 2006. A brain-specific microRNA regulates dendritic spine development. *Nature* 439, 283–289.

Sekino, Y., Kojima, N., Shirao, T., 2007. Role of actin cytoskeleton in dendritic spine morphogenesis. *Neurochem. Int.*, 51, 92–104.

Sheng, M., Hoogenraad, C.C., 2007. The postsynaptic architecture of excitatory synapses: a more quantitative view. *Annu. Rev. Biochem.* 76, 823-847.

Sheng, M., Sala, C., 2001. PDZ domains and the organization of supramolecular complexes. *Annu. Rev. Neurosci.* 24, 1-29.

Shi, Y., Pontrello, C. G., De Fea, K. A., Reichardt, L. F., Ethell, I. M., 2004. Focal adhesion kinase acts downstream of EphB receptors to maintain 102 mature dendritic spines by regulating cofilin activity. *J. Neurosci.* 29, 8129– 8142.

Shim, K. S., Lubec, G., 2002. Drebrin, a dendritic spine protein, is manifold decreased in brains of patients with Alzheimers disease and Down syndrome. *Neurosci. Lett.* 324, 209–212. ‘

Soria-Fregozo, C., Pérez-Vega, M.I., 2012. Actin-binding proteins and signalling pathways associated with the formation and maintenance of dendritic spines. *Neurología*, 27, 421-431.

Südhof, T. C., 2004. The synaptic vesicle cycle. *Annu. Rev. Neurosci.* 27, 509-547.

Takahashi, H., Mizui, T., Shirao, T., 2006. Down-regulation of drebrin A expression suppresses synaptic targeting of NMDA receptors in developing hippocampal neurons. *J. Neurochem.,* 97(Suppl 1), 110–115.

Takahashi, H., Sekino, Y., Tanaka, S., Mizui, T., Kishi, S., Shirao, T., 2003. Drebrin-dependent actin clustering in dendritic filopodia governs synaptic targeting of postsynaptic density-95 and dendritic spine morphogenesis. *J. Neurosci.* 23, 6586-6595.

Tu, J. C., Xiao, B., Yuan, J. P., Lanahan, A. A., Leoffer, T. K., Li, M., Linden, D. J., Worley, P.F., 1998. Homer binds a novel proline-rich mot if and links group 1 metabotropic glutamate receptors with IP3 receptors. *Neuron* 21, 717–726.

Tu. J. C., Xiao, B., Naisbitt, S., Yuan, J. P. Petralia, R.S., Brakeman, P., Doan, A., Aakalu, V.K., Lanahan, A.A., Sheng, M., Worley. P.F., 1999. Coupling of mGluR/Homer and PSD-95 complexes by the Shank family of postsynaptic density proteins. *Neuron* 23, 583–592.

Ueda, S., Negishi, M., Katoh, H., 2013. Rac GEF Dock4 interacts with cortactin to regulate dendritic spine formation. *Mol. Biol. Cell.* 24, 1602–1613.

Valtschanoff, J.G., Weinberg, R.J., 2001. Laminar organization of the NMDA receptor complex within the postsynaptic density. *J. Neurosci.* 21, 1211–1217.

Vigers, A. J., Amin, D. S., Talley-Farnham, T., Gorski, J. A., Xu, B., Jones, K. R., 2012. Sustained expression of brain-derived neurotrophic factor is required for maintenance of dendritic spines and normal behavior. *Neuroscience* 212, 1–18.

Vuori, K., Ruoslahti, E., 1995. Tyrosine phosphorylation of p130Cas and cortactin accompanies integrin-mediated cell adhesion to extracellular matrix. *J. Biol. Chem.* 270, 22259–22262.

Waddington, C. H., 1952. The epigenetics of birds. *CUP Archive.*

Weaver, A. M., Karginov, A. V., Kinley, A. W., Weed, S. A., Li, Y., Parsons, J. T., Cooper, J.A., 2001. Cortactin promotes and stabilizes Arp2/3- induced actin filament network formation. *Curr. Biol.* 11, 370-374.

Westphal, R. S., Tavalin, S. J., Lin, J. W., Alto, N. M., Fraser, I. D., Langeberg, L. K., Sheng, M., Scott, J.D., 1999. Regulation of NMDA receptors by an associated phosphatase-kinase signaling complex. *Scince* 285, 93-96.

Witke, W., Sutherland, J. D., Sharpe, A., Arai, M., Kwiatkowsky, D. J., 2001. Profilin 1 is essential for cell survival and cell divisiön in early mouse development. *Proc. Natl. Acad. Sci. USA* 98, 3832-3836.

Wolven, A. K., Belmont, L. D., Mahoney, N. M., Almo, S. C., Drubin, D. G., 2000. In vivo importance of actin nucleotide exchange catalyzed by profilin. *J. Cell. Biol.* 150, 895-904.

Wyszynski, M., Kharazia, V., Shanghvi, R., Rao, A., Beggs, A. H., Craig, A. M., Weinberg, R., Sheng, M., 1998. Differential regional expression and ultrastructural localization of alpha-actinin-2, a putative NMDA receptoranchoring protein, in rat brain. *J. Neurosci.* 18, 1383–1392.

Xie, Z., Photowala, H., Cahill, M. E., Srivastava, D. P., Woolfrey, K. M., Shum, C. Y., Huganir, R. L., Penzes, P., 2008. Coordination of synaptic adhesion with dendritic spine remodeling by AF-6 and Kalirin-7. *J. Neurosci.* 28, 6079–6091.

Xie, Z., Srivastava, D. P., Photowala, H., Kai, L., Cahill, M. E., Woolfrey, K. M., Shum, C.Y., Surmeier, D.J., Penzes, P., 2007. Kalirin-7 controls activity-dependent structural and functional plasticity of dendritic spines. *Neuron* 56, 640–656.

Xu, X., Kozikowski, A.P., Pozzo-Miller, L., 2014. A selective histone deacetylase-6 inhibitor improves BDNF trafficking in hippocampal neurons from Mecp2 knockout mice: implications for Rett syndrome. *Front. Cell. Neurosci.* 8, 68.

Yamazaki, H., Kojima, N., Kato, K., Hirose, E., Iwasaki, T., Mizui, T., Takahashi, H., Hanamura, K., Roppongi R. T., Koibuchi, N., Sekino, Y., Mori, N., Shirao, T., 2014. Spikar, a novel drebrin-binding protein, regulates the formation and stabilization of dendritic spines. *J. Neurochem.* 128, 507–522.

Yin, H. L., Janmey, P. A., 2003. Phosphoinositide regulation of the actin cytoskeleton. *Annu. Rev. Physiol.* 65, 761–789.

Yoshii, A., Constantine-Paton, M., 2007. BDNF induces transport of PSD-95 to dendrites through PI3K-AKT signaling after NMDA receptor activation. *Nat. Neurosci.* 10, 702-711.

Yoshimura, Y., Sogawa, Y., Yamauchi, T., 1999. Protein phosphatase 1 is involved in the dissociation of Ca2+/calmodulin-dependent protein kinase II from the postsynaptic densities. *FEBS Lett.* 446, 239-242.

Yuste, R., 2010. Motitlity, in: Yuste, R., (Ed.), Dendritic Spines. The MIT Press, England, pp.103-131.

Zhang, H., Webb, D. J., Asmussen, H., Niu, S., Horwitz, A. F., 2005. A GIT1/PIX/Rac/PAK signaling module regulates spine morphogenesis and synapse formation through MLC. *J. Neurosci.* 25, 3379–3388.

Zhou, Z., Hong, E. J., Cohen, S., Zhao, W. N., Ho, H. Y., Schmidt, L., Chen, W.G., Lin, Y., Savner, E., Griffith, E.C., Hu, L., Steen, J.A., Weitz, C.J., Greenberg, M.E., 2006. Brain-specific phosphorylation of MeCP2 regulates activity-dependent Bdnf transcription, dendritic growth, and spine maturation. Neuron 52, 255–269.

Chapter 3

Electrical Properties of Dendritic Spines

Allan T. Gulledge*
Department of Molecular and Systems Biology Geisel School of Medicine at Dartmouth College, Hanover, New Hampshire, USA

Allan T. Gulledge*
Department of Molecular and Systems Biology Geisel School of Medicine at Dartmouth College. Hanover, New Hampshire, USA

Introduction

Dendritic spines are morphological specializations in the dendritic arbor that serve as sites of excitatory (generally glutamatergic) synaptic input. These exquisite and dynamic structures have inspired generations of neuroscientists to wonder whether their biophysical properties influence synaptic transmission and/or contribute to synaptic plasticity. Elucidating a physiological role of dendritic spines has been challenging because they are extremely small (≤ 1 μm) with variable morphology, making them difficult to measure accurately with light microscopy and impossible to access electrically with traditional experimental approaches. Thus, estimation of the biophysical properties of dendritic spines has proven technically challenging, but has recently benefitted from the development of voltage- and ion-sensitive imaging tools, the ability to "uncage" glutamate to activate individual dendritic spines, and the increasing feasibility of electrical recording near or at synaptic spines. Despite a growing literature, consensus on the electrical impact of dendritic spines has been difficult to achieve, in part due to differences in perspective regarding synaptic transmission. From a

In: Dendritic Spines
Editor: Ignacio González-Burgos
ISBN: 979-8-89113-080-7

somatocentric point of view (where the focus is on a synapse's ability to depolarize the axon toward action potential threshold), dendritic spines have negligible impact on individual synaptic responses and are not plausible contributors to synaptic plasticity. Yet from a synaptocentric point of view (where electrical signaling at the synapse is considered important), dendritic spines boost the amplitudes of EPSPs to facilitate recruitment of voltage-sensitive conductances and promote supralinear summation of synaptic input. These two views are complementary, rather than mutually exclusive, as becomes apparent when examining the interplay of spine structural plasticity (i.e., changes in spine morphology) and synaptic plasticity (i.e., changes in synaptic conductance).

This chapter reviews the intrinsic electrical properties of dendritic spines and the increasingly sophisticated approaches used to estimate the electrical coupling between spine and dendrite. Further discussed are the consequences of dendritic spines for synaptic integration and the possible role of spine structural plasticity in stabilizing synaptic integration in dendrites in the face of synapse-specific changes in synaptic efficacy.

The Electrical Properties of Dendritic Spines

Like the rest of the neuron, dendritic spines are bounded by plasma membrane that has the intrinsic electrical properties of capacitance and conductivity. Capacitance results from the hydrophobic fatty acids that form a thin non-conductive barrier separating the aqueous (and conductive) extracellular and cytosolic environments. This generates electrical capacitance that stores charge across the plasma membrane. At the same time, transmembrane proteins form a wide variety of ion channels (including the ionotropic glutamate receptors that typically gate excitatory synaptic transmission in the vertebrate brain) that provide "in parallel" electrical conductance across the plasma membrane. These two electrical properties allow for, and shape, electrical signals in all neurons.

Although there are important membrane specializations in dendritic spines (for review, see Valles and Barrantes, 2021), it is their three-dimensional structure, rather than any particular properties of their plasma membranes, that have the largest impact on synaptic transmission. Consider, for example, a classical mushroom-shaped spine (Figure 1A) that has an excitatory synaptic conductance (i.e., a cluster of AMPA-type glutamate receptors) on an enlarged spine "head" that is connected to its parent dendrite via a narrow spine "neck". When the synapse is activated, AMPA receptors provide the spine head membrane additional conductance (a "synaptic conductance") that in most cases will trigger an inward current (a "synaptic current") that generates an excitatory postsynaptic potential (EPSP)[1]. Once entering the dendritic spine, the synaptic current can follow three possible routes. The electrical properties of the plasma membrane allow for some of the current to be stored (transiently) in the membrane capacitance (C_M), while some current will exit through ion channels that provide electrical conductance across the dendritic spine's plasma membrane (conductance of membranes is often referred to by its reciprocal, membrane resistance, or R_M, with greater conductance corresponding to lower resistance). The remainder of the current will flow through the narrow spine neck and its associated "neck resistance" (R_{neck}) into the dendrite, where it will produce a dendritic EPSP.

In what proportions does the synaptic current distribute across these three alternatives? The key to answering this question is to recognize that the dendritic spine has an extremely small surface area (typically < 2 μm^2) that provides vanishingly small capacitance (~10 femtofarad; capable of storing less than 0.5% of a typical AMPA-mediated synaptic current) and conductance (in the low picosiemen range). Meanwhile, the spine neck has a conductance at least 1,000 times greater, in the nanosiemen range (i.e., $R_{neck} \leq$ 1 GΩ; see below and Table 1 for physiological estimates of R_{neck} and the equivalent conductance). Thus, even in spines having narrow, high resistance spine necks, > 99% of the synaptic current will flow through the spine neck and enter the dendritic shaft (Figure 1A, B; see also Kawato and Tsukahara, 1984; Wilson, 1984; Gulledge et al., 2012). The primacy of spine neck resistance, relative to membrane properties, in regulating EPSPs is revealed in simulations of synaptic transmission. As shown in Figure 1C, adjusting R_M or C_M neuron-wide over a 100-fold range has little impact on the amplitudes or kinetics of EPSPs measured in the spine head (due to the negligible surface area of the dendritic spine), but has large effects on EPSPs measured in dendrites and somata (where membrane surface areas are much larger).

---

[1] The magnitude and direction of synaptic current is determined by the "driving force" for the synapse, which is the difference between the membrane potential (V_M) and the equilibrium potential for the synaptic conductance (E_{syn}). In the case of AMPA-type glutamate receptors that are permeable to both sodium and potassium ions, E_{syn} is ~0 mV. Therefore, for a neuron with a resting membrane potential of -70 mV, the driving force will be -70 mV (i.e., -70 mV - 0 mV) and an inward current (as indicated by the negative value for driving force) will generate a depolarizing excitatory postsynaptic potential, or EPSP.

Table 1. Estimates of neck resistance in dendritic spines

Study	**Neuron type** (preparation; temperature)	**Species** (sex; age or weight)	**Approach**	**Reported spine-neck resistance** (equivalent conductance)
Computational estimates of spine neck resistance based on spine morphologies				
Harris & Stevens (1988, 1989)	CA1 pyramidal neurons (fixed tissue) and Cerebellar Purkinje neurons (fixed tissue)	Long-Evans rats (male; 137 - 310 grams)	Estimates based on measurements of spine morphologies (electron microscopy) in chemically fixed tissue, assuming a cytosolic resistivity of 100 Ω·cm	CA1: 0.9 - 411 MΩ (2 - 1108 nS) Purkinje neuron: 2.6 - 80 MΩ (12 - 383 nS)
Segev et al., (1995)	Barrel Cortex (fixed tissue)	CD/1 mouse (male; 2-month-old)	Estimates based on measurements of spine length (electron microscopy) in chemically fixed tissue, assuming a neck width of 100 nm and cytosolic resistivity of 210 Ω·cm	27 - 540 MΩ (1.9 - 37 nS)
Trommald & Hullenberg (1997)	Dentate granule cells (fixed tissue)	Wistar rats (male; 250 - 280 grams)	Estimates based on measurements of spine morphologies (electron microscopy) in chemically fixed tissue, assuming a cytosolic resistivity of 80 Ω·cm	0.01 - 45 MΩ (22 nS - 100 µS)

In: Dendritic Spines
Editor: Ignacio González-Burgos
ISBN: 979-8-89113-080-7

Study	**Neuron type** (preparation; temperature)	**Species** (sex; age or weight)	**Approach**	**Reported spine-neck resistance** (equivalent conductance)
da Costa (2013)	Visual cortical neurons (fixed tissue)	Cats (NA; 2 to 19 months old)	Estimates based on measurements of spine morphologies (electron microscopy) in chemically fixed tissue, assuming a cytosolic resistivity of 65 Ω·cm	1 - 204 MΩ (4.9 nS - 1 μS) median of 41 MΩ (2.4 nS)
Tønnesen et al., (2014)	CA1 pyramidal neurons (organotypic or acute hippocampal slices; 31.5 °C)	Mice (NA; organotypic - from 5-7 days old, in vitro for 2-4 weeks; acute slices - 4-5 weeks old)	Estimates based on measurements of spine morphologies with super-resolution (stimulated emission depletion) light microscopy and assuming a cytosolic resistivity of 150 Ω·cm	2.1 - 598 MΩ (1.7 - 476 nS) median of 44 MΩ (23 nS)
Eyal et al., (2018)	Layer 2/3 pyramidal neurons in temporal cortex (fixed tissue)	Human (males; 40 or 85 years old; from Benavides-Piccione et al., 2013)	Estimates based on measurements of spine morphologies (from confocal microscopy), assuming a range of cytosolic resistivity (from 100 - 300 Ω·cm)	19 - 128 MΩ (8 - 53 nS)
Tamada et al., (2020)	Unspecified neocortex (fixed or cryopreserved tissue)	Mice (male; 6-10 weeks old)	Estimates based on measurements of spine morphologies (from electron microscopy), assuming a cytosolic resistivity of 109 Ω·cm	Chemically fixed: 1.4 - 206 MΩ (4.8- 712 nS) median of 34 MΩ (30 nS) Cryopreserved: 2.2 - 595 MΩ (1.7- 446 nS) median of 80 MΩ (12.5 nS)

Study	**Neuron type** (preparation; temperature)	**Species** (sex; age or weight)	**Approach**	**Reported spine-neck resistance** (equivalent conductance)
Gemin et al., (2021)	Layer 2/3 neocortical pyramidal neurons (fixed tissue)	Swiss mice (NA; 78 - 129 days old)	Estimates based on measurements of spine morphologies (electron microscopy) in chemically fixed tissue, assuming a cytosolic resistivity of 300 Ω·cm	8 MΩ - 1.45 GΩ (0.7 - 125 nS) median of 188 MΩ (5.3 nS)

Table 1. (Continued)

Study	**Neuron type** (preparation; temperature)	**Species** (sex; age or weight)	**Approach**	**Reported spine-neck resistance** (equivalent conductance)
Ofer et al., (2022)	Layer 3 pyramidal neurons from cingulate cortex (fixed tissue)	Human (males; 40 or 85 years old; from Benavides-Piccione et al., 2013)	Estimates based on measurements of spine morphologies (from confocal microscopy) relative to values reported in mouse cortex (from Cornejo et al., 2022)	~180 MΩ (~5.6 nS)
Estimates of spine neck resistance based on molecular diffusion across spine necks				
Svoboda et al., (1996)	CA1 pyramidal neurons (acute slices; 22-24 °C)	Rats (NA; 2-3 weeks old)	Estimates based on diffusion of fluorescent molecules between spine head and dendrite (fluorescence recovery after photobleaching) using 2-photon microscopy and assuming a cytosolic resistivity of 250 Ω·cm	4-50 MΩ, estimated upper bound of 150 MΩ (20 - 250 nS, lower bound of 6.7 nS)

Study	Neuron type (preparation; temperature)	Species (sex; age or weight)	Approach	Reported spine-neck resistance (equivalent conductance)
Tønnesen et al., (2014)	CA1 pyramidal neurons (organotypic or acute slices; 31.5 °C)	Mice (NA; organotypic - from 5-7 days old, in vitro for 2-4 weeks; acute slices - 4-5 weeks old)	Estimates based on diffusion of fluorescent molecules between spine head and dendrite (fluorescence recovery after photobleaching) and assuming a cytosolic resistivity of 150 Ω·cm	YFP: 9 - 807 MΩ (1.2 - 109 nS) median of 62 MΩ (16 nS) Alexa: 5 MΩ -1.3 GΩ (0.8 - 476 nS) median of 57 MΩ (18 nS)
Acker et al., (2016)	Prefrontal layer 5 pyramidal neurons (acute slices; 22 °C)	CD1 mice (female and male; 17-30 days old)	Estimates based on diffusion (fluorescence recovery after photobleaching) using 2-photon microscopy and assuming a cytosolic resistivity of 150 Ω·cm	52 - 521 MΩ (2 - 19 nS) median of 161.1 MΩ (6.2 nS)
Miyazaki & Ross (2017)	CA1 pyramidal neurons (acute slices; 24 °C)	Sprague Dawley rats (female and male; 2 - 4 weeks old)	Estimates based on the diffusion of sodium from spine heads in response to electrically evoked EPSPs, assuming a cytosolic resistivity of 180 Ω·cm	250 MΩ (4 nS)
Estimates of spine neck resistance based on calcium- or voltage-imaging				
Palmer & Stuart (2009)	Somatosensory layer 5 pyramidal neurons (acute slices; ~22 °C)	Wistar rats (NA; 3-5 weeks of age)	Focal electrical stimulation with voltage imaging of spine and dendrite	No precise range; upper bound of ~500 MΩ (~2 nS)
Harnett et al., (2012)	CA1 pyramidal neurons (acute slices; 34-37 °C)	Rats (male; 7-12 weeks old)	Use calcium transients (from voltage-gated calcium channels) as indicator of spine-head depolarization in response to	~370 - ~830 MΩ (~1.2 - 2.7 nS) mean of 514 MΩ

Study	**Neuron type** (preparation; temperature)	**Species** (sex; age or weight)	**Approach**	**Reported spine-neck resistance** (equivalent conductance)
			glutamate uncaging or local dendritic depolarization.	
Popovic et al., (2015)	Somatosensory layer 5 pyramidal neurons (acute slices; temperature not specified)	Swiss Webster mice (female and male; 18-30 days old)	Use focal or uncaged glutamate combined with voltage imaging of spine and dendrite	1 - 165 MΩ (uncaging) 10 - 70 MΩ (iontophoresis) (6 - 1000 nS)

Table 1. (Continued)

Study	**Neuron type** (preparation; temperature)	**Species** (sex; age or weight)	**Approach**	**Reported spine-neck resistance** (equivalent conductance)
Acker et al., (2016)	Prefrontal layer 5 pyramidal neurons (acute slices; 22 °C)	CD1 mice (female and male; 17-30 days old)	Use glutamate uncaging and voltage imaging of spine head EPSPs together with whole-cell recording of somatic EPSPs. Computational simulations adjusted to fit the data used to estimate spine neck resistances.	23 - 420 MΩ (2 - 43 nS) median of 210.2 MΩ (4.8 nS)
Kwon et al., (2017)	Hippocampal neurons (cell cultures; 21 – 23 °C)	Mice (NA; harvested at P0, imaged after 12-16 days *in vitro*)	Use glutamate uncaging combined with voltage imaging of spine and dendrite	13 - 297 MΩ (3 -77 nS) mean of 95 ± 85 MΩ

Study	**Neuron type** (preparation; temperature)	**Species** (sex; age or weight)	**Approach**	**Reported spine-neck resistance** (equivalent conductance)
Beaulieu-Laroche & Harnett (2018)	Somatosensory layer 5 pyramidal neurons (acute slices; 33 – 37 °C)	Sprague-Dawley rats (male; 4-15 weeks old)	Use calcium transients (from voltage-gated calcium channels) as indicator of spine-head depolarization in response to glutamate uncaging or local dendritic depolarization.	~315 MΩ - ~1.1GΩ (~0.9 - 3.2 nS) mean of 588 MΩ
Cornejo et al., (2022)	Somatosensory layer 2/3 pyramidal neurons (*in vivo*)	Mice (female and male; 4-8 weeks old)	Use *in vivo* 2-photo imaging of an endogenously encoded voltage indicator in response to spontaneous synaptic or optogenetically triggered depolarization of spine heads	0 - 531 MΩ (1.9 nS - μS range) median of 213.7 MΩ (4.7 nS)
Estimates of spine neck resistance based on electrical recordings from spine heads				
Jayant et al., (2017)	Hippocampal neurons (organotypic cultures; 37 °C) or neocortical pyramidal neurons (acute slices; temperature not stated)	Mice (female and male; cultures: 2-4 weeks *in vitro*; acute slices: 7-20 days old)	Use nanoelectrode recordings of spine head (~27 mV) and somatic (~0.5 mV) EPSPs and estimates of dendritic input resistance	250 - 536 MΩ (1.9 - 4 nS) (as disclosed in Kwon et al., 2017)

List of studies estimating neck resistances of dendritic spines from experimental data. Estimated neck resistances are shown together with their equivalent conductances for comparison with typical synaptic conductances (~1 nS) and the much smaller conductance of the spine head plasma membrane (~1 pS).

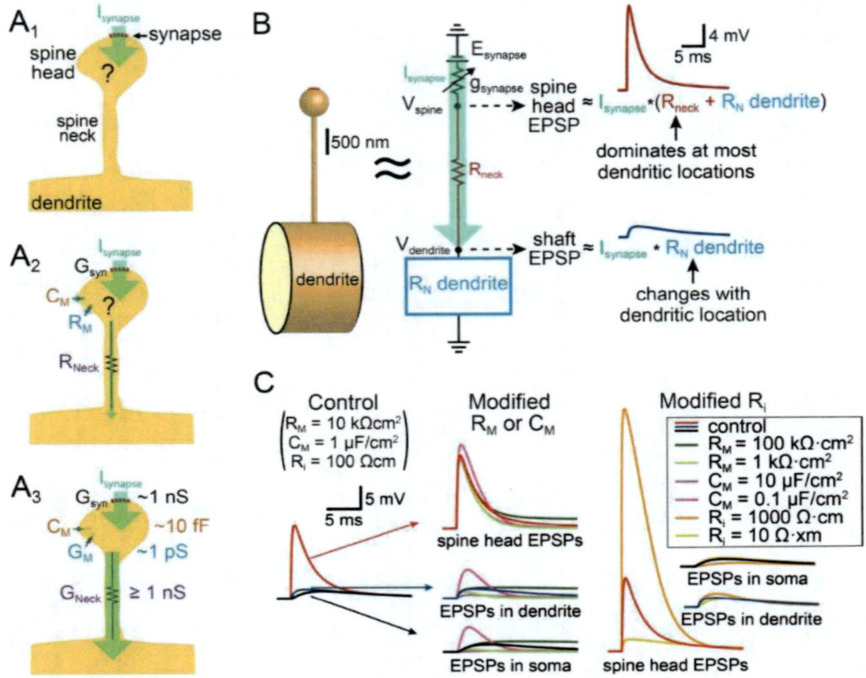

Figure 1. Electrical properties of dendritic spines. ***A_1-A_3***) Cartoon of a dendritic spine with an excitatory synapse placed on a spine head that is connected to the dendrite by a narrower spine neck. Upon synaptic activation of the synaptic conductance (G_{syn}), synaptic current ($I_{synapse}$) enters the spine head (***A_1***), where there exist three avenues available for current flow (***A_2***). Some current will charge the capacitance (C_M) of the plasma membrane that defines the spine head and neck, while some current may escape the spine head and neck across the resistance of the plasma membrane (R_M), and the remaining current will flow across the electrical resistance of the spine neck (R_{neck}) into the dendrite. Given that the total surface area of a spine (head and neck) is on the order of 1 μm^2, and remembering that conductance (G) is the reciprocal of resistance (R), it becomes apparent that the first two routes for current flow are infinitesimal relative to the very large conductance of the spine neck, which typically is also larger than the synaptic conductance (*A_3*; see also Table 1). Thus, almost all synaptic current entering the spine will reach the dendrite, even when spine neck resistances are considered large (i.e, approaching 1 GΩ [1 nS]). *B*) At left is a diagram of the spine structure used in Figures 2 - 4. Assuming a cytoplasmic resistivity of 100 Ω·cm, this spine neck has a resistance of 255 MΩ [~3.9 nS conductance]. To right is a simplified electrical equivalent circuit

In: Dendritic Spines
Editor: Ignacio González-Burgos
ISBN: 979-8-89113-080-7

for the spine, with resistance and capacitance of the plasma membrane omitted due to their inconsequential magnitudes. The synaptic voltage response in the spine head is proportional to the sum of the "in series" neck resistance and input resistance (R_N) of the parent dendrite. Meanwhile because almost all synaptic current passes into the dendrite, the voltage response in the dendrite is determined by the product of synaptic current and the input resistance of the dendrite. Spine neck resistance thus leads to excitatory postsynaptic potentials (EPSPs) being larger in the spine head relative to the parent dendrite. *C*) Traces at left show control EPSPs generated by a 1 nS AMPA-like conductance on the spine shown in *B* placed half-way down a 250 µm dendrite (see Figure 2A for model details) recorded in the spine head (red), the dendritic shaft directly below the spine (blue), and in the soma (black) in a model neuron in which membrane specific resistivity (R_M), capacitance (C_M), and cytosolic resistivity (R_i) are varied across a 100-fold range. Adjusting R_M or C_M (middle traces, color coded as indicated) has very little impact on spine-head EPSPs, but greatly affects dendritic and somatic EPSPs. On the contrary, varying R_i across a similar 100-fold range (which directly scales spine neck resistance) has a large effect on EPSPs in the spine head, but only modest impact on dendritic and somatic EPSPs. (Panel *B* adapted from Gulledge et al., 2012, under the terms of the Creative Commons Attribution License.)

On the other hand, neuron-wide manipulation of cytosolic resistivity (R_i), which directly influences spine neck resistance, has enormous effects on EPSPs in the spine head, but more moderate impact on EPSPs measured in dendrites or somata.

To understand how dendritic spines influence excitatory synaptic transmission, one needs to consider how EPSPs are generated and the factors that determine their amplitudes. When synaptic current enters a neuronal compartment (e.g., a spine head, the dendrite, soma, or axon), the change in membrane potential (V_M) during the EPSP is determined by Ohm's law (voltage = current • resistance), where voltage in this case is the change in membrane potential (i.e., the EPSP amplitude), the current is the synaptic current, and the resistance is the input resistance (R_N) for the neuronal compartment in question (Figure 1B) [1]. Because the R_N of a neuronal compartment is dependent on its three-dimensional structure, including its connectivity to other neuronal compartments, the amplitudes and kinetics of EPSPs are variable and highly dependent on where in the neuron they occur, even when they are generated by identical synaptic currents. This is reflected in the great diversity of EPSP amplitudes that are observed in the dendrites of

---

[1] More precisely, the EPSP amplitude and kinetics are a function of the frequency-dependent input impedance (Z_N) of a given neuronal compartment. Z_N incorporates all membrane properties and connectivity to other compartments. For simplicity, in this chapter R_N is used broadly.

large neurons (for reviews, see Gulledge et al., 2005; Spruston, 2008; see also Figure 3), and holds true at the level of the dendritic spine, where the R_N for the spine head reflects the "in series" (and therefore additive) electrical resistance of the spine neck (R_{neck}) and the R_N of the parent dendrite (R_{dend}). Thus, although the spine head and dendrite experience an almost identical synaptic current (as described above), EPSPs generated in the spine head will be proportionally larger than those measured in the parent dendrite according to the relative contribution that R_{neck} makes toward the total spine head resistance (i.e., R_{neck} + R_{dend}, Figure 1B). Only when the spine neck resistance is zero (i.e., in the absence of a spine) will EPSP amplitudes be identical in the two compartments.

By regulating the amplitude of the EPSP in the spine head, R_{neck} also influences synaptic current. This is because the driving force for the synaptic current (described in Footnote 1, above) changes during the EPSP as the difference between V_M and the synaptic equilibrium potential, E_{syn}, becomes smaller (i.e., driving force is reduced as the membrane depolarizes toward E_{syn} during the EPSP). Figure 2A shows a model neuron of modest proportions and typical membrane and cytosolic properties (see Figure 2 legend for additional model details). One of four morphologically distinct synapses, including three occurring on dendritic spines and one on the dendritic shaft (Figure 2B), is placed half-way down one dendrite and an AMPA-type synaptic conductance (1 nS; Beaulieu-Laroche and Harnett, 2018) is activated to generate synaptic currents (Figure 2C) and EPSPs that are measurable in the spine head, dendrite, and soma (Figure 2D). All three dendritic spines have similar spine heads (500 nm diameter) and spine neck lengths (2 μm), but differ in spine neck structure; one spine has a uniform spine neck diameter (100 nm), while the other two have tapering necks (from 50 - 150 nm) that either narrow or widen toward the spine head. Thus, all three spine necks have identical surface areas, membrane properties, and average spine neck diameters. As shown in Figure 2D, synapses on dendritic spines generate larger local (i.e., "at the synapse") EPSPs than does the identical synaptic conductance placed on the dendritic shaft. Further, spine neck resistance is disproportionately determined by the narrowest part of the spine neck, as dendritic spines with tapering spine necks (but conserved average diameters and surface areas) generate larger local EPSPs than equivalent uniform-diameter spine necks. This has implications for some experimental methodologies that estimate R_{neck} based on approximating spine necks to cylinders (see "Estimates of spine neck resistance," below).

Because synaptic current is determined by driving force, and driving force is reduced to a greater extent during large EPSPs, synapses on spines will generate smaller synaptic currents than those occurring at shaft synapses, even when synaptic conductance is the same (Figure 2C). The overall impact of spine neck resistance on synaptic currents and EPSP amplitudes is shown in Figure 2E; higher values of R_{neck} generate larger local EPSPs despite the resulting decrease in synaptic current. At the same time, EPSPs measured in the dendrite or soma are only modestly affected by spine necks (Figure 2D). Whereas spine head EPSP amplitudes (in this case, 15.3 - 18.1 mV, depending on spine neck geometry) are amplified up to eight-fold relative to the EPSP generated by an identical synapse placed directly on the dendritic shaft (2.3 mV), EPSPs measured in the dendrite (1.9 - 2.3 mV) or soma (1.5 - 1.8 mV) vary by less than 23%, regardless of which spine morphology, if any, is employed. The small decrements in dendritic and somatic EPSPs associated with dendrite spines are due to the small ($\leq$ 24% in this example) decreases in AMPA current associated with dendritic spines, rather than to any direct limitation of current flow through the spine neck (Figure 2F).

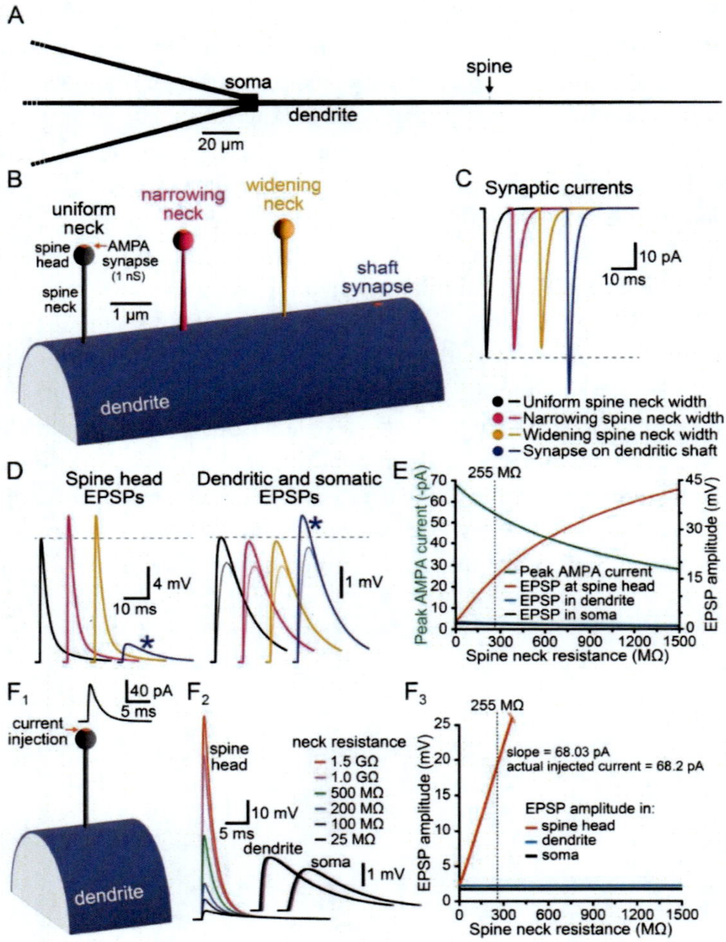

Figure 2. Impact of dendritic spines on synaptic currents and postsynaptic potentials. **A**) Diagram of model neuron having a 10-µm diameter soma and four dendrites (250 um in length, tapering from 3 to 1 µm, only one dendrite shown at full length) with membrane resistivity and capacitance set to 10,000 Ω·cm^2 (equilibrating at -70 mV) and 1 µF/cm^2, respectively, and a cytosolic resistivity of 100 Ω·cm. A dendritic spine is shown halfway along one dendrite. B) Diagram of four synapses tested in the model, including three spine morphologies, each having a 0.5 µm diameter spine head and a 2-µm-long spine neck. One spine neck is of uniform diameter (100 nm diameter; R_{neck} = 255 MΩ [3.9 nS]), while the other two taper (either toward or away from the dendrite) from 150 nm to 50 nm, preserving the surface area and average spine neck diameter. Each spine head contains an AMPA-type synapse with a peak conductance of 1 nS (0.2 ms rise, 2 ms decay). A fourth AMPA-type synapse is placed directly onto the dendritic shaft. C) Synaptic currents generated by each of the

four synapses shown in B (color coded). Note that synaptic current is sensitive to the shape of the spine neck. D) Excitatory postsynaptic potentials (EPSPs; color coded according to B) recorded in the spine head (left), the dendritic shaft directly below the dendritic spine (middle), and in the soma (right); note the different y-axis scales between spine head and dendrite/soma. Asterisks identify the EPSPs generated by the shaft synapse recorded in the spine head and dendritic shaft. While plotted at different scales, they are indistinguishable at the same scale. E) Plot of AMPA currents (left axis) and EPSP amplitudes in the three compartments (right axis) across spine neck resistance for a synapse on a uniform dendritic spine placed half-way down the dendrite (as depicted in A). Dashed line indicates the spine neck resistance of 255 MΩ (3.9 nS). Note that as spine neck resistance increases, the EPSP amplitude in the spine head increases in amplitude while synaptic current is suppressed due to reduced driving force during the depolarizing EPSP. F) Simulations using current injection to the spine head (F_1), rather than a synaptic conductance, demonstrate that the spine neck shapes EPSPs in the spine head only (F_2), and that virtually all the synaptic current is delivered to the dendrite via the spine neck, regardless of R_{neck} (F_2). Traces are color-coded to R_{neck} as indicated and dendritic and somatic EPSPs offset slightly in time to make them visible, as they are otherwise indistinguishable at this scale (note also the different scale for spine-head EPSPs). F_3 shows EPSP amplitudes in response to current injection to the spine head for a range of spine neck resistances. The slope of the IV-I relationship for spine-head EPSPs approximates the synaptic current.

Finally, it is worth noting that dendritic spines, as a population, influence the electrotonic structure of spiny neurons. Even as the surface area of an individual dendritic spine is minute and inconsequential with regard to current flow at the synapse, the high density of dendritic spines in spiny neurons nearly doubles their dendritic surface area (Bannister and Larkman, 1995; Parajuli et al., 2020), increasing total membrane conductance and capacitance and lowering R_{dend} (Holmes, 1989). This population effect of dendritic spines will increase the electrotonic length of dendrites and influence synaptic integration.

Estimates of Spine Neck Resistance

As described above, the electrical impact of dendritic spines depends almost entirely on spine neck resistance. Therefore, to understand how dendritic spines influence synaptic transmission it is necessary to determine the physiological range of spine neck resistances. Although direct measurement of R_{neck} has not been feasible due to the small size of dendritic spines, a number of experimental approaches have been used to estimate neck resistances (Table 1). As discussed below, each of these approaches has advantages and limitations that need to be considered when interpreting results. When evaluated in aggregate, however, these studies point to a range of spine neck resistances that has important consequences for synaptic integration.

Estimates of R_{neck} Based on Spine Morphologies

Given that spine neck resistance is primarily a function of spine neck morphology, R_{neck} can be estimated based on measurements of spine neck dimensions. This is typically done by approximating spine necks as cylinders and estimating R_{neck} based on neck length and cross-sectional area according to the following equation:

$$R_{neck} = \frac{R_i \cdot neck_{length}}{\pi \cdot (neck_{radius})^2}$$

where R_i is the specific intracellular (cytosolic) resistivity in Ω·cm, while necklength and neckradius are the length and radius of the spine neck, respectively, which can be measured using electron microscopy (EM; Harris and Stevens, 1988, 1989; Segev et al., 1995; Trommald and Hulleberg, 1997; da Costa, 2013; Tamada et al., 2020; Gemin et al., 2021) or, with less precision, super-resolution (Tønnesen et al., 2014) or confocal (Eyal et al., 2018; Ofer et al., 2022) microscopy. While EM measurements allow for the greatest precision of measurement, light microcopy approaches allow imaging in live neurons, including longitudinal imaging over time, and avoid structural changes that may occur during chemical fixation (see below).

As listed in Table 1, estimates of spine neck resistance based on spine morphology span a wide range of values, typically ranging from the very low MΩ range to many hundreds of MΩ, with one study (Gemin et al., 2021) reporting spine neck resistances above 1 GΩ. This range of neck resistances can be attributed primarily to the diversity of measured spine morphologies. However, the accuracy of estimates is limited by assumptions for the value of R_i and/or limitations of experimental methodologies. Assumed values for R_i, an attribute that is proportional to neck resistance, have a particularly large impact on estimates of spine neck resistance. Consequently, studies assuming low values for R_i (< 100 Ω·cm; Trommald and Hulleberg, 1997; da Costa, 2013) tend to report lower neck resistances than do studies assuming higher cytosolic resistivity (> 100 Ω·cm; Segev et al., 1995; Tønnesen et al., 2014; Eyal et al., 2018; Tamada et al., 2020; Gemin et al., 2021). Unfortunately, there is no consensus regarding which values for R_i are most appropriate for mammalian neurons at physiological temperatures, or even whether R_i is uniform among and within neurons. Experimentally derived estimates range widely from $< 50\ \Omega\cdot cm$ to several hundred $\Omega\cdot cm$ (Major et al., 1994; Trevelyan and Jack, 2002; Gouwens et al., 2018; Fletcher and Williams, 2019). Thus, estimates of spine neck resistance based purely on morphology (or molecular diffusion; see below) are limited by uncertainty regarding appropriate R_i values.

A further complication for estimating spine neck resistance based on morphology is the assumption that the spine neck is an aqueous cylinder. Spine necks tend to have tortuous morphologies and spine neck resistance is disproportionately impacted by the point in the neck with the smallest cross-sectional area (see Figure 2C - D). Spine necks also may contain membrane bound organelles (e.g., the spine apparatus) that are expected to reduce conductance (i.e., increase resistance) across the spine neck, but which are typically unaccounted for in morphological estimates of R_{neck}. Finally, the process of chemical fixation can induce structural changes that impact estimates of spine neck resistance. Tamada et al., (2020) used electron microscopy to compare spine neck morphologies following chemical or cryo fixation. They report that chemical fixation increases the width of spine necks relative to cryo fixation, leading to lower estimates of spine neck resistance (by about 50%). On the other hand, measurements of spine necks via light microscopy tend to be exaggerated due to limitations in optical resolution, also leading to lower estimates of R_{neck}. Thus, even though R_{neck} is directly determined by dendritic spine structure, the accuracy of estimates of R_{neck} based on morphology is limited by a number of methodological considerations and uncertainty regarding the physiological value of R_i.

Estimates of R_{neck} Based on Molecular Diffusion

An alternative approach for estimating spine neck resistances in living neurons was pioneered by Svoboda et al., (1996). Using intracellular aqueous fluorophores, the diffusion rate across the spine neck can be determined by photobleaching spine heads and measuring the time course of their fluorescence recovery (i.e., fluorescence recovery after photobleaching, or FRAP). The diffusional time constant for the fluorophore can then be used to estimate R_{neck} by the following equation:

$$R_{neck} = \frac{R_i \cdot \tau_{eq} \cdot D_{fluoro}}{V_{head}}$$

where R_i is the specific cytoplasmic resistivity, τ_{eq} is the time constant for equilibration of the fluorophore across the spine neck, D_{fluoro} is the diffusion constant for the fluorophore, and V_{head} is the volume of the spine head (Svoboda et al., 1996). As listed in Table 1, a number of studies have used this approach to estimate spine neck resistance in hippocampal and neocortical pyramidal neurons, finding values spanning a wide range (from the low MΩ to many hundreds of MΩ) that is generally consistent with estimates of R_{neck} based purely on spine neck morphologies (see above). However, there are important limitations to estimates of R_{neck} based on molecular diffusion. First, the use of light microscopy tends to underestimate R_{neck} by biasing sampling toward larger, more identifiable dendritic spines and exaggerating estimates of V_{head}. Second, the process of photobleaching may damage spine structure and impact diffusion characteristics. Finally, as is true for estimates based purely on neck morphologies, calculations of R_{neck} based on molecular diffusion suffer from broadly accepted, but poorly validated, values of R_i. These limitations may contribute to the finding that molecular diffusion time constants are poorly correlated with morphology-based estimates of neck resistance when neck resistance is above ~80 MΩ (Tønnesen et al., 2014).

Estimates of R_{neck} Based on Calcium- or Voltage-Imaging

As an alternative to estimating spine neck resistance based on morphology and/or molecular diffusion, several studies have used optical approaches to measure voltage responses in spines and their parent dendrites following transient depolarization of the spine head (Table 1). This depolarization can be achieved with "minimal" electrical stimulation close to the dendritic spine (Palmer and Stuart, 2009), exogenous glutamate at the spine head via either 2-photon uncaging or iontophoresis (Harnett et al., 2012; Popovic et al., 2015; Acker et al., 2016; Kwon et al., 2017), activation of light-triggered ion channels expressed in the spine (Cornejo et al., 2022), or spontaneous excitatory synaptic transmission (Jayant et al., 2017; Cornejo et al., 2022). Any difference in the voltage response measured in the spine head relative to the dendritic shaft can be attributed to the spine neck resistance, which can be calculated according to the following equation if the dendritic input resistance is known:

$$R_{neck} = (\frac{\Delta V_{spine}}{\Delta V_{dend}} - 1) \cdot R_{dend}$$

where ΔV_{spine} and ΔV_{dend} are the amplitudes of the voltage changes (i.e., EPSP amplitudes) in the spine head and dendrite, respectively, and R_{dend} is the input resistance of the dendrite, which depends heavily on dendritic location but is directly measurable using whole-cell recordings in larger dendrites. Harnett et al., (2012) accomplished this for dendritic spines along the main apical dendrites of CA1 pyramidal neurons, combining dendritic recordings with calcium imaging in the spine head to indirectly measure voltage responses following glutamate uncaging. In these experiments, uncaging produced calcium transients in spine heads that were attributable solely to activation of voltage-gated calcium channels. Because there is no appreciable voltage attention from dendrite to spine head (Nuriya et al., 2006; Palmer and Stuart, 2009; Holthoff et al., 2010; Kwon et al., 2017; Cornejo et al., 2022; and see above), Harnett et al., were able to use current injection in the dendrite to determine the level of depolarization necessary to reproduce the glutamate-evoked calcium signal in the spine head (i.e., to determine the value for ΔV_{spine} generated by glutamate uncaging). The dendritic electrode also recorded dendritic responses to glutamate uncaging (ΔV_{dend}) and directly measured R_{dend}, allowing calculation of spine neck resistances that spanned from ~380 to ~830 MΩ (mean of 514 MΩ). Using a similar approach, Beaulieu-Laroche and Harnett (2018) estimated spine neck resistances ranging from ~315 MΩ to slightly over 1 GΩ (mean of 588 MΩ) in neocortical layer 5 pyramidal neurons.

While the studies described above measured EPSP amplitudes in spine heads indirectly via calcium imaging, others have used voltage-sensitive fluorophores to compare voltage responses in spine heads and dendrites in brain slices (Palmer and Stuart, 2009; Popovic et al., 2015; Acker et al., 2016; Kwon et al., 2017) or in vivo (Cornejo et al., 2022). Most of these studies report larger voltage responses to synaptic activation in spine-heads, suggesting high R_{neck} values (but see Popovic et al., 2015). However, because these studies lack direct measures of dendritic input resistance, their estimates of spine neck resistances (typically in the range of 10 - 500 MΩ) rely on fitting data to computational models or assumed values for R_{dend}. Additional disadvantages of voltage indicators are their suboptimal signal-to-noise ratios at subthreshold voltages, their relatively slow kinetics, and the low temporal resolution (typically < 1 kHz) of imaging in most applications. These limitations will tend to bias calculations of spine neck resistance toward lower values because they may underestimate peak amplitudes of the faster EPSPs expected in high-resistance spine heads. Still, the ranges of spine neck resistances reported, as well as the visualization of isolated depolarization of spine heads during spontaneous synaptic transmission in vivo (Cornejo et al., 2022), suggest that spine neck resistances are generally sufficient to boost EPSP amplitudes in the spine head relative to EPSPs measured in the dendritic shaft.

Estimates of R_{neck} Based on Electrical Recordings from Dendritic Spines

The most direct method for measuring spine neck resistance would be to establish simultaneous electrical recordings of spine heads and their parent dendrites and measure the voltage drop to a given current injection to the spine head. While spine heads are too small for conventional electrical recording, one study developed quantum-dot-coated "nanopipettes" (tip diameters ≤ 30 nm) that allowed for visual targeting of spine heads (Jayant et al., 2017). While the extreme access resistances associated with nanopipettes complicates quantification of voltage signals, Jayant et al., were able to access dendritic spines and record local EPSPs that were quite large (~25 mV or more) relative to somatic responses (< 1 mV). Combining those data with assumptions of dendritic input resistance, Jayant et al., estimated spine neck resistances in the range of hundreds of megaohms (Table 1).

A Range of Physiological Spine Neck Resistances

The studies described above (and listed in Table 1) represent the best experimental attempts at quantifying spine neck resistance. Although limitations in approach, including the need to assume values for specific parameters such as R_i and/or R_{dend}, introduce unsatisfying ambiguity to estimations, precise determinations of R_{neck} may not be particularly useful given the intrinsic diversity and structural plasticity (see below) of spine morphologies. Based on the estimates that do exist (which exhibit considerable overlap in values between 50 and 500 MΩ), the best that can be said with confidence is that many spines have neck resistances sufficient to boost local EPSP amplitudes at the spine head, but which are not high enough to significantly impact the amplitudes of individual EPSPs measured in the dendrite or soma. As described below, by amplifying EPSPs selectively at the site of synaptic input, dendritic spines can influence synaptic integration in spiny neurons.

The Physiological Impact of Dendritic Spines

Based on their negligible impact on somatic EPSPs, it has been argued that dendritic spines are not electrically isolated from the dendrite (e.g., Koch and Poggio, 1983; Turner and Schwartzkroin, 1983; Kawato and Tsukahara, 1984; Koch and Zador, 1993; Svoboda et al., 1996; Popovic et al., 2014). Indeed, not only are somatic EPSPs not affected by the presence of dendritic spines, but dendritic voltage signals, including action potentials, propagate fully into dendritic spines without any voltage attenuation (e.g., Nuriya et al., 2006; Palmer and Stuart, 2009; Holthoff et al., 2010; Jayant et al., 2017; Kwon et al., 2017; Cornejo et al., 2022). Thus, from a somatocentric point of view, dendritic spines may seem of little consequence for synaptic transmission. On the other hand, even moderate spine neck resistances can boost the amplitude of postsynaptic EPSPs in the spine head, suggesting electrical compartmentalization from the dendrite (e.g., Araya et al., 2006; Grunditz et al., 2008; Bloodgood et al., 2009; Harnett et al., 2012). Consistent with this hypothesis is the recent demonstration that synaptically activated dendritic spines cannot be effectively voltage-clamped even under the most ideal experimental conditions with the clamping electrode placed at the base of the dendritic spine (Beaulieu-Laroche and Harnett, 2018). Consequently, from a synaptocentric point of view, dendritic spines appear to be electrically compartmentalized.

Although these two perspectives regarding the electrical isolation of dendritic spines appear to be in conflict, they are not mutually exclusive (Segev et al., 1995). Instead, as discussed below, they reflect complementary attributes of dendritic spines that may provide computational advantages to spiny neurons.

Dendritic Spines Limit Location-Dependent Variability in EPSPs

As described above, EPSPs are shaped by their local environment (i.e., by the R_N of the neuronal compartment in which they occur) and therefore identical synaptic currents will generate EPSPs of variable amplitudes and kinetics, depending on the location of the synapse in the dendritic tree (Figure 3). This effect is especially pronounced in larger neurons, where local R_{dend} varies greatly as one moves from large perisomatic dendrites to the narrow tips of distal dendritic branches that may be many hundreds of micrometers from the soma (for reviews, see Gulledge et al., 2005; Spruston, 2008). Thus, a synaptic conductance (e.g., a 1 nS AMPA-like conductance) placed on the dendritic shaft will generate larger and faster EPSPs at more distal locations along the dendrite (Figure 3A, B). One consequence of dendritic spines is that they provide a more uniform local geometry for excitatory synaptic input, thereby limiting the location-dependent variability of EPSPs measured at the synapse (i.e., in the spine head), even as they have little impact on dendritic or somatic EPSP amplitudes (Gulledge et al., 2012). In Figure 3C are plots of the peak AMPA-mediated synaptic currents and EPSP amplitudes for synapses placed at different locations along the relatively short dendrite (250 um) used in our model neuron. Synapses occurring on dendritic spines (R_{neck} of 255 MΩ) show less variability (as measured by the coefficient of variation, or CV) in synaptic current and EPSP amplitudes measured at the synapse, relative to those generated by synaptic inputs occurring on the dendritic shaft. These effects are magnified in larger neurons such as cortical pyramidal neurons and Purkinje cells (Gulledge et al., 2012), and, as discussed below, may provide several advantages to spiny neurons, including decreasing the need for location-specific synaptic mechanisms.

Dendritic Spines Promote Supralinear Synaptic Summation

While dendritic spines have a negligible effect on individual EPSPs measured at the soma, they are expected to more consistently engage voltage-sensitive mechanisms at the synapse by enabling larger and less variable EPSPs in spine heads. The most common of these mechanisms are NMDA-type glutamate receptors (which, under physiological conditions, require both glutamate release and postsynaptic depolarization to open) and voltage-gated calcium channels (e.g., Siegel et al., 1994; Conti et al., 1999; Parajuli et al., 2012). Because EPSPs in dendritic spines are larger, faster, and less variable across dendritic location than are EPSPs generated by shaft synapses, they more readily and uniformly engage these active synaptic properties (Grunditz et al., 2008; Gulledge et al., 2012; Li and Gulledge, 2021). An example of this is shown in Figure 4, which compares voltage responses to co-activated synaptic inputs occurring on dendritic spines or directly on the dendritic shaft. In a simple "ball and stick" model neuron with synapses containing both AMPA and NMDA receptors, a burst of distributed synaptic input (40 Hz) generates supralinear summation due to voltage-dependent activation of NMDA receptors (Figure 4A). This effect is enhanced in proportion to R_{neck} when inputs occur on dendritic spines, which can facilitate the generation of NMDA "spikes" (Gulledge et al., 2012; Harnett et al., 2012). A similar effect of dendritic spines is shown in a reconstructed layer 5 pyramidal neuron (Figure 4B), where spine-dependent amplification of NMDA currents was sufficient to enhance action potential output in response to spatiotemporal patterns of synaptic input (Figure 4C, see also Gulledge et al., 2012, and Li and Gulledge, 2021). The larger and more uniform EPSPs occurring in dendritic spines promote more uniform activation of NMDA-type glutamate receptors and other active conductances located at the synapse, regardless of where the synapse is placed in the dendritic tree.

It has been proposed in computational studies that dendritic spines may facilitate synaptic initiation of sodium-based action potentials in electrically isolated spine heads (i.e., "active spine heads"; Miller et al., 1985; Shepherd et al., 1985; Segev and Rall, 1988; Tsay and Yuste, 2002), and one experimental study has reported that EPSPs in pyramidal neuron spine heads are amplified by local voltage-gated sodium channels (Araya et al., 2007). However, several other groups have questioned the idea that sodium channels boost EPSPs at the level of the dendritic spine (Bloodgood and Sabatini, 2007; Grunditz et al., 2008; Palmer and Stuart, 2009; Miyazaki and Ross, 2017). Similarly, the limited ultrastructural data available are inconsistent, with one study finding sodium channels localized to dendritic spines (Johnson et al., 2017) and another suggesting a much lower level of expression in spines (Lorincz and Nusser, 2010). If voltage-gated sodium channels are expressed in spine heads, they would be expected to facilitate supralinear summation in a cooperative manner much as observed with NMDA-type glutamate receptors (Segev and Rall, 1988).

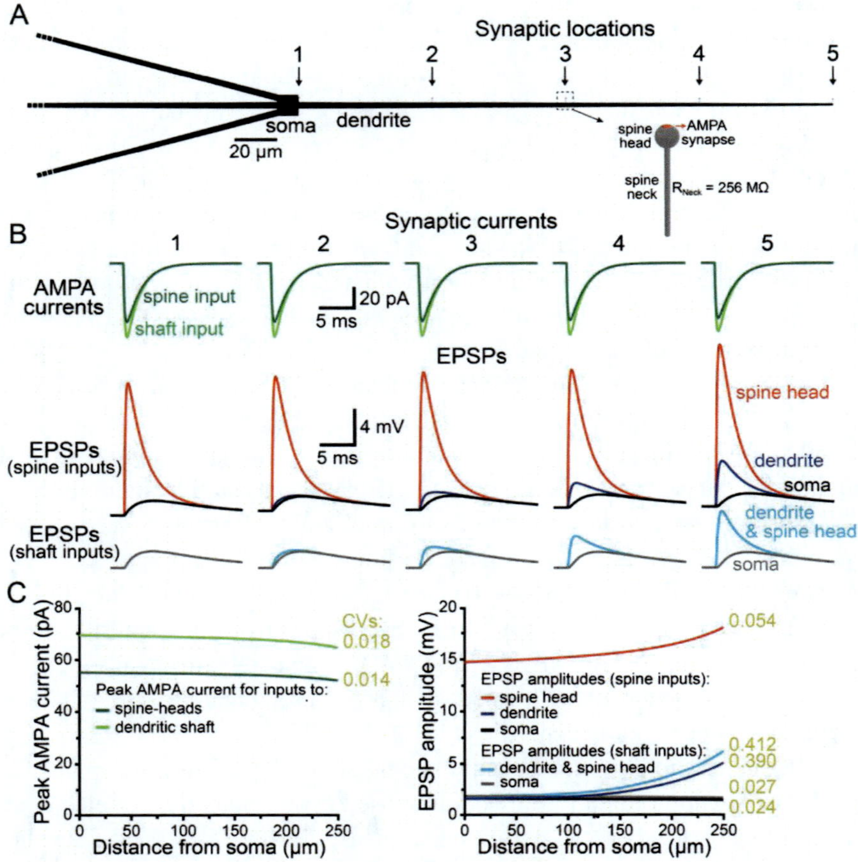

Figure 3. Dendritic spines reduce the location-dependent variability of local (i.e., at the synapse) EPSPs and AMPA current amplitudes, while having little impact on dendritic and somatic EPSPs. A) Diagram of neuron from Figure 2A showing locations of dendritic spine placements (1 through 5). B) Synaptic currents (top traces) and EPSPs (lower traces) recorded in the indicated dendritic compartments for AMPA synapses (1 nS) placed on the spine head or on the dendritic shaft, as indicated. C) Plots of peak AMPA currents (left) and EPSP amplitudes (right) for

AMPA-like synaptic inputs to spine heads or the dendritic shaft, as indicated, across all dendritic locations (1 μm increments). The coefficient of variation (CV; a measure of relative variability) for AMPA currents and EPSP amplitudes across dendritic location are shown in gold. Note that EPSPs generated and measured in spine-heads are much less variable across dendritic location than are dendritic EPSPs.

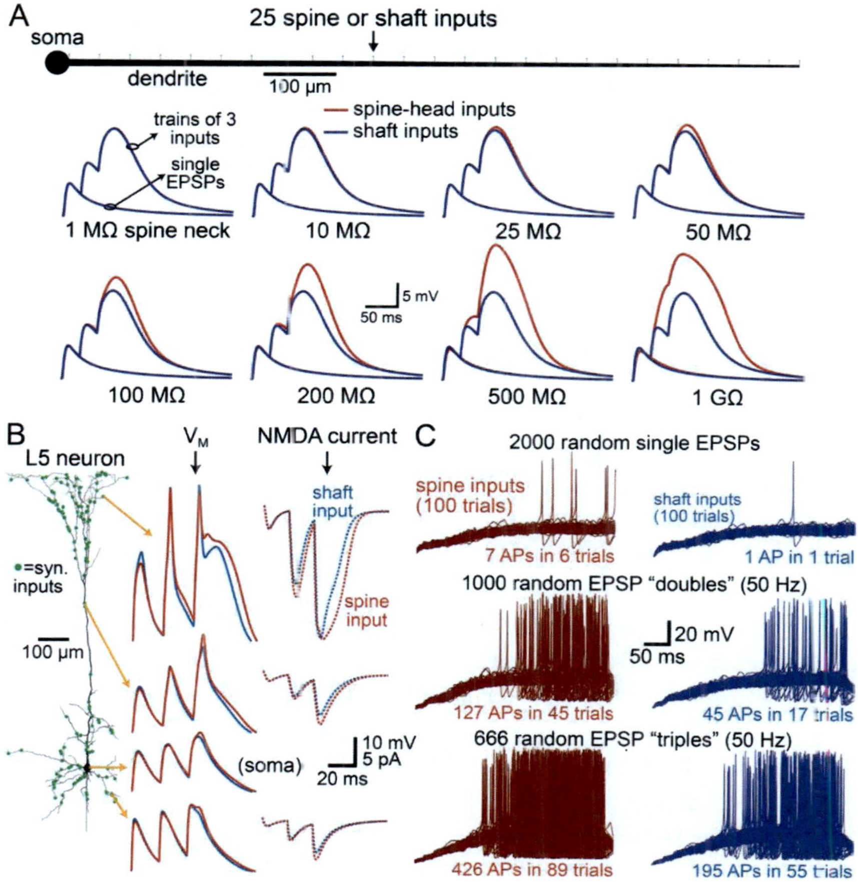

Figure 4. Dendritic spines facilitate NMDA conductances to promote excitability. A) Diagram of a ball-and-stick model neuron with 25 synapses dispersed evenly along a 1 mm long dendrite. Resting membrane potential is -60 mV. Below are somatic responses to single EPSPs or trains of three EPSPs (40 Hz) generated by synaptic inputs (combined AMPA and NMDA conductances) located on spines (red) or on the dendritic shaft (blue) for models with the indicated spine neck resistances. Dendritic spines enhance the NMDA-conductance, increasing synaptic current. B) Left, model layer 5 neuron with synaptic inputs (to spines or dendritic shafts) located at 2-μm

intervals along dendrites. Green dots indicate the locations of 130 randomly selected synaptic inputs. Solid-line traces show voltage responses to trains of 3 AMPA+NMDA EPSPs (50 Hz) delivered simultaneously to selected inputs on spines (200 MΩ; red) or shafts (blue), and recorded at the indicated somatic and dendritic locations. NMDA currents for the indicate synapses are shown with dotted lines. C) In the layer 5 neuron shown in B, randomly selected synapses were activated at random timings over a 400 ms period. Shown are 100 such responses superimposed for inputs to dendritic spines (left; red) or the corresponding shaft locations (right; blue). Each pair of trials (spines and shafts) involved the same input locations and timings. Top traces show responses to 2000 inputs activated once within the 400 ms window (random timings). Middle traces show responses to 1000 inputs activated twice each with 20 ms intervals (total of 2000 synaptic activations). Bottom traces are responses to 666 inputs activated three times with 20 ms intervals (total of 1998 synaptic activations). Inputs onto dendritic spines enhance the likelihood of action potential (AP) generation. (Figure and legend adapted from Gulledge et al., 2012, under the terms of the Creative Commons Attribution License.)

Coordinated Plasticity of Synapses and Spines

The idea that spine necks might serve to restrict electrical flow from synapses to dendrites (as first proposed by Chang, 1952) led to the hypothesis that plasticity of spine morphology "might be used physiologically to change the relative weights of synaptic inputs [and] provide a basic mechanism for learning in the nervous system" (Rall, 1970). Indeed, Bliss and Lømo (1973) considered spine neck plasticity a plausible mechanism for the inducible long-lasting potentiation of synaptic efficacy (now commonly known as long-term potentiation, or LTP) that they discovered in the hippocampus, and Ito (1984) similarly proposed that spine neck plasticity might explain the long-term depression (LTD) of synaptic transmission that he observed in Purkinje cells. In support of the hypothesis that LTP involves structural plasticity of dendritic spines, a series of electron microscopy studies reported that LTP-induction protocols similar to those used by Bliss and Lømo resulted in "swollen" postsynaptic dendritic spines having shorter and wider (and presumably lower resistance) spine necks (Van Harreveld and Fifková, 1975; Fifková and Van Harreveld, 1977; Fifková and Anderson, 1981) and proposed that such changes may be mediated by alteration of the actin cytoskeleton (Fifková and Delay, 1982). Although additional studies reported learning- or activity-dependent plasticity of spine necks in a variety of species and brain areas (e.g., Coss and Globus, 1978; Bradley and Horn, 1979; Coss et al., 1980; Brandon and Coss, 1982; Chang and Greenough, 1984), a growing body of computational studies demonstrated spine neck plasticity to be an implausible mechanism for modulating synaptic efficacy (Kawato and Tsukahara, 1984; Turner, 1984; Harris and Stevens, 1988, 1989; Koch and Zador, 1993; Segev et al., 1995; and see Figure 5A, B), and interest in testing the relationship between spine structural plasticity and synaptic plasticity temporarily waned.

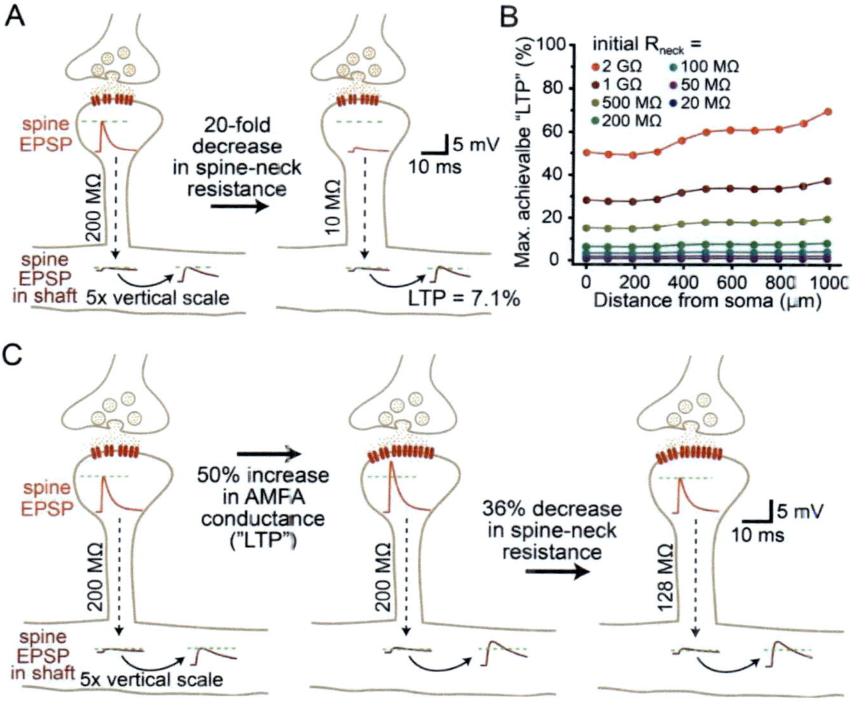

Figure 5. Spine neck plasticity may serve to regulate local (at the synapse) EPSPs, but is not effective in regulating synaptic efficacy measured at the soma. A) Diagram of a spine (R_{neck} = 200 MΩ) located half-way along a 1 mm dendrite in a ball-and-stick model neuron (as in Figure 4A). The spine experiences a reduction in spine neck resistance in the absence of any change in synaptic conductance. In the baseline condition (left), the AMPA conductance (500 pS) generates an EPSP of 7.73 mV in the spine head, and an EPSP of 0.72 mV in the dendritic shaft. Following a 20-fold decrease in spine neck resistance (from 200 to 10 MΩ; right), the same synaptic conductance generates a 1.14 mV EPSP at the spine head, and a 0.77 mV EPSP in the dendritic shaft: a boost of synaptic efficacy of only 7.1%. B) Plots of the maximum possible synaptic potentiation (e.g., LTP) achievable by reducing spine neck resistance to zero vs distance from the soma for a variety of initial spine neck resistances. C) Cartoons depicting a spine experiencing long-term potentiation (LTP). Under baseline conditions (left), a 500 pS AMPA conductance generates a 7.7 mV EPSP in the spine head and a 0.7 mV EPSP in the dendritic shaft. Following

LTP induction (middle), the AMPA conductance is increased to 750 pS (+50% LTP), and the EPSPs in the spine head and dendritic shaft increase to 11.1 mV and 1.04 mV, respectively. If the spine neck resistance is subsequently reduced by 36% (to 128 MΩ; right) the EPSP amplitude in the spine head returns to 7.7 mV, while the shaft EPSP increases marginally to 1.08 mV.

Following recent advances in fluorescence microscopy (including the development of 2-photon glutamate uncaging; Matsuzaki et al., 2001) that allow longitudinal observation of morphological changes in dendritic spines, there has been renewed interest in testing whether spine structural plasticity contributes to LTP and LTD. Although fluorescence microscopy typically lacks the spatial resolution necessary to precisely measure changes in spine neck dimensions (i.e., changes in R_{neck}; but see Tønnesen et al., below), there is now broad consensus that synaptic plasticity is associated with bidirectional structural changes in dendritic spines. Activity patterns that drive LTP (e.g., high frequency electrical stimulation, synaptic activation in the absence of extracellular magnesium, or pairing glutamate-uncaging with postsynaptic action potentials) reliably increase the size of postsynaptic spine heads in a manner that depends on NMDA receptor activation and reorganization of the actin cytoskeleton (Lang et al., 2004; Matsuzaki et al., 2004; Okamoto et al., 2004; Kopec et al., 2006; Harvey and Svoboda, 2007; Honkura et al., 2008; Steiner et al., 2008; Tanaka et al., 2008; Yang et al., 2008; Zhang et al., 2008; Lee et al., 2009; Govindarajan et al., 2011; Bosch et al., 2014; Meyer et al., 2014; Tønnesen et al., 2014). On the other hand, protocols that induce long-term depression (LTD; e.g., prolonged 1-Hz stimulation) lead to NMDA-receptor-dependent spine shrinkage and/or retraction (Nagerl et al., 2004; Okamoto et al., 2004; Zhou et al., 2004; Thomazeau et al., 2021). Although the data describing bidirectional structural plasticity are robust, the limitations of light microscopy and the small dimensions of spine necks make it difficult to interpret how morphological changes might influence the electrical coupling of synapses (i.e., R_{neck}). The exception is a study that used super-resolution (stimulated emission depletion) microscopy with an optical resolution (~50 nm) sufficient to measure spine neck dimensions with some accuracy before and after induction of LTP (Tønnesen et al., 2014). They reported that, in addition to increasing spine head volume, synapse-specific LTP-induction led to wider and shorter spine necks that they estimated would reduce R_{neck} values by ~50%, on average. Further, they found a strong correlation between plasticity-induced increases in spine head volume (which itself is proportional to the magnitude of synaptic potentiation; Matsuzaki et al., 2004; Harvey and Svoboda, 2007) and estimated reductions in R_{neck}.

Consistent with the hypothesis that R_{neck} is reduced during LTP, Araya et al., (2014) reported a correlation between spine neck length and synaptic efficacy, albeit without the resolution necessary to monitor changes in spine neck width.

If modification of R_{neck} is not a plausible mechanism to change synaptic efficacy (see above and Figure 5A, B), are there other electrical advantages of the structural plasticity associated with LTP and LTD? One hypothesis is that plasticity-associated regulation of R_{neck} allows the synapse to maintain a stable postsynaptic EPSP amplitude in the spine head despite changes in synaptic conductance and synaptic efficacy at the soma (Gulledge et al., 2012; Beaulieu-Laroche and Harnette, 2018; see also Figure 5C). Such coordinated structural and synaptic plasticity would maintain consistent postsynaptic activation of voltage-gated processes (e.g., NMDA receptors or voltage-gated ion channels) in the spine head while still allowing synapse-specific modulation of synaptic efficacy (i.e., LTP or LTD).

Similarly, dendritic spines may provide metabolic efficiency for spiny neurons by allowing them to express identical voltage-sensitive ion channels (including NMDA receptors and calcium channels) at all dendritic locations and achieve similar levels of synaptic activation despite distance-dependent changes in R_{dend} and/or activity-dependent changes in synaptic efficacy. While this attractive hypothesis remains to be explicitly tested, a recent electron microscopy study in CA1 pyramidal neurons reported that dendritic spine necks tend to be longer and narrower (and therefore presumably of higher resistance) in large-caliber proximal dendrites relative to spine necks in thinner, more distal dendrites (Parajuli et al., 2020). If dendritic spines do tune R_{neck} to exert homeostatic control of EPSP amplitudes in the spine head, it will be important to determine the signaling pathways and structural mechanisms that trigger and regulate that plasticity.

Conclusion

Despite being first-described well over a century ago (Ramón y Cajal, 1888), dendritic spines continue to mystify and inspire researchers seeking to identify the physiological advantages they provide spiny neurons. Biophysically, dendritic spines are relatively simple structures whose most salient feature is the electrical resistance provided by their spine necks. Increasingly innovative experimental approaches have estimated a range of spine neck resistances, with the bulk of evidence pointing toward moderate (~50 - 500 MΩ) resistances that are sufficient to boost postsynaptic EPSP amplitudes in the spine head, but not high enough to impact dendritic and somatic EPSPs. Spine-dependent boosting of EPSPs at the synapse enhances supralinear summation of synaptic input by facilitating activation of NMDA-type glutamate receptors and voltage-gated calcium channels, processes thought critical for normal synaptic integration in many neurons. Dendritic spines also show structural plasticity that is correlated with long-term changes in synaptic efficacy (e.g., LTP and LTD). Such structural plasticity may allow neurons to tune spine neck resistance with respect to synaptic conductance and/or dendritic location in order to regulate EPSP amplitudes in spine heads and promote more consistent activation of voltage-sensitive conductances. While many mysteries remain with regard to dendritic spines, continuing innovations in microscopy and electrophysiology promise ever-more-precise measurements and manipulations that will refine our understanding of their physiological properties and functional roles in synaptic integration.

References

Acker, C. D., Hoyos, E., Loew, L. M., 2016. EPSPs measured in proximal dendritic spines of cortical pyramidal neurons. *eNeuro 3*.

Araya, R., Jiang, J., Eisenthal, K. B., Yuste, R., 2006. The spine neck filters membrane potentials. *Proc. Natl. Acad. Sci. USA* 103, 17961-17966.

Araya, R., Nikolenko, V., Eisenthal, K.B., Yuste, R., 2007. Sodium channels amplify spine potentials. *Proc. Natl. Acad. Sci. USA* 104, 12347-12352.

Araya, R., Vogels, T.P., Yuste, R., 2014. Activity-dependent dendritic spine neck changes are correlated with synaptic strength. *Proc. Natl. Acad. Sci. USA* 111, E2895-2904.

Bannister, N.J., Larkman, A.U., 1995. Dendritic morphology of CA1 pyramidal neurones from the rat hippocampus: II. Spine distributions. *J. Comp. Neurol.* 360, 161-171.

Beaulieu-Laroche, L., Harnett, M.T., 2018. Dendritic spines prevent synaptic voltage clamp. *Neuron 97*, 75-82 e73.

Bliss, T.V., Lomo, T., 1973. Long-lasting potentiation of synaptic transmission in the dentate area of the anaesthetized rabbit following stimulation of the perforant path. *J. Physiol.* 232, 331-356.

Bloodgood, B. L., Giessel, A. J., Sabatini, B. L., 2009. Biphasic synaptic Ca influx arising from compartmentalized electrical signals in dendritic spines. *PLoS Biol.* 7, e1000190.

Bloodgood, B.L., Sabatini, B.L., 2007. Nonlinear regulation of unitary synaptic signals by CaV(2.3) voltage-sensitive calcium channels located in dendritic spines. *Neuron 53*, 249-260.

Bosch, M., Castro, J., Saneyoshi, T., Matsuno, H., Sur, M., Hayashi, Y., 2014. Structural and molecular remodeling of dendritic spine substructures during long-term potentiation. *Neuron 82*, 444-459.

Bradley, P., Horn, G., 1979. Neuronal plasticity in the chick brain: morphological effects of visual experience on neurones in hyperstriatum accessorium. *Brain Res.* 162, 148-153.

Brandon, J.G., Coss, R.G., 1982. Rapid dendritic spine stem shortening during one-trial learning: the honeybee's first orientation flight. *Brain Res.* 252, 51-61.

Chang, F.L., Greenough, W.T., 1984. Transient and enduring morphological correlates of synaptic activity and efficacy change in the rat hippocampal slice. *Brain Res.* 309, 35-46.

Chang, H.T., 1952. Cortical neurons wiht particular reference to the apical dendrites. *Cold Spring Harbor Symposia on Quantitative Biology* 17, 189-202.

Conti, F., Barbaresi, P., Melone, M., Ducati, A., 1999. Neuronal and glial localization of NR1 and NR2A/B subunits of the NMDA receptor in the human cerebral cortex. Cereb. *Cortex 9*, 110-120.

Cornejo, V.H., Ofer, N., Yuste, R., 2022. Voltage compartmentalization in dendritic spines in vivo. *Science* 375, 82-86.

Coss, R.G., Brandon, J.G., Globus, A., 1980 Changes in morphology of dendritic spines on honeybee calycal interneurons associated with cumulative nursing and foraging experiences. *Brain Res.* 192, 49-59.

Coss, R.G., Globus, A., 1978. Spine stems on tectal interneurons in jewel fish are shortened by social stimulation. *Science* 200, 787-790.

da Costa, N.M., 2013. Diversity of thalamorecipient spine morphology in cat visual cortex and its implication for synaptic plasticity. *J. Comp. Neurol.* 521, 2058-2066.

Eyal, G., Verhoog, M.B., Testa-Silva, G., Deitcher, Y., Benavides-Piccione, R., DeFelipe, J., de Kock, C.P.J., Mansvelder, H.D., Segev, I., 2018. Human cortical pyramidal neurons: From spines to spikes via models. *Front. Cell. Neurosci.* 12, 181.

Fifková E, Van Harreveld, A., 1977. Long-lasting morphological changes in dendritic spines of dentate granular cells following stimulation of the entorhinal area. *J. Neurocytol.* 6, 211-230.

Fifková, E., Anderson, C.L., 1981. Stimulation-induced changes in dimensions of stalks of dendritic spines in the dentate molecular layer. *Exp. Neurol.* 74, 621-627.

Fifková, E., Delay, R.J., 1982 Cytoplasmic actin in neuronal processes as a possible mediator of synaptic plasticity. *J. Cell. Biol.* 95, 345-350.

Fletcher, L.N., Williams, S.R., 2019 Neocortical topology governs the dendritic integrative capacity of layer 5 pyramidal neurons. *Neuron 101*, 76-90 e74.

Gemin, O., Serna, P., Zamith, J., Assendorp, N., Fossati, M., Rostaing, P., Triller, A., Charrier, C., 2021. Unique properties of dually innervated dendritic spines in pyramidal neurons of the somatosensory cortex uncovered by 3D correlative light and electron microscopy. *PLoS Biol*. 19, e3001375.

Gouwens, N. W., Berg, J., Feng, D., Sorensen, S. A., Zeng, H., Hawrylycz, M. J., Koch, C., Arkhipov, A., 2018. Systematic generation of biophysically detailed models for diverse cortical neuron types. *Nat. Commun. 9*, 710.

Govindarajan, A., Israely, I., Huang, S.Y., Tonegawa, S., 2011. The dendritic branch is the preferred integrative unit for protein synthesis-dependent LTP. Neuron 69, 132-146.

Grunditz, A., Holbro, N., Tian, L., Zuo, Y., Oertner, T.G., 2008. Spine neck plasticity controls postsynaptic calcium signals through electrical compartmentalization. *J. Neurosci.* 28, 13457-13466.

Gulledge, A.T., Carnevale, N.T., Stuart, G.J., 2012. Electrical advantages of dendritic spines. *PLoS ONE 7*, e36007.

Gulledge, A.T., Kampa, B.M., Stuart, G.J., 2005. Synaptic integration in dendritic trees. *J. Neurobiol. 64*, 75-90.

Harnett, M. T., Makara, J. K., Spruston, N., Kath, W. L., Magee, J. C., 2012. Synaptic amplification by dendritic spines enhances input cooperativity. Nature 491, 599-602.

Harris, K.M., Stevens, J.K., 1988. Dendritic spines of rat cerebellar Purkinje cells: serial electron microscopy with reference to their biophysical characteristics. *J. Neurosci. 8*, 4455-4469.

Harris, K. M., Stevens, J. K., 1989 Dendritic spines of CA 1 pyramidal cells in the rat hippocampus: serial electron microscopy with reference to their biophysical characteristics. *J. Neurosci. 9*, 2982-2997.

Harvey, C. D., Svoboda, K., 2007. Locally dynamic synaptic learning rules in pyramidal neuron dendrites. *Nature* 450, 1195-1200.

Holmes, W. R., 1989. The role of dendritic diameters in maximizing the effectiveness of synaptic inputs. *Brain Res.* 478, 127-137.

Holthoff, K., Zecevic, D., Konnerth, A., 2010. Rapid time course of action potentials in spines and remote dendrites of mouse visual cortex neurons. *J. Physiol.* 588, 1085-1096.

Honkura, N., Matsuzaki, M., Noguchi, J., Ellis-Davies, G.C., Kasai, H., 2008 The subspine organization of actin fibers regulates the structure and plasticity of dendritic spines. *Neuron 57*, 719-729.

Ito, M., 1984. The modifiable neuronal network of the cerebellum. *Jpn. J. Physiol.* 34, 781-792.

Jayant, K., Hirtz, J. J., Plante, I. J., Tsai, D. M., De Boer, W. D., Semonche, A., Peterka, D.S., Owen, J.S., Sahin, O., Shepard, K.L., Yuste, R., 2017. Targeted intracellular voltage recordings from dendritic spines using quantum-dot-coated nanopipettes. *Nat. Nanotechnol. 12*, 335-342.

Johnson, K. W., Herold, K. F., Milner, T. A., Hemmings, H. C., Jr., Platholi, J., 2017. Sodium channel subtypes are differentially localized to pre- and post-synaptic sites in rat hippocampus. *J. Comp. Neurol.* 525, 3563-3578.

Kawato, M., Tsukahara, N., 1984. Electrical properties of dendritic spines with bulbous end terminals. *Biophys. J.* 46, 155-166.

Koch, C., Poggio, T., 1983. A theoretical analysis of electrical properties of spines. *Proc. R. Soc. Lond. B Biol. Sci.* 218, 455-477.

Koch, C., Zador, A., 1993. The function of dendritic spines: devices subserving biochemical rather than electrical compartmentalization. *J. Neurosci.* 13, 413-422.

Kopec, C.D., Li, B., Wei, W., Boehm, J., Malinow, R., 2006. Glutamate receptor exocytosis and spine enlargement during chemically induced long-term potentiation. *J. Neurosci.* 26, 2000-2009.

Kwon, T., Sakamoto, M., Peterka, D.S., Yuste, R., 2017. Attenuation of synaptic potentials in dendritic spines. *Cell. Rep.* 20, 1100-1110.

Lang, C., Barco, A., Zablow, L., Kandel, E.R., Siegelbaum, S.A., Zakharenko, S.S., 2004. Transient expansion of synaptically connected dendritic spines upon induction of hippocampal long-term potentiation. *Proc. Natl. Acad. Sci. USA* 101, 16665-16670.

Lee, S. J., Escobedo-Lozoya, Y., Szatmari, E. M., Yasuda, R., 2009. Activation of CaMKII in single dendritic spines during long-term potentiation. *Nature* 458, 299-304.

Li, C., Gulledge, A.T., 2021. NMDA receptors enhance the fidelity of synaptic integration. *eNeuro 8.*

Lorincz, A., Nusser, Z., 2010. Molecular identity of dendritic voltage-gated sodium channels. *Science* 328, 906-909.

Major, G., Larkman, A.U., Jonas, P., Sakmann, B., Jack, J.J., 1994 Detailed passive cable models of whole-cell recorded CA3 pyramidal neurons in rat hippocampal slices. *J. Neurosci.* 14, 4613-4638.

Matsuzaki, M., Ellis-Davies, G.C., Nemoto, T., Miyashita, Y., Iino, M., Kasai, H., 2001. Dendritic spine geometry is critical for AMPA receptor expression in hippocampal CA1 pyramidal neurons. *Nat. Neurosci*. 4, 1086-1092.

Matsuzaki, M., Honkura, N., Ellis-Davies, G.C., Kasai, H., 2004. Structural basis of long-term potentiation in single dendritic spines. *Nature* 429, 761-766.

Meyer, D., Bonhoeffer, T., Scheuss, V., 2014. Balance and stability of synaptic structures during synaptic plasticity. *Neuron* 82, 430-443.

Miller, J. P., Rall, W., Rinzel, J., 1985. Synaptic amplification by active membrane in dendritic spines. *Brain Res.* 325, 325-330.

Miyazaki, K., Ross, W. N., 2017. Sodium dynamics in pyramidal neuron dendritic spines: Synaptically evoked entry predominantly through ampa receptors and removal by diffusion. *J. Neurosci.* 37. 9964-9976.

Nagerl, U.V., Eberhorn, N., Cambridge, S.B., Bonhoeffer, T., 2004. Bidirectional activity-dependent morphological plasticity in hippocampal neurons. *Neuron 44*, 759-767.

Nuriya, M., Jiang, J., Nemet, B., Eisenthal, K.B., Yuste, R., 2006. Imaging membrane potential in dendritic spines. *Proc. Natl. Acad. Sci. USA* 103, 786-790.

Ofer, N., Benavides-Piccione, R., DeFelipe, J., Yuste, R., 2022. Structural analysis of human and mouse dendritic spines reveals a morphological continuum and differences across ages and species. *eNeuro 9.*

Okamoto, K., Nagai, T., Miyawaki, A., Hayashi, Y., 2004. Rapid and persistent modulation of actin dynamics regulates postsynaptic reorganization underlying bidirectional plasticity. *Nat. Neurosci.* 7, 1104-1112.

Palmer, L. M., Stuart, G. J., 2009. Membrane potential changes in dendritic spines during action potentials and synaptic input. *J. Neurosci.* 29, 6897-6903.

Parajuli, L. K., Nakajima, C., Kulik, A., Matsui, K., Schneider, T., Shigemoto, R., Fukazawa, Y., 2012 Quantitative regional and ultrastructural localization of the Ca(v)2.3 subunit of R-type calcium channel in mouse brain. *J. Neurosci.* 32, 13555-13567.

Parajuli, L. K., Urakubo, H., Takahashi-Nakazato, A., Ogelman, R., Iwasaki, H., Koike, M., Kwon, H.B., Ishii, S., Oh, W.C., Fukazawa, Y., Okabe, S., 2020. Geometry and the organizational principle of spine synapses along a dendrite. *eNeuro 7.*

Popovic, M. A., Carnevale, N., Rozsa, B., Zecevic, D., 2015. Electrical behaviour of dendritic spines as revealed by voltage imaging. *Nat. Commun.* 6, 8436.

Popovic, M.A., Gao, X., Carnevale, N.T., Zecevic, D., 2014. Cortical dendritic spine heads are not electrically isolated by the spine neck from membrane potential signals in parent dendrites. *Cereb. Cortex* 24, 385-395.

Rall, W., 1970. Cable properties of dendrites and effects of synaptic location, in: Excitatory Synaptic Mechanisms: *Proceedings of the Fifth International Meeting of Neurobiologists, Sandefjord, Norway*, pp. 175-187.

Ramón y Cajal, S., 1888. Estructura de los centros nerviosos de las aves. Con dos laminas litográficas. Revista trimestral de histología normal y patológica 1, 1-10.

Segev, I., Friedman, A., White, E. L., Gutnick, M. J., 1995 Electrical consequences of spine dimensions in a model of a cortical spiny stellate cell completely reconstructed from serial thin sections. *J. Comput. Neurosci.* 2, 117-130.

Segev, I., Rall, W., 1988. Computational study of an excitable dendritic spine. *J. Neurophysiol.* 60, 499-523.

Shepherd, G. M., Brayton, R. K., Miller, J. P., Segev, I., Rinzel, J., Rall, W., 1985. Signal enhancement in distal cortical dendrites by means of interactions between active dendritic spines. *Proc. Natl. Acad. Sci. USA* 82, 2192-2195.

Siegel, S. J., Brose, N., Janssen, W. G., Gasic, G. P., Jahn, R., Heinemann, S.F., Morrison, J.H., 1994. Regional, cellular, and ultrastructural distribution of N-methyl-D-aspartate receptor subunit 1 in monkey hippocampus. *Proc. Natl. Acad. Sci. USA* 91, 564-568.

Spruston, N., 2008 Pyramidal neurons: dendritic structure and synaptic integration. *Nat. Rev. Neurosci.* 9, 206-221.

Steiner, P., Higley, M. J., Xu, W., Czervionke, B. L., Malenka, R. C., Sabatini, B.L., 2008. Destabilization of the postsynaptic density by PSD-95 serine 73 phosphorylation inhibits spine growth and synaptic plasticity. *Neuron 60*, 788-802.

Svoboda, K., Tank, D.W., Denk, W., 1996. Direct measurement of coupling between dendritic spines and shafts. *Science* 272, 716-719.

Tamada, H., Blanc, J., Korogod, N., Petersen, C.C., Knott, G.W., 2020. *Ultrastructural comparison of dendritic spine morphology preserved with cryo and chemical fixation.* Elife 9.

Tanaka, J., Horiike, Y., Matsuzaki, M., Miyazaki, T., Ellis-Davies, G.C., Kasai, H., 2008. Protein synthesis and neurotrophin-dependent structural plasticity of single dendritic spines. *Science* 319, 1683-1687.

Thomazeau, A., Bosch, M., Essayan-Perez, S., Barnes, S.A., De Jesus-Cortes, H., Bear, M.F., 2021. Dissociation of functional and structural plasticity of dendritic spines during NMDAR and mGluR-dependent long-term synaptic depression in wild-type and fragile X model mice. Mol. *Psychiatry* 26, 4652-4669.

Tønnesen, J., Katona, G., Rózsa, B., Nägerl, U.V., 2014. Spine neck plasticity regulates compartmentalization of synapses. *Nat. Neurosci.* 17, 678-685.

Trevelyan, A. J., Jack, J., 2002. Detailed passive cable models of layer 2/3 pyramidal cells in rat visual cortex at different temperatures. *J. Physiol.* 539, 623-636.

Trommald, M., Hulleberg, G., 1997. Dimensions and density of dendritic spines from rat dentate granule cells based on reconstructions from serial electron micrographs. *J. Comp. Neurol.* 377, 15-28.

Tsay, D., Yuste, R., 2002. Role of dendritic spines in action potential backpropagation: a numerical simulation study. *J. Neurophysiol.* 88, 2834-2845.

Turner, D. A., 1984. Conductance transients onto dendritic spines in a segmental cable model of hippocampal neurons. *Biophys. J.* 46, 85-96.

Turner, D. A., Schwartzkroin, P.A., 1983. Electrical characteristics of dendrites and dendritic spines in intracellularly stained CA3 and dentate hippocampal neurons. *J. Neurosci. 3*, 2381-2394.

Valles, A.S., Barrantes, F.J., 2021. Nanoscale sub-compartmentalization of the dendritic spine compartment. *Biomolecules* 11.

Van Harreveld, A., Fifková, E., 1975. Swelling of dendritic spines in the fascia dentata after stimulation of the perforant fibers as a mechanism of post-tetanic potentiation. *Exp. Neurol.* 49, 736-749.

Wilson, C. J., 1984. Passive cable properties of dendritic spines and spiny neurons. *J. Neurosci.* 4, 281-297.

Yang, Y., Wang, X.B., Frerking, M., Zhou, Q., 2008. Spine expansion and stabilization associated with long-term potentiation. *J. Neurosci.* 28, 5740-5751.

Zhang, Y. P., Holbro, N., Oertner, T.G., 2008. Optical induction of plasticity at single synapses reveals input-specific accumulation of alphaCaMKII. *Proc. Natl. Acad. Sci. USA* 105, 12039-12044.

Zhou, Q., Homma, K. J., Poo, M. M., 2004. Shrinkage of dendritic spines associated with long-term depression of hippocampal synapses. *Neuron 44*, 749-757.

Chapter 4

Synaptic Plasticity-Associated Signaling Pathways in Dendritic Spines

Marta Zagrebelsky*
Division of Cellular Neurobiology, Zoological Institute,
Technical University of Braunschweig, Braunschweig, Germany

Introduction

Activity-dependent synaptic plasticity defines processes resulting in the strengthening or weakening of synaptic connections in response to changes in neuronal activity and directly impacts learning and memory formation (Hebb, 1949). Long-term potentiation (LTP; Bliss et al., 2018) is the strengthening of synaptic connections resulting in increased efficiency of synaptic transmission and higher postsynaptic responses and is considered a prominent cellular correlate of learning and memory processes (Morris, 1989; Whitlock et al., 2006). On the contrary, long-term depression (LTD; Collingridge et al., 2010) represents the weakening of neuronal connections and is supposed to play a role in regulating flexibility (Nicholls et al., 2008) and specificity (Escobar and Derrick, 2007) during learning. In LTP, high frequency stimulation of presynaptic excitatory terminals results in opening of α-amino-3-hydroxy-5-methyl-4-isoxazolepropionic acid (AMPA) receptors at the postsynaptic density (PSD) and increased sodium influx, thus depolarizing the postsynaptic terminal enough to remove the magnesium block and allow influx of calcium (Ca^{2+}) through the N-methyl-D-aspartate (NMDA) receptors. Similarly, LTD depends upon the Ca^{2+} influx via the opening of NMDA receptors. However,

* Corresponding Author's Email:

In: Dendritic Spines
Editor: Ignacio González-Burgos
ISBN: 979-8-89113-080-7

since it occurs upon low frequency stimulation of the presynaptic terminal, the postsynaptic depolarization only partially removes the magnesium block resulting in a lower Ca^{2+} influx. The characteristics of the influx of Ca^{2+}, as a second messenger define the specific triggering a series of different downstream signaling pathways: while a large influx mediates the strengthening of synaptic connections or LTP, a lower Ca^{2+} influx results in their weakening or LTD.

The PSD of most excitatory synapses is located on dendritic spines, micron-sized membrane protrusions connected to the dendrite via a thin neck and forming a bulbous head containing the PSD. Dendritic spines come in different shapes and sizes and the size of their head directly correlates with the size of the PSD, the content of AMPA receptors and of the ready-releasable pool of neurotransmitter (Arellano et al., 2007; Ofer et al., 2021) thus, with the strength of synaptic transmission. Moreover, the head of dendritic spines form small biochemical compartments containing an extremely complex network of interconnected signaling proteins (Yasuda, 2017) and confining both biochemical (Bloodgood and Sabatini, 2007) and electric signals (Araya et al., 2006), thereby contributing to the synapse specificity of plasticity and influencing the inter-spine crosstalk processes (Adrian et al., 2014). Importantly, dendritic spines are not static structures. Their shape and size change constantly in an activity-dependent manner reflecting the strength (Arellano et al., 2007) and plasticity of synaptic connections and underlying learning and memory processes (Holtmaat and Svoboda, 2009; Kasai et al., 2010). While LTP is associated with an increase in the amount of AMPA receptors at the PSD as well as of the spine head volume, also named structural LTP (sLTP; Yuste and Bonhoeffer, 2001; Matsuzaki et al., 2004), LTD results in a decrease in the number of AMPA receptors and the shrinkage of the spine head (sLTD; Zhou et al., 2004).

This chapter focuses on the signaling pathways activated upon induction of the NMDA receptor-dependent synaptic plasticity within dendritic spines and resulting in both functional and structural alterations at synapses.

Ca^{2+}-Triggered Signaling Events within Spines upon Induction of Activity-Dependent Plasticity

Influx of calcium ions, via NMDARs is both necessary and sufficient for activity-dependent synaptic plasticity. Indeed, upon synaptic activation Ca^{2+}

plays a crucial role in initiating the downstream signaling (Figure 1A,B; Kennedy et al., 2005) leading to both functional and structural synaptic changes at spines. Two-photon time lapse imaging of calcium dynamics revealed the fast accumulation of Ca^{2+} within single dendritic spine heads upon synaptic activation (Yuste and Denk, 1995). The diffusion dynamics of Ca^{2+} and thus, the duration of calcium transients within spines are predominantly regulated via their binding to calcium buffering proteins, making them more susceptible to spatial constrains including the spine morphology and the neck length and width (Majewska et al., 2000; Sabatini et al., 2002; Noguchi et al., 2005; Tønnesen et al., 2014). Indeed, under physiological conditions the diffusion of Ca^{2+} along the spine neck is extremely low (Svoboda et al., 1996) allowing fast, localized Ca^{2+} buildups within the spine head. Moreover, the kinetic of Ca^{2+} re-uptake into intracellular stores such as the smooth endoplasmic reticulum (SER) influences both the amplitude and the duration of the calcium transients (Sabatini et al., 2002). The compartmentalization of Ca^{2+} within spine heads is critical in modulating the initiation of downstream signaling. Upon synaptic activation short Ca^{2+} pulses lasting in the order of 10-100ms (Sabatini et al., 2002; Noguchi et al., 2005) are translated into signaling events lasting minutes to hours due to their interaction with Ca^{2+}-binding proteins. Among the Ca^{2+}-binding proteins, Calmodulin (CaM) is one of the most relevant due to its ability to interact with several downstream signaling proteins within spines. Upon binding Ca^{2+}, CaM interacts with and activates both CaM-dependent kinases (CaMK; Wayman et al., 2008) as well as the phosphatase Calcineurin (Mulkey et al., 1994; Zhou et al., 2004) thereby initiating respectively the specific signaling events downstream of LTP induction and spine enlargement or LTD induction and spine shrinkage.

CaMKs form a family of multifunctional Ser/Thr protein kinases. Among them CaMKII is particularly abundant in the brain where it phosphorylates and thereby regulates the activity of numerous protein substrates involved in plasticity processes (Figure 1B; Wayman et al., 2008). CaMKII has been shown to be essential for NMDA-dependent LTP induction in particular in the adult hippocampus (Giese et al., 1998; Yasuda et al., 2003) as well as for the maintenance of dendritic spine enlargement upon glutamate uncaging (Matsuzaki et al., 2004). Indeed, some unique functional and structural characteristics of CaMKII make it an ideal candidate to be regulating LTP induction. Particularly, the intramolecular auto-phosphorylation of CaMKII is able to transform a transient increase in Ca^{2+} concentration into long lasting signaling due to its autonomous activity. Active CaMKII subunits autophosphorylate adjacent subunits, thus allowing the enzyme to remain active even after the Ca^{2+}/CaM dissociation. The importance of this phosphorylation event in LTP was shown by mutating the CaMKII autophosphorylation site in mice (Giese et al., 1998). Moreover, the autonomous CaMKII activity was shown to be able to decode the frequency of stimulation by detecting the oscillations in postsynaptic Ca^{2+} concentration.

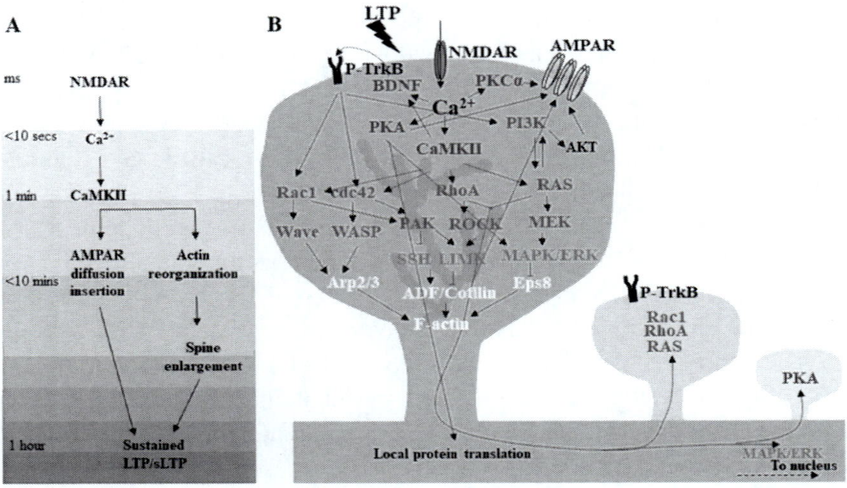

Figure 1. Signaling pathways associated with LTP within the dendritic spine.
A) Temporal event sequence occurring within spines upon synapse stimulation: Activation of NMDARs results in Ca^{2+} influx followed by the activation of CaMKII to modulate AMPAR trafficking and structural plasticity; B) Parallel and interconnected signaling pathways within spines converging on the regulation of AMPAR number at synapses, dynamics of the actin cytoskeleton and local protein synthesis.

The activity profile of CaMKII was measured within individual activated dendritic spines using Fluorescence Resonance Energy Transfer (FRET) sensors. The results showed a very fast activation and subsequent fast inactivation of CaMKII (Takao et al., 2005; Lee et al., 2009). Interestingly, CaMKII activation is necessary only for the first ~1 min of LTP induction (Figure 1; Murakoshi et al., 2011). Its inhibition after LTP induction, in fact does not reverse it suggesting that CaMKII is required for induction, but not for the maintenance, of LTP (Buard et al., 2010). Moreover, CaMKII activation was shown to be restricted within the head of the stimulated spine

(Lee et al., 2009; Yasuda, 2017) and was especially accumulated at the PSD, where upon its activation binds and phosphorylates several proteins in an activity- and NMDA-dependent manner (Wayman et al., 2008). These observations suggest that CaMKII activation may be especially important in determining the synapse specificity of LTP.

Following its interaction with CaM within spines, Ca^{2+} also binds and activates the phosphatase Calcineurin (also known as protein phosphatase 2B), whose activation is crucial in for the induction of LTD and dendritic spine shrinkage, but not LTP (Mulkey et al., 1994; Zhou et al., 2004). FRET imaging studies recording both CaMKII and Calcineurin activities in hippocampal neurons at varying glutamate-uncaging frequencies showed a lower activation threshold of Calcineurin, compared to CaMKII. Indeed, while low glutamate uncaging frequency strongly stimulated Calcineurin, but not CaMKII high frequency stimulation, which induced spine growth, activated both CaMKII and Calcineurin with distinct spatiotemporal kinetics (Fujii et al., 2013). While the activation of both enzyme shows the same temporal pattern, decaying within a few seconds (Fujii et al., 2013; Chang et al., 2017), spatially the activation domains of CaMKII and Calcineurin are clearly separated and associated with distinct molecular complexes thereby independently processing neuronal inputs (Fujii et al., 2013). While activated CaMKII is restricted to the spine head, after its activation Calcineurin spreads over several micrometers (Fujii et al., 2013) possibly mediating the heterosynaptic weakening and shrinkage of nearby unstimulated spines upon high-frequency stimulation of individual spines (Oh et al., 2015).

The short Ca^{2+}-dependent activation of CaMKII and Calcineurin initiates the signaling of several downstream pathways lasting more than tens of minutes and finally resulting in the insertion of new AMPARs at the PSD and thereby synaptic strengthening as well as the reorganization of the actin cytoskeleton within the stimulated dendritic spines and thus, their structural plasticity. Activated CaMKII mediates the phosphorylation of the AMPAR auxiliary protein stargazin, which allows extrasynaptic diffusive receptors to bind to PSD95, thereby fixing them to the PSD (Choquet, 2018). Moreover, while under basal conditions CaMKII associates with F-actin, when Ca^{2+} flows into the spines head, autophosphorylation of CaMKII causes it to dissociate from it allowing the binding of the depolymerizing protein cofilin and other actin binding proteins to remodel the actin cytoskeleton and allow spine head growth. After its dephosphorylation, CaMKII re-associated with the actin filaments thereby stabilizing them (Kim et al., 2015) and contributing to the maintenance of the structural alterations.

Signaling Pathways Activated Downstream of CaMKII Following LTP Induction at Dendritic Spines

The Ras-Raf-MEK-ERK Signaling Pathway

One of the central pathways in regulating synaptic plasticity downstream of the NMDAR-CamKII activation is the Ras–Raf–mitogen-activated protein kinase/ERK kinase (MEK)–extracellular signal regulated kinase (ERK) signaling pathway (Figure 1B). This pathway directly links the activation of NMDAR to LTP and sLTP by controlling both the insertion of AMPARs at the postsynaptic density as well as the spine head enlargement. The Ras family includes several small GTPases whose activity is regulated by a series of guanosine nucleotide exchange factors (GEFs) and GTPase activation proteins (GAPs), respectively activating and inactivating them. Several GEFs and GAPs reside in or near the PSD and are activated during synaptic plasticity processes. Particularly, the Ras GTPase activator SynGAP, which resides in the PSD where it associates with PSD95. Its phosphorylation by CaMKII decreases the activity of SynGAP resulting in the increased activation of Ras (Chen et al., 1998; Oh et al., 2004). Indeed, Ras was shown to be activated in a CaMKII-dependent manner following application of a train of two-photon glutamate uncaging pulses at individual dendritic spines (Harvey et al., 2008). However, activation of Ras also depended on the activity of the phosphatidylinositol 3 kinase (PI3K) and the protein kinase C (PKC; Harvey et al., 2008). Following stimulation, Ras activation peaked within 1 minute after stimulation and returned to baseline after 15 minutes, and was shown to spread out of the activated spine into the underlying dendrite as well as neighboring spines (Harvey et al., 2008; Oliveira and Yasuda, 2013, 2014). The Ras GTPase acts via several downstream effectors, including Raf–MEK–ERK (Thomas and Huganir, 2004) and the phophotydilinositol-3 kinase (PI3K; Qin et al., 2005). The spatiotemporal dynamics of ERK activation were studied using a highly sensitive FRET sensor for ERK activity combined with two-photon fluorescence lifetime imaging (2pFLIM; Tang and Yasuda, 2017). Following glutamate uncaging at singe spines ERK was rapidly activated, in a NMDAR-dependent manner within the stimulated spine as well as in the underling dendrite over at least 12μm. The activation of ERK reached a maximum after 5 minutes and remained stable in parallel with the enlargement of the spine head indicating its involvement in regulating sLTP. The involvement of the Ras-Raf-MEK-dependent ERK activation in LTP and memory processes has been shown in several studies using dominant negative approaches or specific pharmacological inhibitors (Thomas and Huganir, 2004). Expression of constitutively active and dominant negative Ras resulted in increased and decreased synaptic EPSCs respectively by regulating the synaptic delivery of AMPARs thereby occluding or precluding LTP (Zhu et al., 2002). Moreover, PD 98059, a specific MEK blocker, prevented both the increase in ERK activity following high frequency stimulation and the induction of NDMAR-dependent LTP at the hippocampal CA3-CA1 synapse (English and Sweatt, 1996). In addition, the insertion of AMPARs into synapses was blocked by PD 98059 indicating that the activation of ERK downstream of Ras is required for the CaMKII-dependent increase in AMPARs at synapses upon LTP induction (Zhu et al., 2002). Finally, the amplitude of Ras activation upon glutamate uncaging was shown to be proportional to the spine head enlargement and the overexpression of dominant negative Ras (Harvey et al., 2008) as well as the application of the MEK inhibitor PD 98059 (Patterson et al., 2010) blocked sustained structural plasticity upon glutamate uncaging at individual dendritic spines. Taken together, these results indicate that the NMDAR-CaMKII-dependent activation of the Ras–ERK pathway affects the long-term changes in AMPAR response following LTP induction and that ERK is thereby essential for both functional and structural synaptic plasticity. Moreover, ERK activation remains stable over a prolonged period of time and is able to diffuse long distances along the dendritic tree. LTP induction at multiple spines within a short time results in the accumulation of activated ERK within the nucleus, possibly due to its diffusion. In the nucleus, ERK is likely to activate different transcription factors responsible for the maintenance of the late, protein synthesis-dependent phase of LTP (Zhai et al., 2013). Therefore, ERK signaling may be involved in relaying signals from the stimulated spines to the nucleus thereby integrating the activity of individual spines over a wide range of time and space.

The Ras-PI3K-Akt/PKB Pathway

The phosphatidylinositol 3 kinases (PI3Ks) are among the principal downstream effectors of Ras and generate the second messenger phosphatidylinositol-(3,4,5)-trisphosphate (PIP3; Hawkins et al., 2006) following their activity-dependent, Ras mediated activation (Figure 1B; Qin et al., 2005). Upon activation, PI3K triggers the conversion of phosphatidylinositol (3,4)-triphosphate into PIP3 at the plasma membrane. This binds the downstream effector Akt/PKB allowing the phosphoinositide-dependent kinase 1 (PDK1) to phosphorylate and activate it (Alessi et al., 1997). Availability of PIP3 at the post-synapse is critical for LTP induction as well as for maintaining AMPAR clustering at synapses, by regulating their mobility at the spine surface (Arendt et al., 2010) via the interaction with for Phldb2 (pleckstrin homology-like domain, family member 2; Xie et al., 2019). While low levels of Ras activation, drive GluR2-containing AMPARs into the synapse, high levels of Ras activation also promote the insertion of GluR1-containing AMPARs (Zhu et al., 2002). Interestingly, low levels of Ras activation stimulate the Raf-MEK-ERK pathway and induce GluR2-mediated potentiation of synapses. On the other hand, high Ras activity levels facilitate the signaling via the Pi3K-Akt/PKB pathway regulating the synaptic insertion of GluR1-containing AMPARs (Qin et al., 2005). While the activity of both ERK and Akt/PKB are critical to drive AMPARs into synapses, their effect most likely requires other downstream effectors. Both Protein kinase A (PKA) and CaMKII as well as to a lower extent also Protein kinase C (PKC) directly phosphorylate GluR1(Song and Huganir, 2002). However, whether these substrates are activated downstream of ERK and Akt/PKB is not yet clear.

Taken together, the fine-tuned activity levels of Ras control AMPAR insertion at synapses following LTP induction in a subunit specific manner by activating distinct downstream signaling pathways. The role of PI3K and PIP3 in mediating the structural changes at spines following LTP induction are not yet completely clear. However, one recent study showed that upon LTP induction PIP3 increases within dendritic spines where it promotes the formation of spinules (Ueda and Hayashi, 2013). While the function of spinules following LTP induction is not yet known, the accumulation of PIP3 in these structures leads to its decrease within the spine head possibly suggest a role in the sub-spine compartmentalization of the downstream signaling.

The Rho GTPases-Mediated Remodeling of the Actin Cytoskeleton

Dendritic spines contain high amounts of actin organized into a dense a meshwork of connected long- and short-branched filamentous actin (F-actin; Star et al., 2002). The actin cytoskeleton is the principal architectural component found within dendritic spines (Korobova and Svitkina, 2010) and it underlies the maintenance as well as the changing of their size and shape (Hotulainen and Hoogenraad, 2010). In addition, actin filaments form a scaffold for the positioning of channels, receptors as well as different sub-spine structures like the PSD or endosomes (Spence and Soderling, 2015). Therefore, the tightly regulated dynamic remodeling of the actin network is crucial for both the functional and structural plasticity of dendritic spines (Okamoto et al., 2004; Honkura et al., 2008). F-actin forms by polymerization of monomeric globular actin (G-actin) and its dynamics are ensured by treadmilling: a process in which ATP-bound G-actin is added at one end of the actin filament and ADP-bound G-actin is dissociated from the other end. The balance between polymerization and depolymerization of actin filaments is crucial for the plasticity of dendritic spines. Indeed, it was shown that actin polymerization is required for LTP as well as the associated spine head enlargement (Fukazawa et al., 2003; Okamoto et al., 2004). Moreover, while destabilizing the actin cytoskeleton with a low dose of the actin assembly inhibitor LatrunculinA (LatA) interferes specifically with the maintenance of long-term LTP (Krucker et al., 2000) and sLTP, at higher doses LatA prevents both short- and long-term LTP as well as the associated spine head enlargements (Matsuzaki et al., 2004). These observations indicate that the transient and the long-lasting spine enlargements require distinct cytoskeletal organizations and are in agreement with the observation that different F-actin assembly states exists within spines (Honkura et al., 2008).

Several actin-binding proteins (ABPs; Figure 1B) located within spines tightly control the treadmilling of F-actin, by acting on different processes such as actin polymerization, depolymerization, nucleation, branching, capping, cross-linking and trafficking. In this context, a very precise spatiotemporal coordination of the activity of ABPs upon induction of synaptic plasticity is crucial. Indeed, their coordinated activity is thought to underlie both the more dynamic actin network, which is required for its reorganization during the initial phase following LTP induction while as well as the subsequent stabilization of the newly formed network (Okamoto et al., 2004; Kim et al., 2015). Super-resolution imaging studies have shown than the majority of the actin polymerization occurs at the tip and lateral portions of the spine head (Frost et al., 2010). Among the ABPs, three have been primarily involved in promoting actin polymerization during activity-dependent plasticity at spines: the actin-related proteins 2 and 3 complex (Arp2/3), profilin and the formins. Following activation, via one of its nucleation-promoting factors (e.g., N-WASP or Wave1) Arp2/3 binds to the side of an existing actin filament and drives nucleation of a new actin filament to stimulate branched actin polymerization (Rotty et al., 2013). The Arp2/3 complex is essential for dendritic spine maintenance and function such as promoting the tread-milling of the actin filaments and the morphological alterations in response to synaptic activity (Kim et al., 2006; Soderling et al., 2007; Kim et al., 2013). The formins are a family of 15 proteins known to facilitate actin polymerization by promoting the linear growth of the filaments (Harris and Higgs, 2004). Finally, profilin promotes actin polymerization by binding to globular actin (G-actin) thereby facilitating the nucleotide exchange (ADP to ATP) and increasing the availability of G-actin to promote their integration into the growing actin filaments (Harris and Higgs, 2004). Profilin is rapidly recruited to the dendritic spines in an activity-dependent manner (Ackermann and Matus, 2003) and, particularly profilin 2 is required for the enlargement of the dendritic spine head upon LTP induction (Michaelsen-Preusse et al., 2016). Actin polymerization is counterbalanced by its depolymerization, which is primarily regulated by the depolymerizing factors ADF/cofilin. Their function is to sever the actin filaments promoting their disassembly, but also creating new growing filament ends (Bernstein and Bamburg, 2010). Cofilin localizes at the tip of the spine head, below the PSD in areas of high actin dynamics (Racz and Weinberg, 2006). The phosphorylation and dephosphorylation of cofilin tightly regulate its activity and are respectively mediated by the LIMK (Yang et al., 1998) and the phosphatase Slingshot1 (SSH1L; Niwa et al., 2002). Cofilin plays a complex role in regulating functional and structural LTP. Indeed, initially its phosphorylation, resulting in its inactivation was shown to be required for both LTP and spine head enlargement (Fukazawa et al., 2003). However, recently a different study suggested a more complex picture in which cofilin activation is crucial in the early phases following LTP induction, whereas it needs to be deactivated during spine stabilization and LTP consolidation (Bosch et al., 2014). Moreover, cofilin controls activity-induced synaptic availability of glutamate receptors, and exocytosis of synaptic vesicles (Opazo and Choquet, 2011). Elevated ADF/cofilin activity was shown to markedly enhance surface addition of AMPARs after chemically induced long-term potentiation (LTP), whereas inhibition of ADF/cofilin abolished AMPAR insertion at synapses. Moreover, chemically induced LTP elicited a temporal sequence of ADF/cofilin dephosphorylation and phosphorylation underling both AMPAR trafficking and spine enlargement (Gu et al., 2010). Taken together, these results underlie the importance of a very tight spatial and temporal control of the activity of ADF/cofilin within dendritic spines through several upstream regulatory pathways to regulate the different postsynaptic modifications of receptor number and spine size during synaptic plasticity (Rust, 2015). Finally, the actin filaments are stabilized by capping proteins (e.g., Eps8,

CapZ) that, binding to their barbed end prevent further elongation (Menna et al., 2013; Edwards et al., 2014). Eps8, is recruited to the spine head during chemically induced long-term potentiation and the inhibition of its actin-capping activity impairs spine enlargement and plasticity (Menna et al., 2013).

Interestingly, a very precise sub-spine organization of actin fibers has been described and three distinct actin pools have been identified (Honkura et al., 2008). Each of the three actin pools (stable, dynamic and enlargement) are required for the structure and the plasticity of spines and are characterized by different levels of turnover as well as by a distinct sub spine localization underlying the requirement for a very tight spatial and temporal regulation of the actin dynamics. In this context, the coordinated activity of the different functional classes of ABPs and the remodeling of the actin cytoskeleton during synaptic plasticity processes is controlled by the upstream activation of a groups of small GTPase proteins (Figure 1B), including RhoA, Rac1 and cdc42 (Rho GTPases; Harvey et al., 2008; Murakoshi et al., 2011; Bosch et al., 2014; Hedrick et al., 2016). Their activation depends on activated CaMKII and is critical for both the functional and structural aspects of LTP. While the mechanisms mediating the CaMKII-dependent activation of the Rho GTPases are still largely unknown, following LTP induction CaMKII phosphorylates at least two of the GEFs, Kalirin-7 and Trio regulating their activation (Miller et al., 2013; Herring and Nicoll, 2016).

Each of the small GTPases contributes specifically to different aspects of the actin remodeling processes underlying sLTP. RhoA activation controls spine remodeling through the activation of downstream effectors such as the Rho-associated protein kinase (ROCK). Activated ROCK phosphorylates the LIMK, which further phosphorylates, and thereby inactivates ADF/cofilin (Arber et al., 1998). Cdc42 and Rac1 promote actin polymerization and nucleation respectively through the WASP- and Wave-mediated upregulation of Arp2/3 (Hlushchenko et al., 2016). Moreover, Cdc42 and Rac1 stabilize the actin cytoskeleton by promoting the phosphorylation, and thus inhibition of ADF/cofilin-mediated actin depolymerization via the p21-activated kinase (PAK)-mediated activation of the LIMK pathway and the inhibition of the slingshot (SSH) phosphatases pathway (Bosch et al., 2014).

The specific roles of members of the Rho family of small GTPases in regulating the dynamics of the actin cytoskeleton following LTP induction have been recently described in detail using 2-photon glutamate uncaging at single dendritic spines combined to 2pFLIM (Murakoshi et al., 2011). This study recognized the existence of very specific temporal and spatial profiles in the activation of the different Rho GTPases. The activity of RhoA and cdc42 rapidly increased, in the stimulated dendritic spine head and then equally decayed within 5 minutes following LTP induction. Moreover, Rac1 activity rapidly increased upon glutamate uncaging and remained active for up to 30 minutes afterwards (Hedrick et al., 2016). Regarding the spatial pattern of the Rho GTPase activation, while activated RhoA rapidly diffused out of the stimulated spine and spread over ~5 μm along the dendrite, activated Cdc42 remained restricted to the stimulated spine, and exhibited a steep gradient at the spine neck. In addition, after an initial phase in which activated Rac1 showed only a limited diffusion outside the stimulated spine, it then slowly spread ~10 μm into the underlying dendrite (Hedrick et al., 2016). These very specific spatio-temporal patterns of activation of RhoA, Rac1 and cdc42 indicate that they might exert very distinctive roles in regulating the specific processes underlying activity-dependent plasticity at spines. Pharmacological inhibition of RhoA or the downstream ROCK as well as of Rac1 suppressed both the initial transient and the later sustained spine volume change induced by glutamate uncaging (Murakoshi et al., 2011; Hedrick et al., 2016). In contrast, loss of function for cdc42 or the downstream PAK specifically suppressed only the sustained later phase of volume increase (Murakoshi et al., 2011). Taken together, the application of specific loss-of-function approaches for the different Rho GTPases showed that while the activation of RhoA and Rac1 are required both for the initial transient phase of spine head enlargement, the one of cdc42 is essential only for the sustained phase of sLTP. In addition, the specific spatial patterns of activation suggest that while cdc42 is required for the spine specificity of activity-dependent plasticity, the diffusion of RhoA and Rac1 to neighboring, non-activated spines may underlie the phenomenon of synaptic crosstalk in which induction of sLTP at one spine facilitates sLTP at the neighboring ones. Indeed, specifically inhibiting the spreading of RhoA or of Rac1 significantly reduced spine crosstalk without affecting sLTP at the stimulated spine (Hedrick et al., 2016).

PKC

The protein kinase C (PKC) are a family of proteins characterized by several isoforms. These are divided into three classes based on their mode of regulation by secondary messengers: 1) conventional (α, β, γ) which respond to Ca^{2+} and diacylglycerol (DAG), 2) novel (δ, ε, η, θ) which respond to DAG only and 3) atypical (ζ, ι/λ) which are insensitive to Ca^{2+} and DAG and respond to phospholipids.

The classical protein kinase C (PKCα, PKCβ and PKCγ) are typically activated in the presence of Ca^{2+} and of DAG (Lipp and Reither, 2011) at the plasma membrane and are key regulators of plasticity processes and excitatory synapses. While their involvement in regulating LTP is long known (Linden and Routtenberg, 1989; Malinow et al., 1989; Pastalkova et al., 2006; Hongpaisan and Alkon, 2007), the details of their signaling, especially the specific roles of each isozyme are only now being explored (Dell'Acqua and Woolfrey, 2018). A recent study developed novel intermolecular 2pFLIM–Förster resonance energy transfer (FRET) probes to assay the proximity of each PKC isozyme to the plasma membrane, as well as their ability to interact with substrates thereby describing the activation profile of each of the three classical PKC isozymes upon glutamate uncaging at individual spines (Colgan et al., 2018). All PKC isozymes showed a rapid activation (~200ms) and fast decay (~1s), however each of them was activated to different degrees and with distinct kinetics. Particularly, PKCα (Figure 1B) showed the larger peak amplitude and the longer decay time. While all PKC isozymes were only active during the induction phase of LTP, PKCα activity was particularly robust in combination with structural plasticity. Indeed, the specific knockout of PKCα, but not of PKCβ and PKCγ prevented the activity-dependent spine enlargement in CA1 hippocampal neurons as well as the induction of LTP (Colgan et al., 2018). Furthermore, rescue experiments and the analysis of the triple knockout mouse further underlined the specific role of PKCα in the induction of sLTP likely due to the presence of a unique C-terminal PDZ-domain-binding motif featured exclusively in PKCα. Among the PDZ-domain-containing scaffold proteins possibly interacting with PKCα are PICK1, PSD-95 and SAP97. All of them are known to bind to PKCα and regulate AMPA receptor trafficking (Zhu et al., 2016). Moreover, PKCα was recently shown to be involved in regulating the remodeling of the actin cytoskeleton by controlling the activation of Rac1 downstream of both the activation of the NMDARs and the following Ca^{2+} influx and the signaling via the TrkB-PLC to increase DAG in spines (Tu et al., 2020). The very fast activation kinetics of PKC compared to CaMKII suggests that it has a much lower ability to integrate Ca^{2+} signals over time and that the two kinases regulate LTP and sLTP acting on parallel signaling pathways underling the decoding of Ca^{2+} signals at different time scales. Moreover, besides being activated by a fast Ca^{2+} influx via the NMDARs, PKC is also sensitive to DAG produced downstream of the TrkB-PLCγ signaling (Colgan et al., 2018). The ability of activated TrkB to spread over about 10μm along the dendrite (Harward et al., 2016) suggests a possible function of the downstream activation of PKCα in integrating neurotrophic signals as well as the activation of synapses at several nearby dendritic spines (Colgan et al., 2018) to facilitate synaptic plasticity.

The atypical protein kinase M zeta (PKMζ) is a constitutively active PKC isoform due to the lack of the auto-inhibitory regulatory domain of PKC (Schwartz, 1993). PKMζ is elevated during LTP (Sacktor et al., 1993) and has been shown to be necessary and sufficient to maintain synaptic potentiation (Ling et al., 2002; Serrano et al., 2005; Pastalkova et al., 2006) by preventing AMPA receptor endocytosis and promoting stabilization of dendritic spine enlargement (Patel and Zamani, 2021). PKMζ levels increase at synapses following strong, but not weak tetanization in a protein synthesis-dependent manner (Osten et al., 1996). Newly synthesized PKMζ is localized to recently activated synapses by a process called synaptic tag and capture (Sajikumar et al., 2005). Here, a 'tag' is expressed at recently activated synapses that can bind and 'capture' proteins essential for maintaining late-LTP (Frey and Morris, 1997). PKMζ interacts with the ATPase N-ethylmaleimide-sensitive fusion protein (NSF) known to stabilize GluA2-subunit containing AMPA receptors at the post-synaptic membrane (Nishimune et al., 1998; Yao et al., 2008) via its interaction with protein interacting with C kinase-1 (PICK1; Hanley et al., 2002). PICK1 promotes GluA2 receptor endocytosis and stores a pool of extra-synaptic GluA2 receptors (Iwakura et al., 2001). NSF blocks the PICK1-GluA2 interaction, thus preventing endocytosis (Hanley et al., 2002). While the role of PKMζ in sLTP is currently not known, its overexpression was in cultivated hippocampal neurons resulted in an increase in shorter spines, primarily of the stubby type associated with an increase in the size but not the frequency of mEPSCs (Ron et al., 2012) indicating the ability of PKMζ to also regulate dendritic spine structure.

PKA

Cyclic-AMP dependent protein kinase (PKA) is a serine-threonine kinase strongly implicated in the regulation of specific forms of long-term potentiation (LTP). Indeed, its activation depends on the intensity and pattern of electrical stimulation used to induce LTP (Nguyen and Woo, 2003). Glutamate uncaging resulted in a transient, NMDAR-dependent activation of PKA returning to baseline within ~5 min and spreading more than 10 μm along the adjacent dendrite (Figure 1B; Tang and Yasuda, 2017). Furthermore, PKA was shown to be activated by a variety of upstream signals, including

Ca^{2+}, cAMP, adenosine A2 receptors as well as catecholaminergic receptors indicating a possible modulatory role on LTP (Nguyen and Woo, 2003). Indeed, activation of PKA seems to play an important modulatory role on LTP, possibly boosting LTP at single spines. Following the optical coordinated stimulation of glutamatergic and dopaminergic synaptic terminals, activation of PKA by dopamine in the presence of Ca^{2+} promoted spine head enlargement in medium spiny neurons of the striatum possibly via the disinhibition of CaMKII (Yagishita et al., 2014). Once activated, PKA phosphorylates a large number of downstream substrates both in the cytoplasm and in the nucleus making its role in synaptic plasticity extremely complex. PKA directly phosphorylates AMPA receptors and modulate their functions (Banke et al., 2000; Carvalho et al., 2000), thereby rapidly regulating glutamatergic transmission during the early phase of LTP. Furthermore, PKA phosphorylates the GluR1 subunits of AMPA receptors regulating AMPA receptor recycling and their integrations into synapses (Esteban et al., 2003). In addition, PKA also mediates CREB activation and promotes the nuclear translocation of ERK (Nguyen and Woo, 2003) participating in the linking of extracellular signals to the synthesis of new proteins, especially of AMPA receptors, as well as to transcription regulation, thus participating in the late phase of LTP consolidation (Abel et al., 1997). Finally, PKA has also been involved in the regulation of processes dependent on the actin filaments. For instance, PKA phosphorylates the candidate plasticity gene 2 (CPG2), thereby promoting its association with F-actin and contributing in the maintenance of the normal levels of glutamate receptor clathrin-mediated endocytosis (Loebrich et al., 2013). Taken together, the different downstream effectors of PKA account for the variety of its actions during the different phases of LTP, from its induction to its long-term maintenance. Moreover, activated PKA inactivates protein phosphatase 1 (PP1), which dephosphorylates and inactivates CaMKII, an increased presence of cAMP in the spine may prolong the time of CaMKII activation during synaptic plasticity (Blitzer et al., 1998) resulting in an increased degree of sLTP (Luyben et al., 2020), likely by extending the duration of actin cytoskeleton remodeling.

Regulation of Protein Synthesis and Local Translation

While dendritic spine head enlargement can occur in the absence of protein synthesis (Harvey et al., 2008; Harward et al., 2016), some forms of sLTP require the synthesis of new proteins (Tanaka et al., 2008; Govindarajan et al.,

2011; Bosch et al., 2014). Protein synthesis seems to be involved especially in the long-term maintenance of sLTP. Among the molecules that have been clearly involved in protein synthesis-mediated sLTP is PKA. While under normal conditions spine growth returned to baseline levels 2 hours after stimulation, pharmacological activation of PKA maintained the spine growth. This effect of PKA was prevented by blocking protein synthesis (Govindarajan et al., 2011). Local translation of proteins has been shown to occur near the stimulated spines and is able to influence the dynamics of the actin cytoskeleton (Jung et al., 2014; Rangaraju et al., 2017). Several mRNAs of plasticity-related proteins, including β-actin, CaMKII, Arc, PSD-95 reside within the dendrites of hippocampal neurons (Mayford et al., 1996; Tiruchinapalli et al., 2003; Steward et al., 2014) and are regulated in an activity-dependent manner (Sutton et al., 2006; Yoon et al., 2016). Among these, the dendritic localization of the mRNA for CaMKII seems to be important for LTP (Miller et al., 2002). Two different mechanisms have been described by which neuronal activity influences local protein synthesis: 1) by regulating the transport of mRNAs; 2) by locally activating the translation of the mRNAs of plasticity-related proteins which are homogeneously distributed along the entire dendrites. Time lapse imaging of GFP labelled endogenous mRNAs was used to study the transport of RNA granules along dendrites. These studies showed that most of the mRNAs move bidirectionally long microtubules and induction of sLTP by glutamate uncaging stops their movement the base of the stimulated dendritic spines (Buxbaum et al., 2014; Yoon et al., 2016). Moreover, single-molecule in situ hybridization (smFISH) showed that the translation of the mRNA for β-actin is suppressed under basal activity and is transiently unmasked upon chemically induced LTP (Buxbaum et al., 2014).

Activity-regulated synthesis of proteins upon sLTP induction contributes to the modulation of the dynamics of the actin cytoskeleton during structural remodeling of spines. This occurs either directly by the new synthesis of β-actin at stimulated spines or indirectly, for instance by the interaction of the newly synthesized RhoA GTPase (Briz et al., 2015) or Arc (Okuno et al., 2012; Newpher et al., 2018) with CaMKII or other ABPs.

BDNF-TrkB Signaling

In addition to the intracellular pathways activated upon synaptic stimulation, also extracellular factors play a critical role in the regulation of structural plasticity at dendritic spines. In this context, the autocrine signaling of the brain-derived neurotrophic factor (BDNF) via its receptor Tropomyosin receptor kinase B (TrkB) has been suggested to play a crucial role. Indeed. BDNF is an essential regulator of activity-dependent synaptic plasticity (Zagrebelsky and Korte, 2014; Leal et al., 2017) and, its signaling via the TrkB receptor is crucial both for LTP (Korte et al., 1995; Minichiello, 2009) and sLTP (Tanaka et al., 2008; Lai et al., 2012). Interestingly, BDNF is released from dendritic spines of CA1 hippocampal pyramidal neurons upon induction of sLTP and it rapidly activates TrkB receptors in the same spine (Figure 1B). Monitoring TrkB activation using a FRET sensor and 2p-FLIM showed that in response to sLTP induction, TrkB was activated in a rapid (peaking after 1-2 minutes) and sustained (up to 60 minutes) manner (Harward et al., 2016). While initially, activated TrkB was confined within the activated spine, in a later phase (> 30 minutes) it its activation was also seen within the adjacent dendritic segments. TrkB activation was dependent on the NMDA receptors and CamKII signaling as well as on post-synaptically synthetized BDNF. Indeed, imaging a pHluorin-tagged BDNF upon glutamate uncaging showed its rapid release from individually stimulated dendritic spines. The post-synaptic BDNF release and its signaling via TrkB and stimulated spines is necessary for sLTP (Tanaka et al., 2008; Harward et al., 2016). Indeed, scavenging BDNF or blocking its signaling via the TrkB receptor prevented the enlargement of dendritic spines when sLTP was induced by pairing glutamate uncaging with post-synaptic spikes (Tanaka et al., 2008) showing that the post-synaptic autocrine activity of BDNF is able to link NMDAR-CaMKII activation to sLTP. Furthermore, removal of post-synaptic BDNF or its receptor TrkB significantly prevented the activation of Rac1 and cdc42, but not RhoA (Hedrick et al., 2016). Thus, BDNF-TrkB signaling upon sLTP induction affects both the cdc42-mediated spine specific signaling as well as the Rac1-mediated long-range events to the adjacent dendrite and the neighboring spines. This is particularly interesting in the context of inter-synaptic crosstalk. While a weak inhibition of BDNF signaling did not affect sLTP at the suprathreshold-stimulated spines, it attenuated sLTP at the neighboring subthreshold-stimulated ones (Hedrick et al., 2016). Importantly, a weak BDNF-TrkB inhibition reduced Rac1 activity within the dendritic shaft upon induction of inter-synaptic crosstalk, without affecting it in the stimulated spine. Moreover, synaptic crosstalk was prevented by inhibiting the spreading of the activated Rac1. The autocrine BDNF-TrkB loop also plays a role in regulating the protein synthesis-dependent long term phase of sLTP. Indeed, application of low concentrations of exogenous BDNF was able to substitute the post-synaptic spikes and therefore resulted in spine head enlargement upon glutamate uncaging comparable to the one observed by the combined glutamate uncaging and the post-synaptic spikes (Tanaka et al., 2008). This effect of BDNF was blocked by adding the protein synthesis inhibitor anisomycin. While this is possibly due to the activity of BDNF in promoting the local protein synthesis of different plasticity related proteins such as Arc, LIMK1 and CaMKII (Leal et al., 2017), it also seems to be dependent on the stimulation protocol applied since in the study of Harward et al., (2016) the BDNF-mediated spine enlargement did not require protein synthesis. Finally, BDNF application was also shown to promote the delayed capture of PSD scaffolding proteins, like Homer1 underlying the late phase of sLTP (Bosch et al., 2014). Taken together, the autocrine activity of BDNF at individual spines plays a complex role during both the early and the late, protein-synthesis-dependent phases of sLTP. Finally, the activation of PKCα during plasticity requires besides the NMDA receptor-mediated Ca^{2+} influx also the activation of autocrine brain-derived neurotrophic factor (BDNF)–TrkB signaling. These two pathways act on very different spatiotemporal scales of signaling. Thus, by integrating these different signals, PKCα might combine synaptic activation events in time and space, enabling complex processes such as heterosynaptic facilitation of plasticity.

LTD and Spine Shrinkage

The bidirectional functional synaptic plasticity in the form of LTP and LTD has been shown to be closely associated with bidirectional morphological plasticity. While LTP induction is associated with the increase in synaptic strength and the enlargement of the spine head, weakening of synaptic transmission and dendritic spine shrinkage accompany induction of LTD (Nägerl et al., 2004; Zhou et al., 2004). The available knowledge so far indicates that sLTD does not simply depend on the opposite regulation of the same signaling pathways responsible for sLTP, but rather that it depends upon the activation of specific signaling. However, the current understanding on the signaling processes underlying synaptic depression and shrinkage of dendritic spines are still limited possibly due to the only relatively recent development of uncaging protocols for LTD/sLTD induction at single spines (Hayama et al., 2013; Oh et al., 2013). Low-frequency glutamate uncaging was shown to induce shrinkage of individual spines by different mechanisms depending on their original size. Indeed, while in small spines, sLTD only depended on the

NMDAR-mediated Ca^{2+} influx, in bigger spines it additionally required the activation of metabotropic glutamate receptors and of inositol 1,4,5-triphosphate receptors (IP3Rs) leading to the additional Ca^{2+} release from intracellular stores (Oh et al., 2013). The concomitant release of Ca^{2+} from intracellular stores can cause the spread of LTD to neighboring, unstimulated synapses (Nishiyama et al., 2000) in a process known as heterosynaptic LTD. This loss of synapse specificity is normally prevented by the activity of PKC and the PI3 kinase (Daw et al., 2002).

The increase in NMDAR-dependent Ca^{2+} influx upon low-frequency stimulation was shown to result in the CaM-dependent activation of activation of a serine-threonine protein phosphatase cascade including especially the protein phosphatase 2B/calcineurin (CaN) (Malenka and Bear, 2004) followed by dephosphorylation of PKC and PKA substrates (Kameyama et al., 1998; Hrabetova and Sacktor, 2001; van Dam et al., 2002). NMDAR-activation is able to alter the subcellular localization of the serine-threonine protein phosphatase PP1, possibly explaining the specificity of the dephosphorylation for some substrates, but not others (e.g., CaMKII substrate; Morishita et al., 2001).

LTD is associated with alterations in AMPAR function and with their removal from the synapse. While upon LTP induction, phosphorylation has been observed for the ser-831, a substrate of CaMKII and PKC, without a change in ser-845, a PKA substrate, LTD was found to be associated with selective dephosphorylation of ser-845, without any change in ser-831 (Lee et al., 2000). Dephosphorylation of ser-845 was shown to decrease the probability of channel opening for AMPARs, thus contributing to LTD expression (Banke et al., 2000). Moreover, upon low-frequency stimulation AMPARs were rapidly internalized by a dynamin- and clathrin-dependent process (Lin et al., 2000; Man et al., 2000). Indeed, upon NMDAR activation, the clathrin adaptor protein complex AP2 was recruited to the membrane in proximity with the GluR2 C-tail where it initiated clathrin-mediated endocytosis. This process requires the CaN and PP1-mediated dephosphorylation of the phosphatidylinositol 4-phosphate 5-kinase-γ661 (PIP5Kγ661), the major phosphatidylinositol 4,5-bisphosphate ($PI(4,5)P_2$)-producing enzyme (Unoki et al., 2012).

Associated to the internalization of AMPARs, LTD also induces shrinkage of stimulated spines mediated by acting depolymerization resulting in a shift toward a higher G-actin concentration within spines (Okamoto et al., 2004). Interestingly, β-arrestin-2 has been involved in the NMDA-induced remodeling of dendritic spines following chemical induction of LTD via the translocation of the active cofilin upon its CaN- and PI3K-mediated dephosphorylation (Pontrello et al., 2012). Finally, a novel non-ionotropic mediated role of NMDARs has been described in driving spine shrinkage upon LTD induction. While spine shrinkage induced by low-frequency glutamatergic stimulation was blocked by application of competitive glutamate binding site NMDAR antagonists, it persisted in the presence of the glycine/D-serine binding site NMDAR antagonist 7-chlorokynurenate (7-CK) or the pore blocker MK-801 (Stein et al., 2015). Using 2-photon glutamate uncaging combined with different pharmacological approaches showed that the stimulation of the NMDARs drives the spine head shrinkage through NOS1AP (nitric oxide synthase adaptor protein 1)-nNOS (neuronal nitric oxide synthase) interactions. The consequent activity of nNOS, p38MAPK, MK2 (MAPK-activated protein kinase 2) and CaMKII finally promotes actin remodeling by the cofilin-dependent severing of actin filaments (Stein et al., 2020).

It will be very important in future studies to expand the approaches to analyze LTD at single dendritic spines to unravel the specific spatial and temporal profiles of the activation of LTD-related signaling molecules. This will also allow gaining a better understanding of the potential overlap and interactions between LTD- and LTP-associated pathways.

Conclusion

Multiple Synaptic Plasticity-Associated Parallel Pathways and Their Interactions

Being able to obtain a detailed description of the spatial and temporal activation patterns of several signaling molecules and pathways following synapse activation has been a huge step forward in understanding how activity-dependent synaptic plasticity processes, and especially LTP are regulated within dendritic spines. Taken together these studies show that LTP induction activates a series of parallel and interconnected signaling pathways within individual spines following a very specific temporal sequence (Figure 1). Ca^{2+} influx and CaMKII activation are brief and mostly restricted to the dendritic spine, whereas their downstream effectors are often active on longer time scales. Among these, several activated molecules diffuse into and affect the signaling of the underlying dendrite and neighboring spines. Among the small GTPases, while Cdc42 activation is restricted to the activated spine, possibly being involved on ensuring the spine specificity of the LTP, RhoA, Rac1 and RAS diffuse in different manners to the neighboring spines rather mediating the interactions between synapses such as the synaptic crosstalk or the heterosynaptic spine shrinkage. Moreover, the different modes and kinetics of activation of plasticity-related proteins within spines underlies the possibility of integrating different signals, such as synaptic activation and signaling of neurotrophic factors adding an extra level of complexity in the regulation of plasticity.

Each of the signaling pathways involved in LTP at individual spines shares a series of common activators, including NMDAR activation, Ca^{2+} influx, autocrine BDNF signaling and has been shown to converge on common set of mechanisms finally controlling both the changes in synaptic strength and in spine head volume. In fact, all pathways described above cooperate in the very fast regulation of AMPARs trafficking and thereby in regulating the number of AMPARs inserted at synapses and synaptic strength. Moreover, they all also control the dynamics of the actin cytoskeleton by regulating actin polymerization and depolymerization as well as severing and capping of the actin filaments. Finally, all the pathways above are involved in regulating gene transcription and protein translation at a slower time scale. All together, these three main mechanisms downstream the very complex interplay of several signaling pathways underlie the different phases of LTP at spines including its induction and maintenance as well as the interaction with plasticity events occurring at different synapses and their temporal integration.

References

Abel, T., Nguyen, P. V., Barad, M., Deuel, T. A., Kandel, E. R., Bourtchouladze, R., 1997. Genetic demonstration of a role for PKA in the late phase of LTP and in hippocampus-based long-term memory. *Cell* 88, 615-626.

Ackermann, M., Matus, A., 2003. Activity-induced targeting of profilin and stabilization of dendritic spine morphology. *Nat. Neurosci.* 6, 1194-1200.

Adrian, M., Kusters, R., Wierenga, C. J., Storm, C., Hoogenraad, C. C., Kapitein, L. C., 2014. Barriers in the brain: resolving dendritic spine morphology and compartmentalization. *Front. Neuroanat.* 8, 142.

Alessi, D. R., James, S. R., Downes, C. P., Holmes, A. B., Gaffney, P. R., Reese, C. B., Cohen, P., 1997. Characterization of a 3-phosphoinositide-dependent protein kinase which phosphorylates and activates protein kinase Balpha. *Curr. Biol.* 7, 261-269.

Araya, R., Jiang, J., Eisenthal, K. B., Yuste, R., 2006. The spine neck filters membrane potentials. *Proc. Natl. Acad. Sci. USA* 103, 17961-17966.

Arber, S., Barbayannis, F. A., Hanser, H., Schneider, C., Stanyon, C. A., Bernard, O., Caroni, P., 1998. Regulation of actin dynamics through phosphorylation of cofilin by LIM-kinase. *Nature* 393, 805-809.

Arellano. J. I., Benavides-Piccione, R., Defelipe, J., Yuste, R., 2007. Ultrastructure of dendritic spines: correlation between synaptic and spine morphologies. *Front. Neurosci.* 1, 131-143.

Arendt, K. L., Royo, M., Fernández-Monreal, M., Knafo, S., Petrok, C. N., Martens, J. R., Esteban, J. A., 2010. PIP3 controls synaptic function by maintaining AMPA receptor clustering at the postsynaptic membrane. *Nat. Neurosci.* 13, 36-44.

Banke, T., Bowie, D., Lee, H-K., Huganir, R., Schousboe, A., Traynelis, S., 2000. Control of GluR1 AMPA receptor function by cAMP-dependent protein kinase. *J. Neurosci.* 20, 89-102.

Bernstein, B. W., Bamburg, J. R., 2010. ADF/cofilin: a functional node in cell biology. *Trends Cell. Biol.* 20, 187-195.

Bliss, T. V. P., Collingridge, G. L., Morris, R. G. M., Reymann, K. G., 2018. Long-term potentiation in the hippocampus: discovery, mechanisms and function. *Neuroforum* 24, A103-A120.

Blitzer, R. D., Connor, J. H., Brown, G. P., Wong, T., Shenolikar, S., Iyengar, R., Landau, E. M., 1998. Gating of CaMKII by cAMP-regulated protein phosphatase activity during LTP. *Science* 280, 1940-1943.

Bloodgood, B. L., Sabatini, B. L., 2007. Ca2+ signaling in dendritic spines. *Curr. Opin. Neurobiol.* 17, 345-351.

Bosch, M., Castro, J., Saneyoshi, T., Matsuno, H., Sur, M., Hayashi, Y., 2014. Structural and molecular remodeling of dendritic spine substructures during long-term potentiation. *Neuron* 82, 444-459.

Briz, V., Zhu, G., Wang, Y., Liu, Y., Avetisyan, M., Bi, X., Baudry, M., 2015. Activity-dependent rapid local RhoA synthesis is required for hippocampal synaptic plasticity. *J. Neurosci.* 35, 2269-2282.

Buard, I., Coultrap, S. J., Freund, R. K., Lee, Y. S., Dell'Acqua, M. L., Silva, A. J., Bayer, K. U., 2010. CaMKII "autonomy" is required for initiating but not for maintaining neuronal long-term information storage. *J. Neurosci.* 30, 8214-8220.

Buxbaum, A. R., Wu, B., Singer, R. H., 2014. Single β-actin mRNA detection in neurons reveals a mechanism for regulating its translatability. *Science* 343, 419-422.

Carvalho, A. L., Duarte, C. B., Carvalho, A. P., 2000. Regulation of AMPA receptors by phosphorylation. *Neurochem. Res.* 25, 1245-1255.

Chang, J. Y., Parra-Bueno, P., Laviv, T., Szatmari, E. M., Lee, S. R., Yasuda, R., 2017. CaMKII autophosphorylation is necessary for optimal integration of Ca(2+) signals during LTP induction, but not maintenance. *Neuron* 94, 800-808.e804.

Chen, H. J., Rojas-Soto, M., Oguni, A., Kennedy, M. B., 1998. A synaptic Ras-GTPase activating protein (p135 SynGAP) inhibited by CaM kinase II. *Neuron* 20, 895-904.

Choquet, D., 2018. Linking nanoscale dynamics of AMPA receptor organization to plasticity of excitatory synapses and learning. *J. Neurosci.* 38, 9318-9329.

Colgan, L. A., Hu, M., Misler, J. A., Parra-Bueno, P., Moran, C. M., Leitges, M., Yasuda, R., 2018. PKCα integrates spatiotemporally distinct Ca(2+) and autocrine BDNF signaling to facilitate synaptic plasticity. *Nat. Neurosci.* 21, 1027-1037.

Collingridge, G. L., Peineau, S., Howland, J. G., Wang, Y. T., 2010. Long-term depression in the CNS. *Nat. Rev. Neurosci.* 11, 459-473.

Daw, M. I., Bortolotto, Z. A., Saulle, E., Zaman, S., Collingridge, G. L., Isaac, J. T., 2002. Phosphatidylinositol 3 kinase regulates synapse specificity of hippocampal long-term depression. *Nat. Neurosci.* 5, 835-836.

Dell'Acqua, M. L., Woolfrey, K. M., 2018. FRETting over postsynaptic PKC signaling. *Nat. Neurosci.* 21, 1021-1022.

Edwards, M., Zwolak, A., Schafer, D. A., Sept, D., Dominguez R., Cooper, J. A , 2014. Capping protein regulators fine-tune actin assembly dynamics. *Nat. Rev. Mol. Cell. Biol.* 15, 677-689.

English, J. D., Sweatt, J. D., 1996. Activation of p42 mitogen-activated protein kinase in hippocampal long term potentiation. *J. Biol. Chem.* 271, 24329-24332.

Escobar, M. L., Derrick, B., 2007. Long-term potentiation and depression as putative mechanisms for memory formation. *Neural plasticity and memory: From genes to brain Imaging* 15.

Esteban. J. A., Shi, S-H., Wilson, C., Nuriya, M., Huganir, R. L., Malinow, R., 2003. PKA phosphorylation of AMPA receptor subunits controls synaptic trafficking underlying plasticity. *Nat. Neurosci.* 6, 136-143.

Frey, U., Morris, R. G., 1997. Synaptic tagging and long-term potentiation. *Nature* 385, 533-536.

Frost, N. A., Shroff, H., Kong, H., Betzig, E., Blanpied, T. A., 2010. Single-molecule discrimination of discrete perisynaptic and distributed sites of actin filament assembly within dendritic spines. *Neuron* 67, 86-99.

Fujii, H., Inoue, M., Okuno, H., Sano, Y., Takemoto-Kimura, S., Kitamura, K., Kano, M., Bito, H., 2013. Nonlinear decoding and asymmetric representation of neuronal input information by CaMKIIα and calcineurin. *Cell Reports* 3, 978-987.

Fukazawa, Y., Saitoh, Y., Ozawa, F., Ohta, Y., Mizuno, K., Inokuchi, K., 2003. Hippocampal LTP is accompanied by enhanced F-actin content within the dendritic spine that is essential for late LTP maintenance in vivo. *Neuron* 38, 447-460.

Giese, K. P., Fedorov, N. B., Filipkowski, R. K., Silva, A. J., 1998. Autophosphorylation at Thr286 of the alpha calcium-calmodulin kinase II in LTP and learning. *Science* 279, 870-873.

Govindarajan, A., Israely, I., Huang, S-Y., Tonegawa, S., 2011. The dendritic branch is the preferred integrative unit for protein synthesis-dependent LTP. *Neuron* 69, 132-146.

Gu, J., Lee, C. W., Fan, Y., Komlos, D., Tang, X., Sun, C., Yu, K., Hartzell, H. C., Chen, G., Bamburg, J. R., Zheng, J. Q., 2010. ADF/cofilin-mediated actin dynamics regulate AMPA receptor trafficking during synaptic plasticity. *Nat. Neurosci.* 13, 1208-1215.

Hanley, J. G., Khatri, L., Hanson, P. I., Ziff, E. B., 2002. NSF ATPase and alpha-/beta-SNAPs disassemble the AMPA receptor-PICK1 complex. *Neuron* 34, 53-67.

Harris, E. S., Higgs, H. N., 2004. Actin cytoskeleton: formins lead the way. *Curr. Biol.* 14, R520-522.

Harvey, C. D., Yasuda, R., Zhong, H., Svoboda, K., 2008. The spread of Ras activity triggered by activation of a single dendritic spine. *Science* 321, 136-140.

Harward, S. C., Hedrick, N. G., Hall, C. E., Parra-Bueno, P., Milner, T. A., Pan, E., Laviv, T., Hempstead, B. L., Yasuda, R., McNamara, J. O., 2016. Autocrine BDNF-TrkB signaling within a single dendritic spine. *Nature* 538, 99-103.

Hawkins, P. T., Anderson, K. E., Davidson, K., Stephens, L. R., 2006. Signaling through Class I PI3Ks in mammalian cells. *Biochem. Soc. Trans.* 34, 647-662.

Hayama, T., Noguchi, J., Watanabe, S., Takahashi, N., Hayashi-Takagi, A., Ellis-Davies, G. C., Matsuzaki, M., Kasai, H., 2013. GABA promotes the competitive selection of dendritic spines by controlling local Ca2+ signaling. *Nat. Neurosci.* 16, 1409-1416.

Hebb, D. O., 1949. *The organization of behavior: A neuropsychological theory:* Science editions.

Hedrick, N. G., Harward, S. C., Hall, C. E., Murakoshi, H., McNamara, J. O., Yasuda, R., 2016. Rho GTPase complementation underlies BDNF-dependent homo- and heterosynaptic plasticity. *Nature* 538, 104-108.

Herring, B. E., Nicoll, R. A., 2016. Kalirin and Trio proteins serve critical roles in excitatory synaptic transmission and LTP. *Proc. Natl. Acad. Sci. USA* 113, 2264-2269.

Hlushchenko, I., Koskinen, M., Hotulainen, P., 2016. Dendritic spine actin dynamics in neuronal maturation and synaptic plasticity. *Cytoskeleton* (Hoboken) 73, 435-441.

Holtmaat, A., Svoboda, K., 2009. Experience-dependent structural synaptic plasticity in the mammalian brain. *Nat. Rev. Neurosci.* 10, 647-658.

Hongpaisan, J., Alkon, D. L., 2007. A structural basis for enhancement of long-term associative memory in single dendritic spines regulated by PKC. *Proc. Natl. Acad. Sci. USA* 104, 19571-19576.

Honkura, N., Matsuzaki, M., Noguchi, J., Ellis-Davies, G. C., Kasai, H., 2008. The subspine organization of actin fibers regulates the structure and plasticity of dendritic spines. *Neuron* 57, 719-729.

Hotulainen, P., Hoogenraad, C. C., 2010. Actin in dendritic spines: connecting dynamics to function. *J. Cell. Biol.* 189, 619-629.

Hrabetova, S., Sacktor, T. C., 2001. Transient translocation of conventional protein kinase C isoforms and persistent downregulation of atypical protein kinase Mzeta in long-term depression. *Mol. Brain Res.* 95, 146-152.

Iwakura, Y., Nagano, T., Kawamura, M., Horikawa, H., Ibaraki, K., Takei, N., Nawa, H., 2001. N-methyl-D-aspartate-induced alpha-amino-3-hydroxy-5-methyl-4-isoxazoleproprionic acid (AMPA) receptor down-regulation involves interaction of the carboxyl terminus of GluR2/3 with Pick1. Ligand-binding studies using Sindbis vectors carrying AMPA receptor decoys. *J. Biol. Chem.* 276, 40025-40032.

Jung, H., Gkogkas, Ch. G., Sonenberg, N., Holt, Ch. E., 2014. Remote control of gene function by local translation. *Cell* 157, 26-40.

Kameyama, K., Lee, H-K., Bear, M. F., Huganir, R. L., 1998. Involvement of a postsynaptic protein kinase A substrate in the expression of homosynaptic long-term depression. *Neuron* 21, 1163-1175.

Kasai, H., Fukuda, M., Watanabe, S., Hayashi-Takagi, A., Noguchi, J., 2010. Structural dynamics of dendritic spines in memory and cognition. *Trends Neurosci.* 33, 121-129.

Kennedy, M. B., Beale, H. C., Carlisle, H. J., Washburn, L. R., 2005. Integration of biochemical signaling in spines. *Nat. Rev. Neurosci.* 6, 423-434.

Kim, I. H., Racz, B., Wang, H., Burianek, L., Weinberg, R., Yasuda, R., Wetsel W. C., Soderling, S. H., 2013. Disruption of Arp2/3 results in asymmetric structural plasticity of dendritic spines and progressive synaptic and behavioral abnormalities. *J. Neurosci.* 33, 6081-6092.

Kim, K., Lakhanpal, G., Lu, H. E., Khan, M., Suzuki, A., Hayashi, M. K., Narayanan, R., Luyben, T. T., Matsuda T., Nagai, T., Blanpied, T. A., Hayashi, Y., Okamoto, K., 2015. A temporary gating of actin remodeling during synaptic plasticity consists of the interplay between the kinase and structural functions of CaMKII. *Neuron* 87, 813-826.

Kim, Y., Sung, J. Y., Ceglia, I., Lee, K. W., Ahn, J. H., Halford, J. M., Kim, A. M., Kwak, S. P., Park, J. B., Ho Ryu, S., Schenck, A., Bardoni, B., Scott, J. D., Nairn, A. C., Greengard, P., 2006. Phosphorylation of WAVE1 regulates actin polymerization and dendritic spine morphology. *Nature* 442, 814-817.

Korobova, F., Svitkina, T., 2010. Molecular architecture of synaptic actin cytoskeleton in hippocampal neurons reveals a mechanism of dendritic spine morphogenesis. *Mol. Biol. Cell* 21, 165-176.

Korte, M., Carroll, P., Wolf, E., Brem, G., Thoenen, H., Bonhoeffer, T., 1995. Hippocampal long-term potentiation is impaired in mice lacking brain-derived neurotrophic factor. *Proc. Natl. Acad. Sci. USA* 92, 8856-8860.

Krucker, T., Siggins, G. R., Halpain, S., 2000. Dynamic actin filaments are required for stable long-term potentiation (LTP) in area CA1 of the hippocampus. *Proc. Natl. Acad. Sci. USA* 97, 6856-6861.

Lai, K. O., Wong, A. S., Cheung, M. C., Xu, P., Liang, Z., Lok, K. C., Xie, H., Palko, M. E., Yung, W. H., Tessarollo, L., Cheung, Z. H., Ip, N. Y., 2012. TrkB phosphorylation by Cdk5 is required for activity-dependent structural plasticity and spatial memory. *Nat. Neurosci.* 15, 1506-1515.

Leal, G., Bramham, C. R., Duarte, C. B., 2017. BDNF and hippocampal synaptic plasticity. *Vitam. Horm.* 104, 153-195.

Lee, H-K., Barbarosie, M., Kameyama, K., Bear, M. F., Huganir, R. L., 2000. Regulation of distinct AMPA receptor phosphorylation sites during bidirectional synaptic plasticity. *Nature* 405, 955-959.

Lee, S-JR., Escobedo-Lozoya, Y., Szatmari, E. M., Yasuda, R., 2009. Activation of CaMKII in single dendritic spines during long-term potentiation. *Nature* 458, 299-304.

Lin, J. W., Ju, W., Foster, K., Lee, S. H., Ahmadian, G., Wyszynski, M., Wang, Y. T., Sheng, M., 2000. Distinct molecular mechanisms and divergent endocytotic pathways of AMPA receptor internalization. *Nat. Neurosci.* 3, 1282-1290.

Linden, D. J., Routtenberg, A., 1989. The role of protein kinase C in long-term potentiation a testable model. *Brain Res. Brain Res. Rev.* 14, 279-296.

Ling, D. S., Benardo, L. S., Serrano, P. A., Blace, N., Kelly, M. T., Crary, J. F., Sacktor, T.C., 2002. Protein kinase Mzeta is necessary and sufficient for LTP maintenance. *Nat. Neurosci.* 5, 295-296.

Lipp, P., Reither, G., 2011. Protein kinase C: the "masters" of calcium and lipid. *Cold Spring Harbor Perspectives in Biology* 3:a004556.

Loebrich, S., Djukic, B., Tong, Z. J., Cottrell, J. R., Turrigiano, G. G., Nedivi, E., 2013. Regulation of glutamate receptor internalization by the spine cytoskeleton is mediated by its PKA-dependent association with CPG2. *Proc. Natl. Acad. Sci. USA* 110, E4548-4556.

Luyben, T. T., Rai, J., Li, H., Georgiou, J., Avila, A., Zhen, M., Collingridge, G. L., Tominaga, T., Okamoto, K., 2020. Optogenetic manipulation of postsynaptic cAMP using a novel transgenic mouse line enables synaptic plasticity and enhances depolarization following tetanic stimulation in the hippocampal dentate gyrus. *Front. Neural Circuits* 14, 24.

Majewska, A., Brown, E., Ross, J., Yuste, R., 2000. Mechanisms of calcium decay kinetics in hippocampal spines: Role of spine calcium pumps and calcium diffusion through the spine neck in biochemical compartmentalization. *J. Neurosci.* 20, 1722-1734.

Malenka, R. C., Bear, M. F., 2004. LTP and LTD: an embarrassment of riches. *Neuron* 44, 5-21.

Malinow, R., Schulman, H., Tsien, R. W., 1989. Inhibition of postsynaptic PKC or CaMKII blocks induction but not expression of LTP. *Science* 245, 862-866.

Man, H-Y., Lin, J. W., Ju, W. H., Ahmadian, G., Liu, L., Becker, L. E., Sheng, M., Wang, Y. T., 2000. Regulation of AMPA receptor–mediated synaptic transmission by clathrin-dependent receptor internalization. *Neuron* 25, 649-662.

Matsuzaki, M., Honkura, N., Ellis-Davies, G. C. R., Kasai, H., 2004. Structural basis of long-term potentiation in single dendritic spines. *Nature* 429, 761-766.

Mayford, M., Baranes, D., Podsypanina, K., Kandel, E.R., 1996. The 3'-untranslated region of CaMKII alpha is a cis-acting signal for the localization and translation of mRNA in dendrites. *Proc. Natl. Acad. Sci. USA* 93, 13250-13255.

Menna, E., Zambetti, S., Morini, R., Donzelli, A., Disanza, A., Calvigioni, D., Braida, D., Nicolini, C., Orlando, M., Fossati, G., Regondi, M., Pattini, L., Frassoni, C., Francolini, M., Scita, G., Sala, M., Fahnestock, M., Matteoli, M., 2013. Eps8 controls dendritic spine density and synaptic plasticity through its actin-capping activity. *Embo J.* 32, 1730-1744.

Michaelsen-Preusse, K., Zessin, S., Grigoryan, G., Scharkowski, F., Feuge, J., Remus, A., Korte, M., 2016. Neuronal profilins in health and disease: Relevance for spine plasticity and Fragile X syndrome. *Proc. Natl. Acad. Sci. USA* 113, 3365-3370.

Miller, M. B., Yan, Y., Eipper, B. A., Mains, R. E., 2013. Neuronal Rho GEFs in synaptic physiology and behavior. *Neuroscientist* 19, 255-273.

Miller, S., Yasuda, M., Coats, J. K., Jones, Y., Martone, M. E., Mayford, M., 2002. Disruption of dendritic translation of CaMKIIalpha impairs stabilization of synaptic plasticity and memory consolidation. *Neuron* 36, 507-519.

Minichiello, L., 2009. TrkB signaling pathways in LTP and learning. *Nat. Rev. Neurosci.* 10, 850-860.

Morishita, W., Connor, J. H., Xia, H., Quinlan, E. M., Shenolikar, S., Malenka, R. C., 2001. Regulation of synaptic strength by protein phosphatase 1. *Neuron* 32, 1133-1148.

Morris, R. G., 1989. Synaptic plasticity and learning: selective impairment of learning rats and blockade of long-term potentiation in vivo by the N-methyl-D-aspartate receptor antagonist AP5. *J. Neurosci.* 9, 3040-3057.

Mulkey, R. M., Endo, S., Shenolikar, S., Malenka, R. C., 1994. Involvement of a calcineurin/inhibitor-1 phosphatase cascade in hippocampal long-term depression. *Nature* 369, 486-488.

Murakoshi, H., Wang, H., Yasuda, R., 2011. Local, persistent activation of Rho GTPases during plasticity of single dendritic spines. *Nature* 472, 100-104.

Nägerl, U. V., Eberhorn, N., Cambridge, S. B., Bonhoeffer, T., 2004. Bidirectional Activity-Dependent Morphological Plasticity in Hippocampal Neurons. *Neuron* 44, 759-767.

Newpher, T. M., Harris, S., Pringle, J., Hamilton, C., Soderling, S., 2018. Regulation of spine structural plasticity by Arc/Arg3.1. *Semin. Cell Dev. Biol.* 77, 25-32.

Nguyen, P. V., Woo, N. H., 2003. Regulation of hippocampal synaptic plasticity by cyclic AMP-dependent protein kinases. *Prog. Neurobiol.* 71, 401-437.

Nicholls, R. E., Alarcon, J. M., Malleret, G., Carroll, R. C., Grody, M., Vronskaya, S., Kandel, E. R., 2008. Transgenic mice lacking NMDAR-dependent LTD exhibit deficits in behavioral flexibility. *Neuron* 58, 104-117.

Nishimune, A., Isaac, J. T., Molnar, E., Noel, J., Nash, S. R., Tagaya, M., Collingridge, G. L., Nakanishi, S., Henley, J. M., 1998. NSF binding to GluR2 regulates synaptic transmission. *Neuron* 21, 87-97.

Nishiyama, M., Hong, K., Mikoshiba, K., Poo, M-M., Kato, K., 2000. Calcium stores regulate the polarity and input specificity of synaptic modification. *Nature* 408, 584-588.

Niwa, R., Nagata-Ohashi, K., Takeichi, M., Mizuno, K., Uemura, T., 2002. Control of actin reorganization by Slingshot, a family of phosphatases that dephosphorylate ADF/cofilin. *Cell* 108, 233-246.

Noguchi, J., Matsuzaki, M., Ellis-Davies, G. C. R., Kasai, H., 2005. Spine-neck geometry determines NMDA receptor-dependent Ca2+ signaling in dendrites. *Neuron* 46, 609-622.

Ofer, N., Berger, D. R., Kasthuri, N., Lichtman, J. W., Yuste, R., 2021. Ultrastructural analysis of dendritic spine necks reveals a continuum of spine morphologies. *Dev. Neurobiol.* 81, 746-757.

Oh, J. S., Manzerra, P., Kennedy, M. B., 2004. Regulation of the neuron-specific Ras GTPase-activating protein, synGAP, by Ca2+/calmodulin-dependent protein kinase II. *J. Biol. Chem.* 279, 17980-17988.

Oh, W. C., Hill, T.C., Zito, K., 2013. Synapse-specific and size-dependent mechanisms of spine structural plasticity accompanying synaptic weakening. *Proc. Natl. Acad. Sci. USA* 110, E305-312.

Oh, W. C., Parajuli Laxmi, K., Zito, K., 2015. Heterosynaptic structural plasticity on local dendritic segments of hippocampal CA1 neurons. *Cell Reports* 10, 162-169.

Okamoto, K., Nagai, T., Miyawaki, A., Hayashi, Y., 2004. Rapid and persistent modulation of actin dynamics regulates postsynaptic reorganization underlying bidirectional plasticity. *Nat. Neurosci.* 7, 1104-1112.

Okuno, H., Akashi, K., Ishii, Y., Yagishita-Kyo, N., Suzuki, K., Nonaka, M., Kawashima, T., Fujii, H., Takemoto-Kimura, S., Abe, M., Natsume, R., Chowdhury, S., Sakimura, K., Worley, P. F., Bito, H., 2012. Inverse synaptic tagging of inactive synapses via dynamic interaction of Arc/Arg3.1 with CaMKIIβ. *Cell* 149, 886-898.

Oliveira, A. F., Yasuda, R., 2013. An improved Ras sensor for highly sensitive and quantitative FRET-FLIM imaging. *PLoS One* 8, e52874.

Oliveira, A. F., Yasuda, R., 2014. Imaging the activity of Ras superfamily GTPase proteins in small subcellular compartments in neurons. *Meth. Mol. Biol.* 1071, 109-128.

Opazo, P., Choquet, D., 2011. A three-step model for the synaptic recruitment of AMPA receptors. *Mol. Cell. Neurosci.* 46, 1-8.

Osten, P., Valsamis, L., Harris, A., Sacktor, T. C., 1996. Protein synthesis-dependent formation of protein kinase Mzeta in long-term potentiation. *J. Neurosci.* 16, 2444-2451.

Pastalkova, E., Serrano, P., Pinkhasova, D., Wallace, E., Fenton, A. A., Sacktor, T. C., 2006. Storage of spatial information by the maintenance mechanism of LTP. *Science* 313, 1141-1144.

Patel, H., Zamani, R., 2021. The role of PKMζ in the maintenance of long-term memory: a review. *Rev. Neurosci.* 32, 481-494.

Patterson, M. A., Szatmari, E. M., Yasuda, R., 2010. AMPA receptors are exocytosed in stimulated spines and adjacent dendrites in a Ras-ERK-dependent manner during long-term potentiation. *Proc. Natl. Acad. Sci. USA* 107, 15951-15956.

Pontrello, C. G., Sun, M. Y., Lin, A., Fiacco, T. A., DeFea, K. A., Ethell, I. M., 2012. Cofilin under control of β-arrestin-2 in NMDA-dependent dendritic spine plasticity, long-term depression (LTD), and learning. *Proc. Natl. Acad. Sci. USA* 109, E442-451.

Qin, Y., Zhu, Y., Baumgart, J. P., Stornetta, R. L., Seidenman, K., Mack, V., van Aelst, L., Zhu, J. J., 2005. State-dependent Ras signaling and AMPA receptor trafficking. *Genes Dev.* 19, 2000-2015.

Racz, B., Weinberg, R. J., 2006. Spatial organization of cofilin in dendritic spines. *Neuroscience* 138, 447-456.

Rangaraju, V., Dieck, S., Schuman, E. M., 2017. Local translation in neuronal compartments: how local is local? *EMBO Rep.* 18, 693-711.

Ron, S., Dudai, Y., Segal, M., 2012. Overexpression of PKMζ alters morphology and function of dendritic spines in cultured cortical neurons. *Cereb. Cortex* 22, 2519-2528.

Rotty, J. D., Wu, C., Bear, J. E., 2013. New insights into the regulation and cellular functions of the ARP2/3 complex. *Nat. Rev. Mol. Cell Biol.* 14, 7-12.

Rust, M. B., 2015. ADF/cofilin: a crucial regulator of synapse physiology and behavior. *Cell. Mol. Life Sci.* 72, 3521-3529.

Sabatini, B. L., Oertner, T. G., Svoboda, K., 2002. The life cycle of Ca2+ ions in dendritic spines. *Neuron* 33, 439-452.

Sacktor, T. C., Osten, P., Valsamis, H., Jiang, X., Naik, M. U., Sublette, E., 1993. Persistent activation of the zeta isoform of protein kinase C in the maintenance of long-term potentiation. *Proc. Natl. Acad. Sci. USA* 90, 8342-8346.

Sajikumar, S., Navakkode, S., Sacktor, T. C., Frey, J. U., 2005. Synaptic tagging and cross-tagging: the role of protein kinase Mzeta in maintaining long-term potentiation but not long-term depression. *J. Neurosci.* 25, 5750-5756.

Schwartz, J. H., 1993. Cognitive kinases. *Proc. Natl. Acad. Sci. USA* 90, 8310-8313.

Serrano, P., Yao, Y., Sacktor, T. C., 2005. Persistent phosphorylation by protein kinase Mzeta maintains late-phase long-term potentiation. *J. Neurosci.* 25, 1979-1984.

Soderling, S. H., Guire, E. S., Kaech, S., White, J., Zhang, F., Schutz, K., Langeberg, L. K., Banker, G., Raber, J., Scott, J. D., 2007. A WAVE-1 and WRP signaling complex regulates spine density, synaptic plasticity, and memory. *J. Neurosci.* 27, 355-365.

Song, I., Huganir, R. L., 2002. Regulation of AMPA receptors during synaptic plasticity. *Trends Neurosci.* 25, 578-588.

Spence, E. F., Soderling, S. H., 2015. Actin out: Regulation of the synaptic cytoskeleton. *J. Biol. Chem.* 290, 28613-28622.

Star, E. N., Kwiatkowski, D. J., Murthy, V. N., 2002. Rapid turnover of actin in dendritic spines and its regulation by activity. *Nat. Neurosci.* 5, 239-246.

Stein, I. S., Gray, J. A., Zito K., 2015. Non-ionotropic NMDA receptor signaling drives activity-induced dendritic spine shrinkage. *J. Neurosci.* 35, 12303-12308.

Stein, I. S., Park, D. K., Flores, J. C., Jahncke, J. N., Zito, K., 2020. Molecular mechanisms of non-ionotropic NMDA receptor signaling in dendritic spine shrinkage. *J. Neurosci.* 40, 3741-3750.

Steward, O., Farris, S., Pirbhoy, P. S., Darnell, J., Driesche, S. J., 2014. Localization and local translation of Arc/Arg3.1 mRNA at synapses: some observations and paradoxes. *Front. Mol. Neurosci.* 7, 101.

Sutton, M. A., Ito, H. T., Cressy, P., Kempf, C., Woo, J. C., Schuman, E. M., 2006. Miniature neurotransmission stabilizes synaptic function via tonic suppression of local dendritic protein synthesis. *Cell* 125, 785-799.

Svoboda, K., Tank, D. W., Denk, W., 1996. Direct measurement of coupling between dendritic spines and shafts. *Science* 272, 716-719.

Takao, K., Okamoto, K., Nakagawa, T., Neve, R. L., Nagai, T., Miyawaki, A., Hashikawa, T., Kobayashi, S., Hayashi, Y., 2005. Visualization of synaptic Ca2+ /calmodulin-dependent protein kinase II activity in living neurons. *J. Neurosci.* 25, 3107-3112.

Tanaka, J., Horiike, Y., Matsuzaki, M., Miyazaki, T., Ellis-Davies, G. C., Kasai, H., 2008. Protein synthesis and neurotrophin-dependent structural plasticity of single dendritic spines. *Science* 319, 1683-1687.

Tang, S., Yasuda, R., 2017. Imaging ERK and PKA activation in single dendritic spines during structural plasticity. *Neuron* 93, 1315-1324. e1313.

Thomas, G. M., Huganir, R. L., 2004. MAPK cascade signaling and synaptic plasticity. *Nat. Rev. Neurosci.* 5, 173-183.

Tiruchinapalli, D. M., Oleynikov, Y., Kelic, S., Shenoy, S. M., Hartley, A., Stanton, P. K., Singer, R. H., Bassell, G. J., 2003. Activity-dependent trafficking and dynamic localization of zipcode binding protein 1 and beta-actin mRNA in dendrites and spines of hippocampal neurons. *J. Neurosci.* 23, 3251-3261.

Tønnesen, J., Katona, G., Rózsa, B., Nägerl, U. V., 2014. Spine neck plasticity regulates compartmentalization of synapses. *Nat. Neurosci.* 17, 678-685.

Tu, X., Yasuda, R., Colgan, L. A., 2020. Rac1 is a downstream effector of PKCα in structural synaptic plasticity. *Sci. Rep.* 10, 1777.

Ueda, Y., Hayashi, Y., 2013. PIP_3 regulates spinule formation in dendritic spines during structural long-term potentiation. J. *Neurosci,* 33, 11040-11047.

Unoki, T., Matsuda, S., Kakegawa, W., Van Ngo Thai, B., Kohda, K., Suzuki, A., Funakoshi, Y., Hasegawa, H., Yuzaki, M., Kanaho, Y., 2012. NMDA receptor-mediated PIP5K activation to produce PI(4,5)P2 is essential for AMPA receptor endocytosis during LTD. *Neuron* 73, 135-148.

van Dam, E. J., Ruiter, B., Kamal, A., Ramakers, G. M., Gispen, W. H., de Graan, P. N., 2002. N-methyl-D-aspartate-induced long-term depression is associated with a decrease in postsynaptic protein kinase C substrate phosphorylation in rat hippocampal slices. *Neurosci. Lett.* 320, 129-132.

Wayman, G. A., Lee, Y. S., Tokumitsu, H., Silva, A. J., Soderling, T. R., 2008. Calmodulin-kinases: modulators of neuronal development and plasticity. *Neuron* 59, 914-931.

Whitlock, J. R., Heynen, A. J., Shuler, M. G., Bear, M. F., 2006. Learning induces long-term potentiation in the hippocampus. *Science* 313, 1093-1097.

Xie, M. J., Ishikawa, Y., Yagi, H., Iguchi, T., Oka, Y., Kuroda, K., Iwata, K., Kiyonari, H., Matsuda, S., Matsuzaki, H., Yuzaki, M., Fukazawa, Y., Sato, M., 2019. PIP(3)-Phldb2 is crucial for LTP regulating synaptic NMDA and AMPA receptor density and PSD95 turnover. *Sci. Rep.* 9, 4305.

Yagishita, S., Hayashi-Takagi, A., Ellis-Davies, G. C., Urakubo, H., Ishii, S., Kasai, H., 2014. A critical time window for dopamine actions on the structural plasticity of dendritic spines. *Science* 345, 1616-1620.

Yang, N., Higuchi, O., Ohashi, K., Nagata, K., Wada, A., Kangawa, K., Nishida, E., Mizuno, K., 1998. Cofilin phosphorylation by LIM-kinase 1 and its role in Rac-mediated actin reorganization. *Nature* 393, 809-812.

Yao, Y., Kelly, M. T., Sajikumar, S., Serrano, P., Tian, D., Bergold, P. J., Frey, J. U., Sacktor, T. C., 2008. PKM zeta maintains late long-term potentiation by N-ethylmaleimide-sensitive factor/GluR2-dependent trafficking of postsynaptic AMPA receptors. *J. Neurosci.* 28, 7820-7827.

Yasuda, H., Barth, A. L., Stellwagen, D., Malenka, R. C., 2003. A developmental switch in the signaling cascades for LTP induction. *Nat. Neurosci.* 6, 15-16.

Yasuda, R., 2017. Biophysics of biochemical signaling in dendritic spines: Implications in synaptic plasticity. *Biophys. J.* 113, 2152-2159.

Yoon, Y. J., Wu, B., Buxbaum, A. R., Das, S., Tsai, A., English, B. P., Grimm, J. B., Lavis, L. D., Singer, R. H., 2016. Glutamate-induced RNA localization and translation in neurons. *Proc. Natl. Acad. Sci. USA* 113, E6877-e6886.

Yuste, R., Denk, W., 1995. Dendritic spines as basic functional units of neuronal integration. *Nature* 375, 682-684.

Yuste, R., Bonhoeffer, T., 2001. Morphological changes in dendritic spines associated with long-term synaptic plasticity. *Annu. Rev. Neurosci.* 24, 1071-1089.

Zagrebelsky, M., Korte, M., 2014. Form follows function: BDNF and its involvement in sculpting the function and structure of synapses. *Neuropharmacology* 76 Pt C, 628-638.

Zhai, S., Ark, E. D., Parra-Bueno, P., Yasuda, R., 2013. Long-distance integration of nuclear ERK signaling triggered by activation of a few dendritic spines. *Science* 342, 1107-1111.

Zhou, Q., Homma, K. J., Poo, M-m., 2004. Shrinkage of dendritic spines associated with long-term depression of hippocampal synapses. *Neuron* 44, 749-757.

Zhu, J., Shang, Y., Zhang, M., 2016. Mechanistic basis of MAGUK-organized complexes in synaptic development and signaling. *Nat. Rev. Neurosci.* 17, 209-223.
Zhu, J. J., Qin, Y., Zhao, M., Van Aelst, L., Malinow, R., 2002. Ras and Rap control AMPA receptor trafficking during synaptic plasticity. *Cell* 110, 443-455.

Chapter 5

Dendritic Spine Plasticity and the Processing of Cognitive Synaptic Information

Ignacio González-Burgos[1,*]
and Nallely Vázquez-Hernández[2]
[1]Laboratorio "Burgos" de Investigaciones Psicobiológicas, Zapopan, Jalisco, Mexico
[2]Universidad Tecnológica de México (UNITEC), Campus Guadalajara, Guadalajara, Jalisco, México

Introduction

The brain is capable of expressing organizational changes at different levels of the biological organization of nerve tissue. Such changes affect the physiology of levels both further up and further down. They are generically recognized by the term *neuroplasticity*. In the best case, these modifications tend to adapt to changing circumstances in the internal and external environments, though they may have the opposite effect of fostering aberrant connections. In the latter case, they contribute to establishing psychoneurobiological conditions of a pathological nature.

The synapse is the basis of all neural communication. Because the afferent information that reaches neurons varies widely, its processing at the synaptic level must be highly plastic in order to generate, subsequently, coherent behavioral responses to environmental demands. In reality, several synaptic mechanisms support such differential processing and, hence, effectuate changes in the cytoarchitecture of neurons. These include dendritic spines, the excitatory postsynaptic contact sites in those neurons that contain them. Based

* Corresponding Author's Email:

In: Dendritic Spines
Editor: Ignacio González-Burgos
ISBN: 979-8-89113-080-7

on the characteristics of the synaptic inputs they receive, dendritic spines differentially process afferent information to postsynaptic neurons in relation to, among other factors, the characteristic functional activity of the different geometric shapes that have been identified to date. Dendritic spines present plastic modifications that include *de novo* spinogenesis, interconversion of one type to another, redistribution along the dendritic trunk, and reabsorption or pruning. All these events are related to either the processing of information with cognitive or non-cognitive content, or that which is homeostatic in nature.

The plasticity of these spines is not only expressed under normal conditions, for they are particularly sensitive to events that lead to pathology. Under such circumstances, it has been difficult to distinguish between plasticity as an adaptive response to inflicted damage or a result of the damage itself. Even so, the structural modifications of the spines seem to be conducive to functional readaptation, though this does not necessarily occur, due to the fact that the plastic changes mediated by dendritic spines are insufficient to reverse the functional defects associated with the morphophysiological alterations that underlie the many psychophysiopathological conditions where studies have verified their involvement.

The Concept and Study of Neuroplasticity

Individuals' ability to adapt efficiently to the environment with which they interact depends on the normal expression of various fundamental neuropsychological processes. The cerebral organization of these processes is carried out through the formation of neural circuits that are interrelated as functional systems by means of highly-specific synaptic contacts (Luria, 1970, 1982). Information from the environment –internal or external– is transmitted in an integrative way through these neural circuits, but its psychoneural processing may vary due to modifications of previous reference schemes (experience) or the occurrence of neuropsychopathological events. Thus, under conditions both normal and atypical, the contacts among the neurons involved are usually modified to produce other conditions of brain functioning that, at least in theory, tend toward adaptation, though they can also produce alterations that lead to aberrant behaviors. Whatever the case, nerve tissue expresses a wide range of specific biological phenomena at different levels of organization –from the molecular to the systemic– that tend to maintain the balance between the activity of individuals and the demands of their environment.

The ability of nerve cells to react –effectively or not– to novel or adverse conditions is known as neuroplasticity, a term that the World Health Organization formally defined (in 1982) as "*the ability of nervous system cells to reorganize anatomically and functionally after being subjected to developmental, environmental, or pathological influences –including trauma and disease– which could result in an adaptive (or maladaptive) response to functional demand.*" In a broader perspective, the ability to adapt underlies the series of psychobiological qualities that allow the acquisition of the concurrent, eminently sensory, information that enters from the subject's internal and external environments, its selective "filtering," analyses of components that are significant in some sense, and the structuring of eventual behavioral responses in accordance with environmental demands. The behavioral responses generated not only adjust to the needs raised by the environment in this interactive process, but are also capable of modifying previous environmental conditions. This dialectical interaction process favors adaptation, since an individual interacts with his environment actively. This entails deploying activities that go beyond simple survival to transform the environment. The ability to adapt depends on the pressure the environment exerts on the individual and his ability to handle that pressure in terms of the adequacy between environmental demands and the consequent behavior emitted in response, whatever the situation and level of neural processing of the information involved. If we accept this understanding, it is clear that attempts to adapt may, or may not, be successful (Figure 1). Adaptive plastic changes in the brain may be related with neurodevelopment, biological cycles, homeostasis, and experience (learning/memory), whilst inadaptation are commonly related with neuropathological damage and transitory immediate plastic changes.

The neural systems that support manifestations of the psychological processes that help individuals adapt to their environment are structured by multiple, highly-organized synaptic contacts between neurons in different brain regions that become functionally related. The content of the information that neural circuits process includes the biological, psychological, and social reference schemes that underlie an individual's adaptive capacity.

Experimental studies of nervous system plasticity have established that the organization of neuropsychological processes is closely related to structural modifications of the neuronal cytoarchitecture supported, in turn, by highly dynamic changes in the molecular components of the microenvironment of the synapses involved. Under pathological conditions, however, these cognitive capacities may be altered, though prior to this the plastic responses of the nerve cells involved in their psychoneural organization show a certain morphofunctional correspondence.

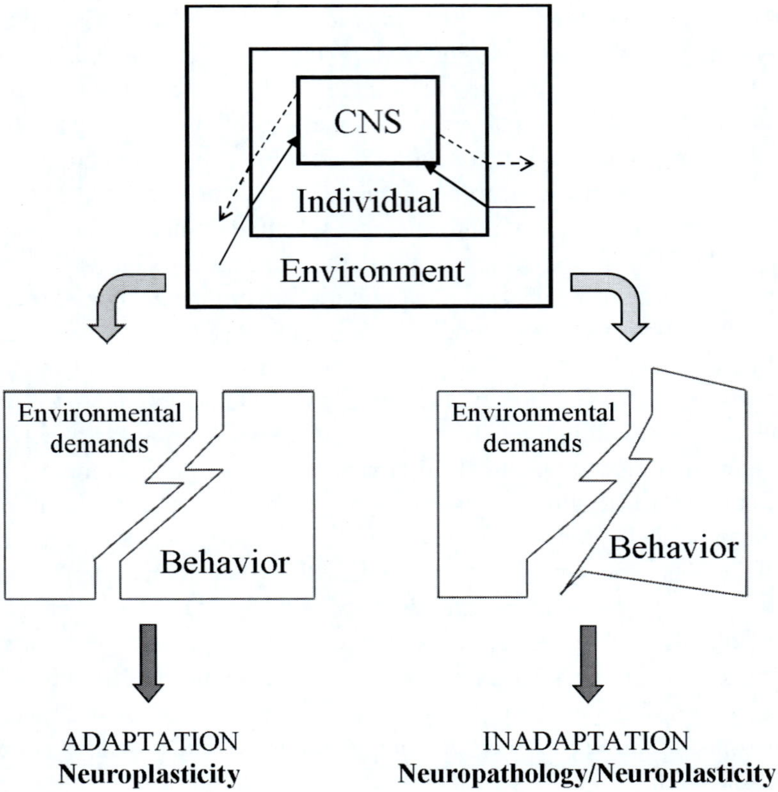

Figure 1. Explanatory diagram on the generation of plastic changes in the brain. Under normal conditions, the perception of sensory stimuli (and thought schemes) are translated into adaptive behavioral patterns with respect to the demands of the environment. Under pathological conditions, such perceptual schemes are not in accordance with the demands of the environment and then the plastic changes that occur in the brain are aberrant or, in the best of cases, adaptive but transitory.

The scientific literature identifies three basic theoretical-methodological approaches of research on neuroplasticity that, ultimately, are complementary.

1. Developmental neurobiology is based on the fact that during the neurogenesis and synaptogenesis of the nervous system both the formation of the various neural regions and the establishment of

synaptic contacts between the neurons that constitute them are ongoing processes that greatly favor the modification of the genetic plan that is in progress. Thus, in the early stages of brain development, emerging plastic events under both normal conditions (e.g., learning) and those forced by a pathology-inducing incident correspond to genetically active capacities whose expression translates into a likely adaptive advantage; such as children's remarkable ability to learn a foreign language or the probable functional recovery after a temporary, transitory brain disorder. In this line of thinking, an adult brain shows a lower plastic capacity that tends to decrease with age because the processes of brain development and differentiation are completed upon reaching biological maturity. This occurs because the gene expression of the characteristic events of development is relatively "turned off." If they eventually gain expression as a result of atypical conditions, they do so in isolation and out of context from their joint expression with other typical events of early development, so their rehabilitative capacity is limited or even aberrant.

2. The study of biological cycles or rhythms. A biological cycle can be understood as the regular repetition of a series of characteristic, related events over time. Typical examples of biological rhythms are the sleep-wake cycle, hormone secretion, and menstruation. Many of the events that underlie the expression of biological cycles are regulated by brain activity that, in turn, is characterized by the expression of plastic phenomena. This is the case, for instance, of the variable efficiency in the cognitive performance of fertile women across the stages of the menstrual cycle. In this case, variations are directly associated with circulating hormone concentrations that are known to modulate the synaptic transmission of cognition-related information in a plastic manner (González-Flores et al., 2015).
3. Designing experimental models that reproduce, at scale, features of certain psychoneuropathologies. This approach to the study of plasticity phenomena has one clear advantage over the others, for it provides the possibility to establish closer causal relations between a condition and its underlying psychobiological phenomena; that is, it increases the possibility of attributing plastic biological responses more specifically to the disorders that researchers attempt to reproduce experimentally.

Rigidity and Neural Dynamism

With the exception of the neurons of the subgranular zone of the dentate gyrus of the hippocampus and the subventricular zone of the lateral ventricles (Lledo et al., 2006), nerve cells do not show mitotic capacity, a fact that has an apparently paradoxical significance and transcendence. The persistence of the neurons that make up nerve tissue suggests a biological system that is rigid in terms of both its structure and its functional activity. However, throughout an individual's life –more so during early stages of postnatal development– it is clear that the brain expresses a high degree of adaptive versatility that is plastic in nature, under normal conditions and even after the infliction of damage or injury. In addition, it has been proposed that the vast amount of information that the brain processes is organized through various forms of intrinsic, plastic homeostatic processes; that is, neural activity which tends to both tune synaptic connections after their stimulation-mediated activity and maintain those connections within the correct dynamic range of synaptic firing (Turrigiano and Nelson, 2004). The expression of natural plasticity (like that which mediates learning processes, is induced by injury, or is homeostatic) suggests that the circuits that sustain the organization of the many neuropsychological processes are not fixed or rigid but, rather, dynamically modifiable and highly specific and efficient due to their ability to transform the information that the individual receives from the environment with which he interacts permanently.

As a derivation of his Theory of Neuronism based on the extensively detailed description of the structure of nerve cells in practically all regions of the vertebrate brain, Cajal (Ramón and Cajal, 1909) proposed that learning could consist in plastic changes in certain properties of neurons or their interconnections, similar to the phenomena that occur during synapse formation in the embryonic stage. This concept was reinforced by the proposal that learning might involve persistent chain activity by interconnected neurons (Forbes, 1922). Later, based on electrophysiological studies, Hebb (1949) proposed one of the most widely accepted theories of synaptic transmission between neurons; one that, in fact, provides solid support for some of the postulates about the biological bases of learning. Hebb posited that although interneuronal circuits are established genetically, the strength or efficiency of certain connections are not fully determined, so the properties of these circuits may well be modified as a result of changes in the activity of neurons. These postulates lead to the deduction that the modification of synaptic activity associated with structural changes in the contacts between the neurons involved represents one of the most effective neurobiological processes that fosters adjustments in individual behavior that, at least in theory, favor more efficient adaptive measures or likelihood of rehabilitation. But this is not always the case.

Although plastic changes can, indeed, favor adaptation under normal conditions, various factors come to hinder this process under pathological or harmful circumstances. While it is true that this results in the expression of plastic phenomena, these phenomena may be aberrant and oblige us to speak of maladaptive plasticity.

Since the pioneering reports outlined above, studies of brain functions have been closely linked to a localizationist notion which holds that there are more-or-less circumscribed regions in the brain that participate in organizing certain psychological processes. Today, reductionist approaches to brain activity have been diluted by more functionalist positions that have spurred the development of the concept of "systems." This means that the expression of a certain psychic quality depends on the concerted activity of various brain regions because they are made up of various psychoneural components.

Brain Plasticity

Plasticity phenomena in nerve tissue are expressed at distinct levels of biological organization. Clearly, modifications of the neural circuits that support the organization and expression of behavior occur as a function of gradual, successive changes in underlying biological systems. The accumulation of changes in the structure and function of the circuits involved in the organization and behavioral expression of a certain psychoneural process leads to changes in the activity of the brain regions involved, changes that can be recorded by various scientific devices. In this regard, the advent of non-invasive techniques for studying brain activity has made it possible to detect changes in the activity of certain brain regions that may be associated with certain functional states, whether normal or atypical. Such functional modifications have been interpreted as plastic responses by the neural circuitry in question. Based on the neural activity recorded, various theoretical constructs have been proposed to characterize and explain the neural phenomena that underlie these systematic plastic responses: vicariousness, redundancy, functional substitution, diaschisis, and compensation (Duffau, 2006; Brailowsky et al., 1998).

Vicariousness

Vicariousness is a phenomenon of plasticity expressed as a result of a circumscribed lesion in a certain region of the brain such that other areas take over the functional role of the damaged part. The expression of vicarious plastic responses assumes that neurons are equipotential; that is, they have the capacity to assume functions that, in principle, are foreign to their natural activity, be it sensory, motor, or of some other kind.

Redundancy

Redundancy refers to the ability of brain structures secondarily related to a damaged area to restore lost functions. This type of plastic response is observed in patients who, after suffering a serious brain lesion in a circumscribed region, do not present functional deficits –at least observably– because a secondary brain region expresses an apparent capacity to perform the same functional activities as those of the damaged area.

Functional Substitution

Another plastic process that explains the recovery of functions lost due to some type of injury is functional substitution. This consists in the start-up of a part of the brain that, in principle, is not functionally related to the injured area. This type of plastic process has been attributed to the genetic reprogramming of the functional capacities of a region to allow it to "replace" the damaged area.

Diaschisis

The state of shock that produces transient or permanent functional sequelae depending on the severity of the lesion in a region distant from the one directly affected due to some type of neural connection, is called diaschisis. In this process, it is assumed that a mechanism of active inhibition of the regions that could assume the functions of the damaged area is set in motion, and that it determines the duration of the functional blockage caused by the lesion.

Compensation

Compensation is a plastic process more closely related to the implementation of cognitive strategies capable of resolving deficiencies in behavioral performance caused by a lesion in a unimodal associative region. The original function is not recovered, but the activation of heteromodal association areas significantly increases their functions, allowing them to compensate for the deficiencies.

Presynaptic Neuronal Plasticity

The fact that living matter is organized in continuous levels obliges us to consider that the changes that occur at a certain level could impact, to some degree, the activity of lower and/or higher levels. This, in turn, leads to the suggestion that the plasticity recorded in broad neural circuits is supported by changes that may occur at the level of the neurons that constitute them, and *vice versa*. At the presynaptic level, plastic events are diverse and, in some cases, characteristic of the developing brain. However, events of this kind can also occur in adult brains and, regardless of their cause, show phenomenological features similar to those that occur in early stages of brain development. Among others, *axonal retraction* due to competition is a plastic phenomenon typical of synaptogenesis that occurs during the normal ontogenetic development of the nervous system. When an injury occurs, *axonal sprouting* is usually observed first in those nerve fibers that remain viable and functional. The resulting sprouts tend to assume the presynaptic role played by those that degenerated as a result of either programmed competitive retraction (Llorens et al., 2011) or inflicted damage (Kaas et al., 2008). However, an axon that is damaged irreversibly at a certain level does not regenerate, and the neuron usually dies from retrograde degeneration. With the emergence of axonal regrowth, *reactive synaptogenesis* can occur secondarily, but may be synaptically functional or ineffective; that is, it may establish aberrant connections that potentially compromise the functionality of the system involved. In the case of a damaged –or even normal– adult brain, presynaptic plasticity is limited by the inhibitory action of two groups of proteins: myelin-associated inhibitory and extracellular matrix-associated inhibitory proteins (Akbik et al., 2012).

Denervation hypersensitivity is a postsynaptic plastic response –in principle compensatory– caused by a decrease in afferent stimulation due to a lesion of the afferent axonal terminals or a blocking of synaptic activity. This plastic process increases the number of receptor sites and/or their affinity for their ligand (neurotransmitter) (Soria-Fregoso et al., 2008), which consequently leads to an increase in neuronal response capacity (López-Vázquez et al., 2006). The degree of severity of the lesion can lead to denervation and cause non-synaptic diffusion of a neurotransmitter whose activity, in this case, would be subject to extrasynaptic regulation mechanisms (Bach and Rita, 1990).

Based on the foregoing, it can be deduced that, generically, presynaptic events that occur under certain conditions, whether normal or atypical, affect the translation of afferent information to postsynaptic neurons. The neurochemical and neurophysiological characteristics of –especially– excitatory synaptic stimuli play a decisive role in the expression of the plastic changes that dendritic spines could express since they underlie the translation of afferent information.

Bioelectric Bases of Synaptic Stimulation

Bioelectric potentials are the means through which information is communicated between neurons by allowing differential encoding of the afferent information that reaches neurons from distinct brain regions (Olvera-Cortés et al., 2015). In neurons, these potentials can be grouped as passive or active, according to their characteristics. Active potentials constitute frequency codes that are translated in the synaptic terminal and then received, and integrated into, the postsynapse after the release of the corresponding neurotransmitter. The most representative form is called the *action potential.*

Action Potential

Action potentials (AP) are all-or-nothing explosive events. They are not attenuated by distance, have similar amplitudes in many nerve cells, and vary mainly in their conduction velocity (Kuffler and Nichols, 1976). APs spread along all cell processes. They are characterized by a depolarizing phase, in which the membrane potential becomes positive until membrane polarity is reversed, when a hyperpolarizing phase begins in which the membrane potential returns to its resting value (Lagerlund, 2002). An excitable cell (such as a neuron) responds to stimulation with a change in its membrane potential proportional to the intensity and duration of the stimulus. When the change is depolarizing (i.e., reduces the membrane potential difference) and sufficiently strong, it can bring the membrane potential to a threshold, triggering an AP. This occurs at the axon hillock and is propagated in a self-regenerative manner with no net loss of current through the entire axon. The conduction of APs occurs point-by-point across both myelinated and unmyelinated axons. In the latter, small segments of the axon that undergo transmembrane potential reversal transmit this depolarization to the next segment from the currents generated in the segment with the reversed polarity (local circuit currents). It is these currents that lead the neighboring segment of the axon to reach the threshold and, hence, to the inversion of polarity typical of the AP and so on, segment-by-segment, until the AP travels across the entire axon and reaches the terminal button (Lagerlund, 2002). Inactivation of voltage-gated sodium channels gives the AP its directionality toward the terminal button. The propagation speed of myelinated fibers is faster than that of unmyelinated fibers because the myelin sheath decreases membrane capacitance that, in turn, reduces the time constant and increases the length constant of the segment of axon between areas devoid of myelin (Lagerlund, 2002), which are known as nodes of Ranvier.

Electrotonic Potentials

The main features of passive, or electrotonic, potentials are (i) their graduated character and (ii) the fact that they do not propagate over distances greater than 1 mm since they disappear due to severe attenuation. They occur in specialized regions of neurons, where responses are initiated or inhibited. Electrotonic potentials play an important role in propagating nerve impulses. At synapses, they are called either excitatory or inhibitory postsynaptic potentials (EPSP or IPSP) (Kuffler and Nichols, 1976). These potentials of membranes result from their passive electrical properties. Because cells are made up of a lipid bilayer surrounded by intra- and extracellular fluid, they constitute capacitors (that is, two conducting plates separated by dielectric material) that can store an electrical charge (Lagerlund, 2002). In addition to capacitance, the membrane has the property of resistance to the passage of charged ions. In the case of axons and dendrites –whose cylindrical shape mirrors that of a cable– a kind of axial resistance occurs that corresponds to the passage of current through their interior (Lagerlund, 2002). Thus, the capacitance and resistance of membranes as electrical material determine many of the responses to current inputs to cells that occur from synaptic contacts, whether excitatory or inhibitory.

A potential change generated at one point on the dendrite (EPSP or IPSP) does not propagate through the entire dendritic tree but, rather, decays with distance from the current entry site. Therefore, large potential changes in a neuron are due to numerous synaptic inputs integrated in time and space, as in the case of the summative effect of synaptic stimulation on the dendritic trunk (inhibitory synapses) or on dendritic spines (excitatory synapses) (Harris and Kater, 1994).

Electrotonic potentials are also involved in propagating APs and influence the conduction velocity. If internal resistance and capacitance increase, the conduction velocity decreases, and *vice versa*. The diameter of the fibers is a determining factor in the conduction speed of unmyelinated fibers, while the myelin in myelinated axons acts as a layer that thickens the membrane, thus increasing its resistance and lowering its capacitance by acting as an insulator. Furthermore, current flow occurs only at the nodes of Ranvier, so only a small proportion of the membrane participates in the nerve impulse and relatively few charges are required to shift the membrane potential. As a result, the impulse propagates rapidly from node-to-node. This is why impulse conduction in myelinated fibers is also called saltatory conduction (Kuffler and Nichols, 1976). One type of passive or electrotonic potential is known as synaptic potentials. They are produced because the release of a neurotransmitter from a presynaptic terminal produces a change in the conductance of the postsynaptic neuron; that is, it modifies its permeability to ions, allowing them to flow, and generate changes in the potential around the area of the synapse.

Synaptic Integration

When two or more synapses are activated simultaneously, distance and time are crucial. Spatial summation refers to the possibility of two or more spatially separated synapses being activated at the same time, thus adding their EPSPs to integrate a larger response. The spatial summation of synaptic potentials is not linear. In fact, it depends on the electrotonic distance between the synaptic inputs and the net driving force during the synaptic potential (Johnston and

Wu, 1997a, 1997b). The electrotonic distance depends on the capacitance and resistance of the membrane, two features that influence the synapses' capacity for spatial integration and, hence, the integration of information by the neurons.

The summation of synaptic potentials can also occur with respect to the time that synapses are activated. This temporal summation depends on the course of decay of the synaptic response over time that, in turn, is directly related to the interval between successive stimulations of the synapse. Temporal summation is also non-linear (Johnston and Wu, 1997a, 1997b). Synaptic potentials thus allow the integration of the large number of eventual synaptic inputs to neurons, such that the firing of a given neuron results from the integration of all the afferent information gathered in time and space. In reality, most of the excitatory synapses in the central nervous system (CNS) are integrative in nature.

Stimulation-Dependent Synaptic Plasticity

This modality refers to the change in the strength of the synaptic connection depending on the previous use or activity of the synapse. In addition to being integrative, the synaptic changes that induce APs may signify plastic changes that consist in variations of synaptic strength resulting from the history of activity between an afferent pathway and a population of target neurons (Olvera-Cortés et al., 2015).

Synaptic plasticity can be short- or long-term. Facilitation, post-tetanic potentiation, and depression are short-term forms of synaptic plasticity, but there are only two main forms of long-term synaptic plasticity: long-term potentiation and long-term depression.

Long-Term Potentiation (LTP)

When a synapse is stimulated by an individual pulse of a certain strength, the response it generates will be greater if high-frequency stimulation was previously applied to the same pathway. LTP is a long-term change in synaptic strength (potential amplitude associated with stimulation of a synapse or pathway) observed under certain conditions in integrative-type synapses. It is characterized by an increase in the amplitude of the EPSP after a brief train of stimuli that decays over several hours, though under certain conditions it can last for months or more. LTP involves an increase in the quantum release of neurotransmitters at the presynaptic level with changes at the postsynaptic level that include the recruitment of AMPA receptors and increases in the GluN1 and GluN2A subunits of NMDA glutamate receptors in the postsynapse (Baez et al., 2013; Squire and Kandel, 1999). LTP is the best-profiled synaptic mechanism for supporting memory (Johnston and Wu, 1997c). It was first described by Bliss and Lømo (1973) in the hippocampus, but has been observed in most excitatory synapses of the CNS. LTP represents a set of mechanisms responsible for increasing and maintaining the strength of synaptic connections. The initial part of LTP has been called early LTP or short-term potentiation (lasting from a few to about 30 minutes).

According to the various processes that occur to establish it, LTP has two phases: induction and maintenance. Induction occurs during, and shortly after, the high-frequency stimulation that induces it. It consists of the steps and mechanisms that lead to the long-lasting changes associated with LTP. The maintenance phase is the period that follows induction, when the change in synaptic efficacy is expressed (Johnston and Wu, 1997c). LTP can be induced under various conditions, some forms through the convergence of depolarization of the postsynaptic cell with stimulation of presynaptic fibers. This is known as associative LTP (Johnston and Wu, 1997d). One key feature of LTP is that the increase in the amplitude of the EPSP is almost always restricted to the pathways that are stimulated; that is, it is specific to such synapses. For this reason, we understand that LTP does not promote a generalized change in postsynaptic activity. This type of LTP is called homosynaptic. Another feature of LTP is cooperativity. A stimulation intensity threshold exists for LTP induction, such that a minimum number of concurrently activated input fibers is required for induction to occur, so we assume that some kind of cooperative interaction among different fibers is required. Extremely high stimulation intensities, or after repeated induction of LTP with a not-so-high stimulation intensity, produce a maximum amount of potentiation called saturation. LTP can be induced by stimulating a group of postsynaptic neurons via two independent but simultaneous synaptic inputs. This feature of LTP is known as associativity (Bliss and Lynch, 1988).

Long-Term Depression (LTD)

LTD was initially described in the cerebellum (Ito and Kano, 1982). Simultaneous stimulation of the parallel and climbing fibers of the cerebellum leads to depression of synaptic efficacy at the synapse between the parallel fibers and the Purkinje cells. The trigger for LTD is an increase in Ca^{+2} in the postsynaptic neuron through the aperture of voltage-gated Ca^{2+} channels that open after depolarization of the Purkinje cells (Linden et al., 1991). The transient increase in internal Ca^{2+} is sufficient to induce LTD (Konnerth et al. 1992). Small increases in Ca^{2+} have been shown to induce LTD, while large increases may induce LTP (Johnston and Wu, 1997d).

Excitatory Synaptic Transmission

Excitatory transmission in the CNS is mediated mainly by the neurotransmitter glutamate through activation of ionotropic and metabotropic receptors; that is, receptors associated with ion channels, or those coupled to second messenger systems and linked by G proteins at the intracellular level, respectively (Meldrum, 2000).

Glutamate Receptors

Glutamate receptors come in two types: ionotropic and metabotropic. Glutamate-mediated rapid chemical excitation occurs at 90% of ionotropic receptors, which are classified in three groups: N-methyl-D-aspartate (NMDA), alpha-amino-3-hydroxy-5-methyl-isoxazolepropionic acid (AMPA), and kainate (KA) receptors (Salinska et al., 2005). NMDA and AMPA receptors are deeply involved in excitatory transmission related to synaptic plasticity processes mediated by dendritic spines. NMDA receptors are cation channels that allow the passage of Na^{+}, K^{+}, and Ca^{2+} ions. They have a high permeability to Ca^{2+} ions (Ascher and Nowak, 1988) and in a resting state are blocked by physiological concentrations of Mg^{2+}, which means that they are voltage dependent (Mayer and Westbrook, 1987; Nowak et al., 1984). AMPA receptors, meanwhile, participate in a coordinated manner with NMDA receptors in rapid transmission and in phenomena related to synaptic plasticity (Fox, 1996).

There are eight subtypes of metabotropic receptors, classified in three groups: the first includes the mGluR1 and mGluR5 subtypes, which are coupled to the phosphatidylinositol pathway, while the members of the second –mGluR2, mGluR3– and third groups (mGluR4, mGluR6, mGluR7, mGluR8), are negatively coupled to adenylyl cyclase. In a typical excitatory synapse, ionotropic receptors outnumber metabotropic ones and are strongly involved in rapid impulse transmission; whereas metabotropic receptors regulate the activity of the former and mediate the slow effects of glutamate (Cooper et al., 2003). Their location on the periphery of the synapses of dendritic spines contributes strongly to this effect, mainly in large spines (i.e., mushroom) (Luján and col., 1996).

In general, both ionotropic and metabotropic receptors have been related to transient and long-lasting synaptic plasticity and, especially, to the induction and maintenance of LTP (Bennett, 2000; Bortolotto et al., 2005). mGluR1 receptors are located mainly on the margins of synapses, a location that favors modulation of the activity of AMPA and NMDA ionotropic receptors (Bessis et al., 2001) and, therefore, of the excitability of postsynaptic neurons (Niswender and Conn, 2010). mGluR2 and mGluR3 are located both pre- and postsynaptically, while those of the third group (mGluR4, mGluR6, mGluR7, mGluR8) are located pre-synaptically and act as autoreceptors with inhibitory activity (Niswender and Conn, 2010).

Spine-Mediated Glutamatergic Transmission

Given that a high percentage of excitatory synaptic contacts are carried out on dendritic spines, these structures possess all the metabolic machinery necessary to process afferent information even that which comes from glutamate receptors. As a result, they are considered quasi-independent compartments of the progenitor dendrite that regulate the flow of afferent excitatory information (Koch and col., 1992).

In broad terms, the glutamate-mediated information transmission process begins with the transmission of an EPSP to the axon terminal. The glutamate concentration in a synaptic vesicle is approximately 100 mmol/l. The release of its contents primarily activates AMPA-type ionotropic receptors (Meldrum, 2000). The AMPA receptor located on the outer surface of the dendritic spine is permeable to the reciprocal and counter-gradient flow of Na^+ and K^+. The small increase in potential difference generated by this ionic current flow produces microenvironmental conditions that favor the removal of the Mg^{2+}

ion from the interior of the NMDA receptor channel and its consequent opening, since it becomes voltage-sensitive. The NMDA receptor is also permeable to Na^+ and K^+, in the same way as AMPA and is permeable to Ca^{2+} ions (Bear and col., 1998).

Under specific conditions, such as the stimulation of mushroom spines, metabotropic receptors are activated secondarily (Luján et al., 1996), causing the release of calcium from the spine apparatus (formed by lamellae of the existing smooth endoplasmic reticulum) inside these spines (Spacek and Harris, 1997). This is related to the transmission of calcium-mediated signals from the spine apparatus to the nucleus of the neuron (Chanadai and Kavalali, 2022). The smallest spines (thin type) have Ca^{2+} chelating systems made up of "trapping" molecules, which also regulate the concentrations of the ion inside the spine (Ellisman et al., 1990). This process is reversible, so the systems involved in the release of Ca^{2+} are the same ones that "capture" it to avoid damage due to an abnormal intracellular increase in its concentration (Ellisman et al., 1990).

Thus, a regulation system exists for the concentration of Ca2+ inside the spine that naturally protects the cell from possible excitotoxic damage mediated by neurotransmitter activity and the abnormal elevation of the Ca2+ concentration, not only in the spine but also in the progenitor dendrite (Kostyuk, 1998). This is further favored by the presence of Ca2+ pumps in the spines, which participate in extruding those ions (Majewska et al., 2000) to prevent an increase in their concentration in the progenitor dendrite as they diffuse toward it (Volfovsky et al., 1999). Upon termination of its neurotransmitter effects, glutamate is recaptured from the intersynaptic gap through a process mediated by transporter proteins located in either the presynaptic terminal or glial cells (Danblot, 2001).

Synaptic Plasticity and Memory

Memory formation consists of several phases. Efficient information *acquisition* and *consolidation* normally require activity-dependent synaptic modifications. Another important aspect of memory formation is the suppression of previously established memories when they become obsolete, an essential process for behavioral adaptations (Bouton, 2004; Myers et al., 2006). The association of pre- and postsynaptic activity between two neurons evokes change in one, or both, neurons, so that the synaptic connection between them is strengthened. In other words, the same force of stimulation induces a greater postsynaptic response (increased synaptic strength or efficacy) (Hebb, 1949). Theoretically, bidirectional regulation of synaptic strength may be functionally involved in dynamic behavioral changes during different stages of memory processing. (Braunewell and Manahan-Vaughan, 2001; Diamond et al., 2005; Silva, 2003).

LTP and Memory

It is now widely accepted that synaptic plasticity underpins memory processes. In the hippocampus (Bin Ibrahim et al., 2022), cerebral cortex (Meunier et al., 2017), and amygdala (Chen et al., 2017) both LTP- and LTD-mediated modifications of synaptic strength are associated with learning. This may be prevented when LTP saturation occurs (Moser et al., 1998), for research has shown that when LTP is produced to its maximum level (saturation) by stimulation protocols, test animals trained on a learning task fail to learn. Thus, LTP could be the mechanism through which learning produces the reinforcement of synaptic strength; that is, the synaptic response may be increased at specific synapses that are activated during learning, and this modification may persist beyond the behavioral event experienced.

One of the most efficient ways to induce hippocampal LTP is through theta burst stimulation. It has been suggested that theta activity –which predominates in the hippocampus during behaviors associated with information processing (Eichenbaum et al., 1992; Vinogradova, 1995)– acts as a natural tetanizer for LTP induction. Supporting this, some increases in the expression of theta activity are associated, for example, with spatial learning that depends on the hippocampus, though not with other types of learning in which this brain region does not participate (Olvera-Cortes et al., 2002; Olvera-Cortes et al., 2004; Olvera-Cortes et al., 2012). Evidence suggests that both the facilitation of learning and impairments to it observed due to the modification of theta activity could be mediated by its consequences for LTP induction.

According to duration, memory has classically been divided into *short-* (lasting minutes to hours) and *long-term* types (ranging from days to years) (Scoville and Milner, 1957). However, memory is a complex process with features that go well beyond its duration. For this reason, studies of memory now characterize it as a "system" (González-Burgos, 2015a) (Figure 2).

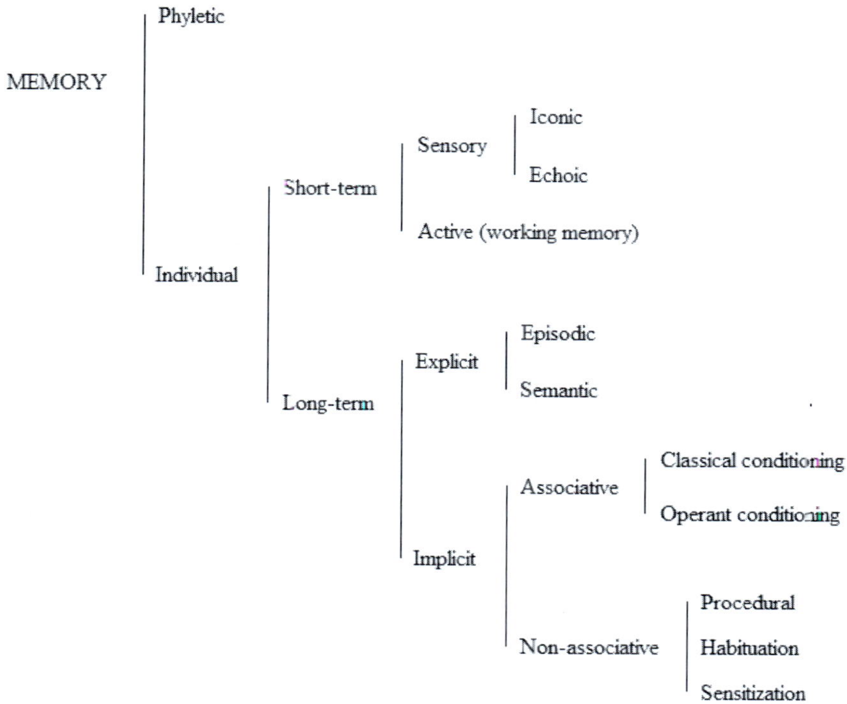

Figure 2. Chategorization of the memory systems. Modified from González-Burgos, 2022.

It has been proposed that LTP consists of two different temporal phases, each requiring a distinct type of stimulation to be triggered. Short-duration LTP (1-3 hours) can be induced by a single high-frequency (100 Hz, 1 s) stimulation train, while long-duration LTP (>4 hours) requires repeated stimulating trains (3-4 trains with separations of 5-10 minutes) (Huang and Kandel, 1994). The early phase of short- and long-term LTP may depend on post-translational modifications of pre-existing proteins and the incorporation of AMPA-type glutamate receptors into the post-synaptic density of the dendritic spines of neurons. In contrast, the late phase of LTP could depend on the protein synthesis that underlies the consolidation of long-term memory (Squire and Kandel, 1999).

It follows, then, that maintenance of LTP requires protein synthesis, which leads to changes in the synapses and a modification of the synaptic strength between them. The presence of LTP and memory formation have been associated with morphological changes in hippocampal synapses

(Geinisman, 2000; O'Connell et al., 1997; Rose, 1995). The increase in protein synthesis that accompanies LTP contributes to the establishment of morphological changes that have been reported in association with LTP, such as increases in the surface area of the postsynaptic zone (Desmond and Levy, 1988; Harris et al., 2003), in the number of spines and their stabilization from nascent short-lived ones to long-lived types (Hill and Zito, 2013; Lee et al., 1980), and in the area of spines and the number of long types (thin spines) (Desmond and Levy, 1988; Fifkova and Van Harreveld, 1977). More recently, LTP has been found to occur at individual hippocampal synapses during spatial learning through activation of a signaling cascade associated with LTP and mediated by the protein cofilin (Fedulov et al., 2007; Hongpaisan and Alkon, 2007). In fact, in one study, rats characterized by exploring behavior showed 30% more dendritic spines with CA1-phosphorylated cofilin expression than non-exploring rats. In addition, their synapses were larger (mushroom type) (González-Ramírez et al., 2014) than neighboring spines that did not show phosphorylated cofilin (Fifkova and Van Harreveld, 1977).

LTD and Memory

Although LTD reduces the strength of glutamatergic synapses, it shares certain properties with LTP, such as cooperativity and associativity. This positions it as a potential Hebbian mechanism for information storage. Pioneering studies proposed LTD as the mechanism that underlies cerebellar-mediated motor learning, as in the case of vestibulo-ocular reflex adaptation and the establishment of conditioned responses (Thompson, 1990), memory consolidation in models of spatial learning (Ge et al., 2010), and exposure to novel objects, which facilitates the induction of LTD (Manahan-Vaughan and Braunewell, 1999). In free-roaming rodents, both novel spatial learning and the learning of groups of objects were observed to trigger LTD expression in the CA1 field of the hippocampus, while preventing LTP induction. Thus, it was suggested that LTD is a key mechanism for this type of learning (Goh and Manahan-Vaughan, 2012).

The importance of LTD in memory processes involves the amygdala and ventromedial prefrontal cortex, interconnected areas involved in regulating both memory consolidation (Izquierdo et al., 2007) and extinction (Milad et al., 2007). This pathway (prefrontal-amygdala) normally generates LTD, but when the stimulation that induces LTD is applied under stressful conditions, LTP is produced. Based on this, it has been suggested that LTD could support the learning of safe situations, while LTP may participate in the memory of aversive events (Maroun, 2006). Both LTP and LTD allow the association and permanence of information in memory traces (González-Burgos, 2015a). At present, they are the most intensively studied phenomena, but could be part of a broader range of plastic responses. In this regard, plastic responses of inhibitory synapses have recently been identified and are now being explored –including both LTP (inhibitory LTP, iLTP) and LTD (inhibitory LTD, iLTD)– in the synapses between GABAergic neurons and principal projection neurons (Chapman et al., 2022; Nufer et al., 2023).

Dendritic Spine Plasticity and Cognitive Processing

Neurons establish synaptic contacts among their three structural components: soma, axon, and dendrites. In general, the vast majority of the information that a neuron receives is "captured" in synaptic contacts on its dendrites, then conducted to the soma for integration. The result of this integrative process is the emission of an action potential that is transmitted along the axon from the soma to the axonal terminal of the neuron. Though not exclusively, in the vast majority of cases the presynaptic component is represented by either the terminal buttons of the axon or the contacts that pass through the axon itself, while the postsynaptic sites are located mainly in dendrites, though also in the soma and even in the axon. Dendrites constitute the neuronal region with which most synaptic contacts are established, whether inhibitory or excitatory in nature. The former represent the ability to "slow down" the transmission of information in a neural circuit. The synaptic transmission that exerts this effect occurs in contacts between axonal buttons and postsynaptic sites generally located on the surface of the dendritic trunk of neurons, though some are present on the surface of both the soma and axons. Excitatory synaptic contacts, in contrast, occur between presynaptic axonal elements and the dendritic trunk of neurons, though the vast majority occur on dendritic spines.

These spines are cytoplasmic protrusions 0.1-2 μm long. They can be seen in the soma (Mugnaini et al., 1967) and the axon hillock of neurons (Westrum, 1970), but most are found in dendrites, which is why they are generically known as "dendritic spines." These spines protrude more or less perpendicularly to the longitudinal axis and correspond to the progenitor dendrites. Most excitatory synaptic contacts occur on these spines. The number of dendritic spines in neurons varies depending on the density of the afferent excitatory innervation to them (Bannister and Larkman, 1995).

Observations of neurons with spines show that they congregate preferentially in dendritic regions distal to the soma (Edwards, 1995). The finding that the synaptic contacts established with spines are excitatory means that they could be related to the fact that inhibitory synapses occur predominantly in the dendritic area close to the soma (Edwards, 1995). Hence, it has been proposed that the excitatory activity that enters from the distal portions through synaptic contacts mediated by spines may be regulated by the influx of inhibitory information during its transit toward the neuronal soma for its subsequent integration (Pérez-Vega et al., 2000). In addition, afferent excitatory information to a spiny neuron is regulated by differences in density (number of spines per unit of dendritic length), distribution (typical presence in specific regions of the dendritic tree), and the proportional density with which different types of spines are present (Bannister and Larkman, 1995).

The proportional density of the different types of spines (thin, mushroom, stubby, wide, branched, double) also varies, though the thin and mushroom types usually predominate, followed by stubby ones and, to a much lesser extent, wide, and then the branched and double kinds (González-Burgos, 2009, 2012, 2022; Lee et al., 2005). Since the thin, mushroom, and stubby spines are the most abundant types, and the ones that show greater variability under diverse environmental circumstances, these three forms are considered critical for integrating the synaptic information they mediate (Pérez-Vega et al., 2000; Kasai et al., 2003).

It is important to highlight that, while the different "shapes" of spines described to date are based on more or less specific electrophysiological characteristics –which has earned them their morphological and functional distinction– they are by no means rigid or fixed, much less immutable. In fact, one could argue that they form a morphofunctional continuum based on their plastic capacity to transform from one type to another (Arellano et al., 2007; Gipson and Olive, 2017; Sala and Segal, 2014), though such events are highly specific and modulated by various molecular and neurochemical factors internal and external to the spines themselves (González-Burgos, 2022). Dendritic spines are not, therefore, immutable entities in terms of their geometric structure. Quite to the contrary, they can present structural modifications with a dynamism on the order of minutes (Muller et al., 2000) in terms of interconversions between their various geometric shapes.

The different shapes of dendritic spines determine differential effects mediated by the excitatory synaptic drive itself which, under certain stimulation conditions, can induce modifications in the geometric structure of the spines (Harris and Kater, 1994). After the action of a synaptic stimulus, these spines can undergo changes in their geometric structure in a matter of minutes (Muller et al., 2000). When such modifications occur, the distribution of glutamate receptors is also modified, such that if the head of the spine is enlarged, the area occupied by the ionotropic receptors expands with respect to that occupied by metabotropic ones, though the area they occupy in the periphery also increases, together with their number (Luján et al., 1996).

LTP induces structural changes in the geometry of dendritic spines. On the one hand, spines that are enhanced during the induction phase usually transform from the thin to the mushroom type (Lang et al., 2004) until LTP consolidation is reached (Yang et al., 2008); on the other, mushroom spines may show "perforations" in the outer region of the postsynaptic density and transform into the branched form (Muller et al., 2000) until, eventually, two thin spines are produced. The ephemeral nature of these processes could help explain the low proportional density of the branched type.

In the early 1990s, this morphological characterization showed that the narrowness of the neck of the spines is directly related to their resistance to the passage of electrical current mediated by Ca^{2+} ions. This suggested that spines might exert a modulating effect on the synaptic currents related to afferent excitatory information to neurons (Koch et al., 1992; Koch and Zador, 1993). Based on this, and a highly varied corpus of complementary evidence, we now know that the role of each type of spine in synaptic impulse transmission is distinct, and that there is a close relation between spine shape and differential processing of the excitatory synaptic information afferent to them, their plastic capacity, and the transmission of synaptic information involved in the processing of cognitive (Kasai et al., 2010a, 2010b) and non-cognitive information (González-Burgos et al., 2017).

Both ionotropic and metabotropic glutamatergic receptors on dendritic spines are involved in short- (transient) and long-term (persistent) synaptic plasticity (Bennett, 2000; Bortolotto et al., 2005). If synaptic stimulation is intense, metabotropic receptors are secondarily activated because, among other reasons, they are predominantly located on the periphery of the synapse (Luján et al., 1996). As part of the process of nerve impulse transmission at the synaptic level, the flow of Ca^{2+} ions toward the progenitor dendrite encounters greater physical and electrical resistance in spines with narrower necks (Volfovski et al., 1999; Majewska et al., 2000). The electrical resistance of the neck depends on its length, the transverse space it occupies, its cytoplasmic resistivity, and the activity of the organelles inside it (Wickens, 1988). Of the total Ca^{2+} that enters a typical spine, 95% is captured by intracellular buffer systems. The remaining 5% operates freely in the interior to exert its signaling effects. Of this 5%, around 70% is expelled from the spine by the action of Ca^{2+} pumps in the lateral zones of the membrane. The other 30% is captured by the smooth endoplasmic reticulum. For signaling purposes, only 0.01% of total Ca^{2+} flows into the parental dendrite through the neck (Nimchinsky et al., 2002).

The concentration of Ca^{2+} inside the spine is a determining factor in its eventual plastic geometric transformation (Koch and Zador, 1993; Kovalchuk et al., 2000). Moderate increases in Ca^{2+} levels within normal limits cause elongation of the spines, but excessive increases trigger a contraction of the spine, or even its disappearance (Segal et al., 2000). Consistent with this, studies have reported that small pulses of glutamate stimulation elongate spines, whereas larger ones induce their retraction or disappearance (Korkotian and Segal, 1999). Consequently, excessive excitatory stimulation of synapses with dendritic spines can decrease their density (Jiang et al., 1998), while weak stimulation may augment their number (Kirov and Harris, 1999).

In adults these spines are less plastic, and plastic for shorter periods, than in younger individuals (Grutzendler et al., 2002). During brain development, the proportion of spines that are capable of expressing plastic transformations decreases gradually. These "transient" spines are predominantly of the thin type (Holtmaat et al., 2005) which, as the most efficient ones in the transmission of synaptic impulses (Koch et al., 1992), allow adjustments in information processing. The proportion of persistent spines (i.e., those less plastic than transitory ones) ranges from 35% in juvenile stages to 73% in mature adults (Holtmaat and col., 2005).

Cognitive Information Processing

The interconversions of dendritic spines constitute the morphofunctional substrate that sustains the capacity for synaptic plasticity which underlies phenomena as complex as learning and memory (González-Burgos, 2012; González-Burgos et al., 2012), or even neural activity unrelated to cognition (González-Burgos et al., 2017). Available evidence shows that the morphophysiological activity of thin spines is related to the acquisition of information –learning, for example– while those in the mushroom type are involved in information storage; that is, memory (Bourne and Harris, 2007; Kasai et al., 2003; Matsuzaki et al., 2004; Medvedev et al., 2010). Regardless of the specific process that creates them (González-Burgos and Velázquez-

Zamora, Chapter 1 in this book), neckless stubby spines have been considered the initial stage of dendritic spine maturation (Harris et al., 1992; Velázquez-Zamora et al., 2011), while more recent work shows that they participate in regulating neuronal excitability (Flores-Soto et al., 2021; Romero-Guerrero et al., 2022). In addition to being mediated by excitatory glutamatergic transmission, cognitive performance mediated by the plastic activity of dendritic spines is modulated by various tissue components. Among other elements, estrogens play a preponderant role (Arevalo et al., 2015) in modulating cognitive activity in various physiological conditions, such as the menstrual cycle and menopause (Cervantes-Alfaro et al., 2015; González-Flores et al., 2014).

Studies related to differential cognitive performance in the distinct phases of the estrous cycle of rats have shown a higher proportion of thin spines and fewer mushroom spines in the pyramidal neurons of the CA1 field of the rat hippocampus in the estrus stage compared to proestrus, with a concomitantly higher density of total spines in the latter (González-Burgos et al., 2005). Despite the greater associative capacity of synapses (Harris and Kater, 1994; Leuner and Shors, 2004; Leuner et al., 2003) and estrogenic activity (Woolley et al., 1990) that, hypothetically, would facilitate cognitive activity (Luine and Frankfurt, 2020) during proestrus, in this phase of the cycle a less efficient performance of spatial memory has been observed compared to the estrus stage, when behavioral performance was enhanced (Warren and Jurasca, 1997). The enhanced cognitive efficiency observed during estrous could be due to the higher proportion of thin spines, due to their immediate plasticity (Fischer et al., 1998) and ability to quickly process and integrate synaptic information (Bortolotto et al., 2005), associated with LTP induction (Chidambaram et al., 2019) and the acquisition of novel information in the short term (Hayashi-Takagi et al., 2015; Nabavi et al., 2014); that is, learning (Bourne and Harris, 2007).

Experimental studies related to the cessation of estrogenic activity show that estrogen replacement therapy induces a recovery in dendritic spine density in pyramidal neurons of the CA1 field of the hippocampus, with increases in the number of thin, mushroom, and stubby spines (Velázquez-Zamora et al., 2012a). Similarly, replacement therapy with selective estrogen receptor modulators (SERMs), like tamoxifen or raloxifene, increases the density of dendritic spines, especially thin ones, in layers II/III of the pyramidal neurons in the prefrontal cortex. This underlies improved performance of the allocentric working memory (Velázquez-Zamora et al., 2012b). Likewise, the densities of total spines and of the thin, mushroom, stubby, and wide types increased in hippocampal neurons of male rodents. In addition to a neuroprotective effect, this finding suggests the occurrence of plastic events related to better task performance after the reduction of the hormonal activity that occurs with age (González-Burgos et al., 2012).

Since it derives from hormonal activity, the organization of sexual behavior is also subject to the expression of plastic changes mediated by dendritic spines. Type II neurons of the ventromedial nucleus of the hypothalamus (Pfaff, 1980; Sá and Madeira, 2005) are known to participate in the organization of motor patterns associated with estradiol-mediated mating behavior in females (Griffin and Flanagan-Cato, 2011). In these neurons, a reduction in dendritic spines was observed in the metestrus stage (low estrogen) compared to the other stages (diestrus, proestrus, estrus), with an increase in thin and mushroom spines in proestrus (high estrogen) and a decrease in these spines in estrus. These findings suggest that a more efficient processing capacity operates during the stage of sexual receptivity (González-Burgos et al., 2015). Recent studies also show that sexual experience in male rats is associated with a lower proportion of thin spines and a higher proportion of mushroom spines in both prefrontal projection neurons and median spiny neurons in the shell region of the nucleus accumbens. This evidence seems to support the consolidation not only of motivational and attentional information but also of that related to decision-making and consequent action, all of which are necessary for successful mating in relation to sexual experience (Hernández-González et al., 2023).

According to the most widely accepted characterization of the normal development of aging, this process can occur successfully or be associated with some kind of pathology (Depp and Jeste, 2006; Roberson et al., 2012). Normal aging presents an involutionary phase of the brain, which begins around the seventh decade of life. However, due to the great compensation capacity (plasticity) and reserves of the Nervous System, behavioral alterations are more evident between 80 and 90 years of age. Pot-mortem studies carried out in humans show that from the age of 90 the number of neurons is reduced by up to 45% in some areas of the brain although, from the age of 50, the hippocampus shows a neuronal reduction of 5% during each subsequent decade (Selkoe, 1992).

The areas that lose fewer neurons during the course of age are those included in the temporal lobes (hippocampus), particularly involved in the memory acquired during individual development (episodic and semantic memory), while the most affected correspond to cortical regions of association, such as the frontal, parietal, and temporal cortex (Cerella, 1990).

When this neuronal loss exceeds a certain critical limit, the cognitive-behavioral alterations typical of normal aging appear (Aveleyra-Ojeda et al., 2007). This is consistent with reports showing that during successful aging both learning and the performance of spatial reference memory (González-Burgos, 2015a) occur more slowly than in young individuals, though with equal efficiency (González-Ramírez et al., 2015). These findings may be related to plastic events at the level of the dendritic spines. A higher total spine density was observed in the pyramidal neurons of the hippocampal CA1 field of elderly subjects trained with the Morris maze (González-Burgos, 2015b), but a decrease in thin spines also occurred. Although a greater proportional density of mushroom spines was seen (that, in theory, could facilitate behavioral performance), they may have been dysfunctional due to abnormal activity of the cofilin protein observed in senile rats (Cichon et al., 2012). Consequently, the resulting slowness in solving cognitive tasks during successful aging could reflect compensatory plastic phenomena, though these eventually prove inadequate to compensate for events related to the natural aging process.

Prefrontal and cerebellar activity is also affected during successful aging (Vázquez-Hernández et al., 2021). Though at an underlying level, dendritic spines are clearly involved in the characteristics of behavioral performance associated with working memory and motor coordination, compared to mature young subjects, aged animals show impaired allocentric working memory performance, and balance and motor coordination tests show erratic performance. A reduced density of dendritic spines has been observed in both the pyramidal neurons of layers II/III of the prefrontal cortex and Purkinje neurons in the cerebellum, associated with a proportional decrease in thin, mushroom, stubby, and wide types. This loss of spines suggests a decrease in the associative and integrative capacity of synaptic inputs (Harris and Kater, 1994; Leuner and Shors, 2004; Leuner et al., 2003) that, together with the homogeneous reduction of each type of spine, could help explain both poor cognitive performance and the erratic locomotor behavior that –though not pathology-related– constrain the full functional performance of elderly individuals.

Locomotor activity *per se* causes plastic modifications in the regions involved in its organization. It has been reported that mild, moderate, and severe motor training condition the differential expression of plastic changes by the dendritic spines in the pyramidal neurons of layer V of the motor cortex (Vázquez-Hernández et al., 2017) and in Purkinje neurons in the cerebellum (González-Burgos et al., 2011), similar to the way that motor learning does

(González-Tapia et al., 2015). Motor training has been shown to induce an increase in the density of dendritic spines that translates into an increase in thin, mushroom, stubby and wide types. This demonstrates that spinogenesis occurs, and suggests the acquisition and consolidation of information required for differential performance in the three training schemes, along with a physiological damping effect in the face of synaptic overstimulation mediated by forced or extreme motor performance (González-Burgos, 2022; González-Tapia et al., 2017; Yu and Zuo, 2011). Throughout the training period of motor learning processes, there is an increase in thin spines, accompanied by a reduction in mushroom spines in layer V of the prefrontal cortex, suggestive of an active process of rectification of the motor activity associated with gradually-acquired motor learning (González-Tapia et al., 2016).

Various pathological alterations –paralysis, paresthesia, hyperreflexia, for example– are related to damage in the motor region of the cerebral cortex (Napieralski et al., 1998; Nudo et al., 2001; Oudega and Perez, 2012; Tan et al., 2013; Welniarz et al., 2017). Recent experimental studies show that after a pharmacological lesion of the corticospinal pathway, both gait and motor coordination are impaired, concomitantly with an increase in dendritic spines and a greater proportional density of the thin and stubby types in thoraco-lumbar motoneurons, as well as an increase in the expression of spinophyllin (related to spine density), synaptophysin (related to ionotropic receptor anchoring), and beta III-tubulin (related to axonal regrowth) (Martínez-Torres et al., 2021a). These findings are consistent with the functional recovery that occurs spontaneously after spinal cord injury (Franz et al., 2014), though this is unsustainable over time (Jacobson et al., 2021). The plastic events underlying this spontaneously incipient recovery could be related to the increase in the proteins evaluated, especially the thin spines which, thanks to their attributes of mediating rapid neurotransmission and processing novel information through their associative and integrative capacities, could provide the opportunity to influence early rehabilitation processes. Following this, experimental rehabilitation after lesions to the motor cortex has resulted in an increase in the density of thin, mushroom, stubby, and wide spines in spinal motoneurons (Vázquez-Hernández et al., 2022) that suggests facilitation of the plastic processes that underlie the spontaneous recovery initially observed post-lesion. When a contusive injury is induced in the spinal cord, an increase in thin, stubby, and wide spines is also observed, with the appearance of filopodia (related to spinogenesis) and other spiny, but atypical, structures (Portilla-de Buen et al., 2017).

Taken together, the evidence related to both normal and pathological motor performance reveals the occurrence of plastic events mediated by dendritic spines. These initial plastic processes should be considered an opportunity to intervene and foster the rehabilitation of patients affected by disorders of the psychomotor system, rather than as a natural compensatory response that is clearly insufficient and somewhat misleading.

Generically, alterations in glutamate-mediated neurotransmission can affect cognitive performance. Neonatal administration of monosodium glutamate reduces the number of pyramidal neurons in the CA1 field of the hippocampus, while also increasing the thin and mushroom (González-Burgos et al., 2009) that underlie impairments in spatial learning and memory (Olvera-Cortés et al., 2005). Since the activity of the NMDA receptors is directly involved in transmission mediated by dendritic spines (Velázquez-Zamora et al., 2011), findings like these could be related to the over-expression of subunits of that ionotropic receptor caused by increased glutamatergic activity (Beas-Zárate et al., 2001, 2002; Yan et al., 2002). Although the increase in spines related to cognitive processing suggests a compensatory plastic response that, theoretically, could favor behavioral performance, evidence shows that plastic modifications of dendritic spines do not necessarily support functional rehabilitation. Hence, it is tempting to think that, under atypical conditions, plasticity mediated by dendritic spines is not clearly distinguishable from synaptic changes of a pathological nature (McEachern and Shaw, 1999).

One pathological condition related to motor skills is attention deficit hyperactivity disorder (ADHD). Intracisternal neonatal application of 6-hydroxydopamine to rats produced an increase in motor activity in the juvenile stage with a reduction in dendritic spines in the median spiny neurons of the core of the nucleus accumbens (related to the expression of motor patterns). There was no detriment to the different types of spines, though the thin ones were larger than normal (González-Burgos et al., 2010). The decrease in the number of spines could be due to over-stimulation that resulted from the dopaminergic dysregulation (Spencer et al., 1998) that occurred during the development of the nucleus accumbens under the effects of the lesion inflicted (Antonopolus et al., 2002). The elongation of the thin spines could be interpreted as a compensatory effect that tended toward a better processing of the incoming novel information to structure motor patterns according to environmental demands, before the general pruning of the spines. After treatment with atomoxetine [a drug used to treat ADHD (Aman et al., 2014; Fredriksen and Peleikis, 2016)], motor activity returned to normal values. In the neurons of layer V of the prefrontal cortex, the proportion of thin spines was lower, while the number of mushroom spines increased concomitantly with a reversal of allocentric working memory impairment (Martínez-Torres et al., 2018) that is known to occur in individuals with ADHD (Ramos et al., 2020). Although the increase in spine density related to the storage and consolidation of acquired information (mushroom spines) could point to better cognitive performance and be considered a compensatory effect, the reduction of the spines related to the rapid processing of incoming information and, consequently, the acquisition of novel information (thin spines), could have prevented the normalization of cortical processing of information related to the adequate execution of working memory in hyperactive subjects.

Currently, a wide variety of pathological conditions occur that involve alterations of the neuronal cytoarchitecture, including the density and geometric structure of the dendritic spines of the affected neurons. These disorders include global cerebral ischemia (García-Chávez et al., 2008), Fetal Alcohol Syndrome (González-Burgos et al., 2006), Alzheimer's (Walker and Herskowitz, 2021), Parkinson's disease (Nishihima et al., 2018), schizophrenia (Martínez-Torres et al., 2021b; Reza-Zaldivar et al., 2020), depression (Holmes et al., 2023), autism spectrum disorder (Nishiyama, 2019), and malnutrition (discussed in the last chapter of this book) which, by the way, occur in both underdeveloped and highly-developed, industrialized countries.

Conclusion

Plastic events that occur at different levels of biological organization underlie, in a coordinated manner, the behavioral expression of such cognitive processes as learning and memory. Scientific evidence shows various, highly dynamic plastic changes in neurons at both the pre- and postsynaptic level, due to neurotransmitter-mediated stimulation. Specifically, glutamate-mediated synaptic stimulation generates modifications in the dendritic spines of neurons that modulate postsynaptic responses to inputs related to the processing of information with cognitive content.

Until the end of the last century, the study of plasticity mediated by dendritic spines was limited almost exclusively to eventual variations in their density. Over the last two decades, however, these descriptions have been extended to variations in the shape of the spines, an advance that has amply improved our understanding of their participation in organizing cognitive processes under both normal conditions during the various stages of development, and under various atypical conditions derived from either neuropathological damage or psychopathological disorders. The study of plastic changes in the morphophysiology of dendritic spines provides consistent evidence of the differential participation of the distinct types of spines in various components of cognitive activity. Today, determining the directionality of such events and how they can be manipulated to foster the rehabilitation of behavior related to cognition constitutes a scientific challenge that requires new, more advanced research paradigms.

References

Akbik, F., Cafferty, W. B., Strittmatter, S. M., 2012. Myelin associated inhibitors: a link between injury-induced and experience-dependent plasticity. *Exp. Neurol.* 235, 43-52

Aman, M. G., Smith, T., Arnold, L. E., Corbett-Dick, P., Tumuluru, R., Hollway, J. A., Hyman, S. L., Mendoza-Burcham, M., Pan, X., Mruzek, D. W., Lecavalier, L., Levato, L., Silverman, L. B., Handen, B., 2014. A review of atomoxetine effects in young people with developmental disabilities. *Res. Dev. Disabil.* 35, 1412-1424.

Antonopolus, J., Dori, J., Dinopoulos, A., Chiotelli, M., Parnavelas, J. G., 2002. Postnatal development of the dopaminergic system of the striatum in the rat. *Neuroscience* 110, 245–256.

Arellano, J. I., Benavides-Piccione, R., Defelipe, J., Yuste, R., 2007. Ultrastructure of dendritic spines: correlation between synaptic and spine morphologies. *Front. Neurosci.* 1, 131-143.

Arevalo, M. A., Azcoitia, I., Gonzalez-Burgos, I., Garcia-Segura, L. M., 2015. Signaling mechanisms mediating the regulation of synaptic plasticity and memory by estradiol. *Horm. Behav.* 74, 19-27.

Aveleyra-Ojeda, E., Gómez-Pérez, M. E., Ostrosky-Shejet, F., 2007. Cambios neurofisiológicos, cognoscitivos y neuroendócrinos durante el envejecimiento [Neurophysiological, cognitive and neuroendocrine changes during aging], in: Guevara-Pérez, M. A., Hernández-González, M., Arteaga-Silva, M., Olvera-Cortés, M. E. (Eds.), *Aproximaciones al studio de la Funcionalidad cerebral y el Comportamiento [Approaches to the study of brain function and behavior].* Universidad de Guadalajara, Guadalajara, pp. 251-286.

Bach-y-Rita, P., 1990. Receptor plasticity and volume transmission in the brain: emerging concepts with relevance to neurologic rehabilitation. *J. Neurol. Rehabil.* 4, 121–128.

Baez, M. V., Oberholzer, M. V., Cercato, M. C., Snitcofsky, M., Aguirre, A. I., Jerusalinsky, D. A., 2013. NMDA receptor subunits in the adult rat hippocampus undergo similar changes after 5 minutes in an open field and after LTP Induction. *PLoS One* 8, e55244.

Bannister, N. J., Larkman, A. U., 1995. Dendritic morphology of CA1 pyramidal neurons from the rat hippocampus: II. Spine distributions. *J. Comp. Neurol.* 360, 161-171.

Bear, M. F., Connors, B. W., Paradiso, M. A., 1998. Neurociencia. Explorando el Cerebro [Neuroscience. Exploring the Brain]. MASSON-Williams & Wilkins, Barcelona.

Beas-Zárate, C., Flores-Soto, M. E., Armendariz-Borunda, J., 2002. NMDAR-2C and 2D subunits gene expression is induced in brain by neonatal exposure of monosodium L-glutamate to adult rats. *Neurosci. Lett.* 321, 9-12.

Beas-Zárate, C., Rivera-Huizar, S. V., Martinez-Contreras, A., Feria-Velasco, A., Armendariz-Borunda, J., 2001. Changes in NMDA-receptor gene expression are associated with neurotoxicity induced neonatally by glutamate in the rat brain. *Neurochem. Int.* 39, 1-10.

Bennett, M. R., 2000. The concept of long-term potentiation of transmission at synapses. *Prog. Neurobiol.* 60, 109–137.

Bessis, A. S., Acher, F., Pin, J. P., 2001. Metabotropic glutamate receptor: exciting possibilities in excitatory transmission. *Cell transmissions*. 17, 3-9.

Bin Ibrahim, M. Z., Benoy, A., Sajikumar, S., 2022. Long-term plasticity in the hippocampus: maintaining within and 'tagging' between synapses. *FEBS J.* 289, 2176-2201.

Bliss, T. V., Lømo, T., 1973. Long-lasting potentiation of synaptic transmission in the dentate area of the anaesthetized rabbit following stimulation of the perforant path. *J. Physiol.* 232, 331-356.

Bliss, T. V. P., Lynch, M. A., 1988. Long-term potentiation of synaptic transmission in the huppocampus: properties and mechanisms, in: Landfield, P. W., Deadwyler, S. (Eds.), *Long term potentiation: From biophysiscs to behavior.* Liss, New York, pp. 3-72.

Bortolotto, Z. A., Collett, V. J., Conquet. F., Jia, Z., van der Putten, H., Collingridge, G. L., 2005. The regulation of hippocampal LTP by the molecular switch, a form of metaplasticity, requires mGlu5 receptors. *Neuropharmacology* 49, 13-25.

Bourne, J., Harris, K. M., 2007. Do thin spines learn to be mushroom spines that remember? *Curr. Op. Neurobiol.* 17, 381-386.

Bouton, M. E., 2004. Context and behavioral processes in extinction. *Learn. Mem.* 11, 485-494.

Brailowsky, S., Stein, D. G., Will, B., 1998. *El cerebro averiado. Plasticidad cerebral y recuperación functional [Damaged brain. Brain plasticity and functional recovery].* Fondo de Cultura Económica [Fund of Economic Culture], 2ª Ed., México.

Braunewell, K. H., Manahan-Vaughan, D., 2001. Long-term depression: a cellular basis for learning? *Rev. Neurosci.* 12, 121-140.

Cerella, J., 1990. Aging and information processing rate, in: Birren, J., Schaie, K. (Eds.), *Handbook of Psychology and Aging.* Academic Press, New York, pp. 201-221.

Cervantes, M., Velázquez-Zamora, D. A., González-Burgos, I., 2015. Modulación estrogénica de las funciones cognitivas durante el ciclo menstrual y la menopausia [Estrogenic modulation of cognitive functions during the menstrual cycle and menopause], in: González-Burgos, I. (Ed.), *Psicobiología de la Memoria. Un Enfoque Interdisciplinario [Psychobiology of Memory. An Interdisciplinary Approach].* Bios-Médica, Guadalajara, pp. 263-290.

Chanaday, N. L., Kavalali, E. T., 2022. Role of the endoplasmic reticulum in synaptic transmission. *Curr. Op. Neurobiol.* 73, 102538.

Chapman, C. A., Nuwer, J. L., Jacob, T. C., 2022. The Yin and Yang of GABAergic and glutamatergic synaptic plasticity: opposites in balance by crosstalking mechanisms. *Front. Synaptic Neurosci.* 14, 911020.

Chen, A., Hu, W. W., Jiang, X. L., Potegal, M., Li, H., 2017. Molecular mechanisms of group I metabotropic glutamate receptor mediated LTP and LTD in basolateral amygdala in vitro. *Psychopharmacology* 234, 681-694.

Cichon, J., Sun, Ch., Chen, B., Jiang, M., Chen, X. A., Sun, Y., Wang, Y., Chen, G., 2012. Cofilin aggregation blocks intracellular trafficking and induces synaptic loss in hippocampal neurons. *J. Biol. Chem.* 287, 3919-3929.

Cooper, J. R., Bloom, F. E., Roth, R. H., 2003. The biochemical basis of neuropharmacology. 8th ed. Oxford University Press, New York.

Danbolt, N. C., 2001. Glutamate uptake. *Prog. Neurobiol.* 65, 1-105.

Depp, C. A., Jeste, D. V., 2006. Definitions and predictors of successful aging: a comprehensive review of larger quantitative studies. *Am. J. Geriatr. Psychiatry* 14, 6-20.

Desmond, N. L., Levy, W. B., 1988. Synaptic interface surface area increases with long-term potentiation in the hippocampal dentate gyrus. *Brain Res.* 453, 308-314.

Diamond, D. M., Park, C. R., Campbell, A. M., Woodson, J. C., 2005. Competitive interactions between endogenous LTD and LTP in the hippocampus underlie the storage of emotional memories and stress-induced amnesia. *Hippocampus.* 15, 1006-1025.

Duffau, H., 2006. Brain plasticity: from patophysiological mechanisms to therapeutic applications. *J. Clin. Neurosci.* 13, 885-897.

Edwards, F. A., 1995. Anatomy and electrophysiology of fast synapses lead to a structural model for long-term potentiation. *Physiol. Rev.* 75, 759-787.

Eichenbaum, H., Otto, T., Cohen, N. J., 1992. The hippocampus--what does it do? *Behav. Neural Biol.* 57, 2-36.

Ellisman, M. H., Deering, T. J., Ouyang, Y., Beck, C. F., Tanksley, S. J., Walton, P. D., Airey, J. A., Sutko, J. L., 1990. Identification and localization of ryanodine binding proteins in the avian central nervous system. *Neuron.* 5, 135-146.

Fedulov, V., Rex, Ch. S., Simmons, D. A., Palmer, L., Gall, Ch. M., Lynch, G., 2007. Evidence that long-term potentiation occurs within individual hippocampal synapses during learning. *J. Neurosci.* 27, 8031-8039.

Fifkova, E., Van Harreveld, A., 1977. Long-lasting morphological changes in dendritic spines of dentate granular cells following stimulation of the entorhinal area. *J. Neurocytol.* 6, 211-230.

Fischer, M., Kaech, S., Knutti, D , Matus, A., 1998. Rapid actin-based plasticity in dendritic spines. *Neuron* 20, 847-854.

Flores-Soto, M., Romero-Guerrero, C., Vázquez-Hernández, N., Tejeda-Martínez, A., Martín-Amaya-Barajas, F. L., Orozco-Suárez, S., González-Burgos, I., 2021. Pentylenetetrazol-induced seizures in adult rats are associated with plastic changes to the dendritic spines on hippocampal CA1 pyramidal neurons. *Behav. Brain Res.* 28,113198.

Forbes, A., 1922. The interpretations of spinal reflexes in terms of present knowledge of nerve conduction. *Physiol. Rev.* 2, 361-414.

Fox, K., 1996. The role of excitatory amino acid receptors in developmental plasticity of sensory neocortex, In: Conti, F., Hicks, Ph. (Eds.), *Excitatory Amino Acids in the Cerebral Cortex.* MIT, USA, pp. 229-244.

Franz, S., Ciatipis, M., Pfeifer, K., Kierdorf, B., Sandner, B., Bogdahn, U., Blesch, A., Winner, B., Weidner, N., 2014. Thoracic rat spinal cord contusion injury induces remote spinal gliogenesis but not neurogenesis or gliogenesis in the brain. *PLoS One* 9, e102896.

Fredriksen, M., Peleikis, D. E., 2016. Long-term pharmacotherapy of adults with attention deficit hyperactivity disorder: A literature review and clinical study. *Basic Clin. Pharmacol. Toxicol.* 118, 23-31.

García-Chávez, D., González-Burgos, I., Letechipía-Vallejo, G., López-Loeza, E., Moralí, G., Cervantes, M., 2008. Long-term evaluation of cytoarchitectonic characteristics of prefrontal cortex pyramidal neurons, following global cerebral ischemia and neuroprotective melatonin treatment, in rats. *Neurosci. Lett.* 448, 148-152. 2008.

Ge, Y., Dong, Z., Bagot, R. C., Howland, J. G., Phillips, A. G., Wong, T. P., Wang, Y. T., 2010. Hippocampal long-term depression is required for the consolidation of spatial memory. *Proc. Natl. Acad. Sci. U S A*. 107, 16697-16702.

Geinisman, Y., 2000. Structural synaptic modifications associated with hippocampal LTP and behavioral learning. *Cereb. Cortex.* 10, 952-962.

Gipson, C. D., Olive, M. F., 2017. Structural and functional plasticity of dendritic spines - root or result of behavior? *Genes Brain Behav.* 16, 101-117.

Goh, J. J., Manahan-Vaughan, D., 2013. Spatial object recognition enables endogenous LTD that curtails LTP in the mouse hippocampus. *Cereb. Cortex* 23, 1118-1125.

González-Burgos, I., 2009. Dendritic spines plasticity and learning / memory processes: Theory, evidence and perspectives, in: Baylog, L. R. (Ed.), *Dendritic Spines. Biochemistry, Modelling and Properties.* Nova Science Publishers, Inc., New York, pp. 163-186.

González-Burgos, I., 2012. From synaptic transmission to cognition: an intermediary role for dendritic spines. *Brain Cogn.* 80, 177-183.

González-Burgos, I., 2015a. Neuroanatomía functional de los Sistemas de Memoria [Functional Neuroanatomy of Memory Systems], in: González-Burgos, I. (Ed.), *Psicobiología de la Memoria. Un Enfoque Interdisciplinario [Psychobiology of Memory. An Interdisciplinary Approach].* Bios-Médica, Guadalajara, pp. 15-50.

González-Burgos, I., 2015b. Estudio experimental del aprendizaje y la memoria en animales de laboratorio. Fundamentos teóricos, metodológicos y técnicos [Experimental study of learning and memory in laboratory animals. Theoretical, methodological and technical foundations], in: González-Burgos, I. (Ed.), *Psicobiología de la Memoria. Un Enfoque Interdisciplinario [Psychobiology of Memory. An Interdisciplinary Approach].* Bios-Médica, Guadalajara, pp. 595-624.

González-Burgos, I., 2022. *Psychobiological Principles of the Memory Process.* Nova Science Publishers, Inc. New York. 2022.

González-Burgos, I., Alejandre-Gómez, M., Cervantes, M., 2005. Spine-type densities of hippocampal CA1 neurons vary in proestrus and estrus rats. *Neurosci. Let.* 379, 52-54.

González-Burgos, I., Alejandre-Gómez, M., Olvera-Cortés, M. E., Pérez-Vega, M. I., Evans, S., Feria-Velasco, A., 2006. Prenatal-through-postnatal exposure to moderate levels of ethanol leads to damage on the hippocampal CA1 field of juvenile rats. A stereology and Golgi study. *Neurosci. Res.* 56, 400-408.

González-Burgos, I., García-Martínez, S., Velázquez-Zamora, D. A., Ponce-Rolón, R., 2010. Cytoarchitectural impairments in the medium spiny neurons of the Nucleus Accumbens core of hyperactive juvenile rats. *Int. J. Dev. Neurosci.* 28, 475-480.

González-Burgos, I., González-Tapia, D., Velázquez Zamora, D. A., Feria-Velasco, A., Beas-Zárate, C., 2011. Guided motor training induces dendritic spine plasticity in adult rat cerebellar Purkinje cells. *Neurosci. Lett.* 491, 216-220.

González-Burgos, I., Rivera-Cervantes, M. C., Velázquez-Zamora, D. A., Feria-Velasco, A., García-Segura, L. M., 2012. Selective estrogen receptor modulators regulate dendritic spine plasticity in the hippocampus of male rats, in special issue: Dendritic Spine Plasticity and Cognition. González-Burgos, I., Nikonenko, I., Korz, V. (Eds.). *Neural Plast.* Vol. 2012, Article ID 309494, 6 pages.

González-Burgos, I., Velázquez-Zamora, D. A., González-Tapia, D., Cervantes, M. A., 2015. Golgi study of the plasticity of dendritic spines in the hypothalamic ventromedial nucleus during the estrous cycle of female rats. *Neuroscience* 298, 74-80.

González-Burgos, I., Velázquez-Zamora, D. A., González-Tapia, D., Vázquez-Hernández, N., Martínez-Torres, N. I., 2017. Plasticity of dendritic spines. Not only for cognitive processes, in: Heinbockel, Th. H. (Ed.), *Synaptic Plasticity*, InTech, Croacia, pp. 153-166.

González-Flores, O., Domínguez-Ordóñez, R., González-Burgos, I., 2015. The cellular and molecular mechanisms driving the estrogenic modulation of synaptic activity in mnemonic information processing, in: González-Burgos, I. (Ed.), *Synaptic Fundamentals of Memory Performance.* Nova Science Publishers, Inc., New York, pp. 33-58.

González-Flores, O., Hoffman, K. L., Melo, A. I., Beyer, C., 2014. Effects of estrogen fluctuations during the menstrual cycle on cognitive functions: Cellular systems and brain áreas involved, in: González-Burgos, I. (Ed.), *Estrogens and Cognition. Psichobiological and Clinical Aspects.* Research Signpost, India, pp. 71-96.

González-Ramírez, M. M., Velázquez-Zamora, D. A., Olvera-Cortés, M. E., González-Burgos, I., Changes in the plastic properties of hippocampal dendritic spines underlie the attenuation of place learning in healthy aged rats. *Neurobiol. Learn. Mem.* 109, 94-103.

González-Tapia, D., Martínez-Torres, N. I., Hernández-González, M., Guevara, M. A., González-Burgos, I., 2016. Plastic changes to dendritic spines on layer V pyramidal neurons are involved in the rectifying role of the prefrontal cortex during the fast period of motor learning. Behav. *Brain Res.* 298, 261-267.

González-Tapia, D., Vargas-Genis, C., González-Burgos, I., 2017. Psychobiological and clinical relationships in motor rehabilitation, in: Bell, K., Rusell, C., (Eds.), *Motor learning. A Review and Directions for Research.* Nova Science Publishers, Inc., New York, pp. 86-108.

González-Tapia, D., Velázquez-Zamora, D. A., Olvera-Cortés, M. E., González-Burgos, I., 2015. The motor learning induces plastic changes in dendritic spines of Purkinje cells from the neocerebellar cortex of the rat. *Restor. Neurol. Neurosci.* 33, 639-645.

Griffin, G. D., Flanagan-Cato, L. M., 2011. Ovarian hormone action in the hypothalamic ventromedial nucleus: remodelling to regulate reproduction. *J. Neuroendocrinol.* 23, 465-471.

Grutzendler, J., Kasthuri, N., Gan, W.-B., 2002. Long-term dendritic spine stability in the adult cortex. *Nature* 420, 812-816.

Harris, K. M., Jensen, F. E., Tsao, B., 1992. Three-dimensional structure of dendritic spines and synapses in rat hippocampus (CA1) at postnatal day 15 and adult ages: implications for the maturation of synaptic physiology and long-term potentiation. *J. Neurosci.* 12, 2685-2705.

Harris, K. M., Kater, S. B., 1994. Dendritic spines: Cellular specializations imparting both stability and flexibility to synaptic function. *Ann. Rev. Neurosci.* 17, 341-371.

Hayashi-Takagi, A., Yagishita, S., Nakamura, M., Shirai, F., Wu, Y. I., Loshbaugh, A. L., Kuhlman, B., Hahn, K. M., Kasai, H., 2015. Labelling and optical erasure of synaptic memory traces in the motor cortex. *Nature* 525, 333-338.

Hebb, D. O., 1949. *The organization of behavior*. Wiley, New York.

Hernández-González, M., Barrera-Cobos, F. J., Hernández-Arteaga, E., González-Burgos, I., Flores-Soto, M., Guevara, M. A., Cortes, P. M., 2023. Sexual experience induces a preponderance of mushroom spines in the medial prefrontal cortex of male rats. *Behav. Brain Res.* 447, 114437.

Holmes, S. E., Abdallah, Ch., Esterlis, I., 2023. Imaging synaptic density in depression. *Neuropsychopharmacology* 48, 186-190.

Holtmaat, A. J. G. D., Trachtenberg, J. T., Wilbrecht, L., Shepherd, G. M., Zhang, X., Knott, G. W., Svoboda, K., 2005. Transient and persistent dendritic spines in the neocortex in vivo. *Neuron* 45, 279-291.

Hongpaisan, J., Alkon, D. L., 2007. A structural basis for enhancement of long-term associative memory in single dendritic spines regulated by PKC. *Proc. Natl. Acad. Sci. USA* 104, 19571-19576.

Huang, Y. Y., Kandel, E. R., 1994. Recruitment of long-lasting and protein kinase A-dependent long-term potentiation in the CA1 region of hippocampus requires repeated tetanization. *Learn. Mem.* 1, 74-82.

Ito, M., Kano, M., 1982. Long-lasting depression of parallel fiber-Purkinje cell transmission induced by conjunctive stimulation of parallel fibers and climbing fibers in the cerebellar cortex. *Neurosci. Lett.* 33, 253-258.

Izquierdo, L. A., Barros, D. M., da Costa, J. C., Furini, C., Zinn, C., Cammarota, M., Bevilaqua, L. R., Izquierdo, I., 2007. A link between role of two prefrontal areas in immediate memory and in long-term memory consolidation. *Neurobiol. Learn. Mem.* 88, 160-166.

Jacobson, P. B., Goody, R., Lawrence, M., Mueller, B. K., Zhang, X., Hooker, B. A., Pfleeger, K., Ziemann, A., Locke, Ch., Barraud, Q., Droescher, M., Bernhard, J., Popp, A., Boeser, P., Huang, L., Mollon, J., Mordashova, Y., Cui, Y.-F., Savaryn, J. P., Grinnell, Ch., Dreher, I., Gold, M., Courtine, G., Mothe, A., Tator, Ch. H., Guest, J. D., 2021. Elezanumab, a human anti-RGMa monoclonal antibody, promotes neuroprotection, neuroplasticity, and neurorecovery following a thoracic hemicompression spinal cord injury in non-human primates. *Neurobiol. Dis.* 155, 105385.

Jiang, M., Lee, C. L., Smith, K. L., Swann, J. W., 1998. Spine loss and other persistent alterations of hippocampal pyramidal cell dendrites in a model of early-onset-epilepsy. *J. Neurosci.* 18, 8356-8368.

Johnston, D., y Wu, S. M., 1997a. Functional properties of dendrites Foundations of cellular neurophysiology. MIT Press. Massachusetts, pp. 55-120.

Johnston, D., Wu, S. M., 1997b. Synaptic transmission III: Postsynaptic mechanisms Foundations of cellular neurophysiology. MIT Press. Massachusetts, pp. 357-422.

Johnston, D., Wu, S. M., 1997c. Synaptic transmission I: Presynaptic mechanisms Foundations of cellular neurophysiology. MIT Press. Massachusetts, pp. 287-322.

Johnston, D., Wu, S. M., 1997d. Cellular Neurophysiology of learning and memory Foundations fo cellular neurophysiology. MIT Press. Massachusetts, pp. 441-480.

Kaas, J. H., Qi, H. X., Burish, M. J., Gharbawie, O. A., Onifer, S. M., Massey, J. M., 2008. Cortical and subcortical plasticity in the brains of humans, primates, and rats after damage to sensory afferents in the dorsal columns of the spinal cord. *Exp. Neurol.* 209, 407-416.

Kasai, H., Fukuda, M., Watanabe, S., Hayashi-Takagi, A., Noguchi, J., 2010a. Structural dynamics of dendritic spines in memory and cognition. *TINS* 33, 121-129.

Kasai, H., Hayama, T., Ishikawa, M., Watanabe, S., Yagishita, S., Noguchi, J., 2010b. Learning rules and persistence of dendritic spines. *Eur. J. Neurosci.* 32, 241-249.

Kasai, H., Matsuzaki, M., Noguchi, J., Yasumatsu, N., Nakahara, H., 2003. Structure-stability-function relationships of dendritic spines. *Trends Neurosci.* 26, 360-368.

Kirov, S. A., Harris, K. M., 1999. Dendrites are more spiny on mature hippocampal neurons when synapses are inactivated. *Nat. Neurosci.* 2, 878-883.

Koch, Ch., Zador, A., 1993. The function of dendritic spines: devices subserving biochemical rather than electrical compartmentalization. *J. Neurosci.* 13, 413-422.

Koch, Ch., Zador, A., Brown, T. H., 1992. Dendritic spines: convergence of theory and experiment. *Science* 256, 973-974.

Konnerth, A., Dreessen, J., Augustine, G. J., 1992. Brief dendritic calcium signals initiate long-lasting synaptic depression in cerebellar Purkinje cells. *Proc. Natl. Acad. Sci. U S A.* 89, 7051-7055.

Korkotian, E. y Segal, M., 1999. Bidirectional regulation of dendritic spine dimensions by glutamate receptors. *Neuroreport* 10, 2875-2877.

Kostyuk P., 1998. *Plasticity in Nerve Cell Function.* Oxford University Press, New York.

Kovalchuk, Y., Eilers, J., Lisman, J., Konnerth, A., 2000. NMDA receptor-mediated subthreshold Ca(2+) signals in spines of hippocampal neurons. *J. Neurosci.* 20, 1791-1799.

Kuffler, S. W., Nichols, J. G., 1976. *From Neuron to Brain.* Sunderland, massachusetts: Sinauer Associates Inc.

Lagerlund, T. D., 2002. Electrophysiologic generators in clinical neurophysiology. In D. J. R. (Ed.), *Clinical Neurophysiology.* Oxford University Press. Oxford, Second ed., pp. 53-62.

Lang, C., Barco, A., Zablow, L., Kandel, E. R., Siegelbaum, S. A., Zakharenko, S. S., 2004. Transient expansion of synaptically connected dendritic spines upon induction of hippocampal long-term potentiation. *Proc. Natl. Acad. Sci. USA* 101, 16665-16670.

Lee, K. J., Kim, H., Rhyu, I. J., 2005. The roles of dendritic spine shapes in Purkinje cells. *The Cerebellum.* 4, 97-104.

Leuner, B, Falduto, J., Shors, T. J., 2003. Associative memory formation increases the observation of dendritic spines in the hippocampus. *J. Neurosci.* 23, 659-665.

Leuner, B., Shors, T. J., 2004. New spines, new memories. *Mol. Neurobiol.* 29, 117-130.

Linden, D. J., Dickinson, M. H., Smeyne, M., Connor, J. A., 1991. A long-term depression of AMPA currents in cultured cerebellar Purkinje neurons. *Neuron.* 7, 81-89.

López-Vázquez, M. A., Olvera-Cortés, M. E., González-Burgos, I., 2006. Multiunitary activity of prefrontal piramidal neurons increases Turing spatial working memory performance, after serotonin depletion. *Int. J. Devl. Neuroscience.* 24, 502-503.

Luine, V., Frankfurt, M., Estrogenic regulation of memory: The first 50 years. *Horm. Behav.* 121, 104711.

Luján, R., Nusser, Z., Roberts, D. B., Shigemoto, R., Somogyi, P., 1996. Perisynaptic location of metabotropic glutamate receptors mGluR1 and mGluR5 on dendrites and dendritic spines in the rat hippocampus. *Eur. J. Neurosci.* 8, 1488-1500.

Luria, A. R., 1970. *The functional organizatiopn of the brain.* 222, 66-72.

Luria, A. R., 1982. *Introducción Evolucionista a la Psicología [Evolutionary Introduction to Psychology].* Fontanella, Spain.

Lledo, P. M., Alonso, M., Grubb, M. S., 2006. Adult neurogenesis and functional plasticity in neuronal circuits. *Nat. Rev. Neurosci.* 7, 179-193.

Llorens, F., Gil, V., del Río, J. A., 2011. Emerging functions of myelin-associated proteins during development, neuronal plasticity, and neurodegeneration. *FASEB J.* 25, 463-475.

Majewska, A., Brown, E., Ross, J., Yuste, R., 2000. Mechanisms of calcium decay kinetics in hippocampal spines: role of spine calcium pumps and calcium diffusion through the spine neck in biochemical compartmentalization. *J. Neurosci.* 20, 1722-1734.

Maroun, M., 2006. Stress reverses plasticity in the pathway projecting from the ventromedial prefrontal cortex to the basolateral amygdala. *Eur. J. Neurosci.* 24, 2917-2922.

Martínez-Torres, N. I., González-Tapia, D., Flores-Soto, M., Vázquez-Hernández, N., Salgado-Ceballos, H., González-Burgos, I., 2021a. Spinogenesis in spinal cord motor neurons following pharmacological lesions to the rat motor cortex. *Neurología* 36, 119-126.

Martínez-Torres, N. I., González-Tapia, D., Vázquez-Hernández, N., González-Burgos, I., 2018. Atomoxetine prevents working memory loss in hyperactive rats, mediating plastic changes in prefrontal cortex piramidal neurons. *Pharmacol. Biochem. Behav.* 175, 116-122.

Martínez-Torres, N. I., Vázquez-Hernández, N., Martín-Amaya-Barajas, F. L., Flores-Soto, M., González-Burgos, I., 2021b. Ibotenci acid induced lesions impair the modulation of dendritic spine plasticity in the prefrontal cortex and amygdala, a phenomenon that undelies working memory and social behavior. *Eur. J. Pharmacol.* 896, 173883.

Matsuzaki, M., Honkura, N., Ellis-Davies, G. C. R., Kasai, H., 2004. Structural basis of long-term potentiation in single dendritic spines. *Nature.* 429, 761-766.

Mayer, M. L., Westbrook, G. L., 1987. Permeation and block of N-methyl-D-aspartic acid receptor channels by divalent cations in mouse cultured central neurons. *J. Physiol.*, 394, 501-527.

McEachern, J. C., Shaw, C. A., 1999. The plasticity-pathology continuum: defining a role for the LTP phenomenon. *J. Neurosci. Res.* 58, 42-61.

Medvedev, N. I., Popov, V. I., Rodriguez Arellano, J. J., Dallérac, G., Davies, H. A., Gabbott, P. L., Laroche, S., Kraev, I. V., Doyère, V., Stewart, M. G., 2010. The N-methyl-D-aspartate receptor antagonist CPP alters synapse and spine structure and impairs long-term potentiation and long-term depression induced morphological plasticity in dentate gyrus of the awake rat. *Neuroscience* 165, 1170-1181.

Meldrum, B., 2000. Glutamate as a neurotransmitter in the brain: Review of physiology and pathology. *J. Nutr.* 130, 1007-1015.

Meunier, C. N. J., Chameau, P , Fossier, Ph. M., 2017. Modulation of synaptic plasticity in the cortex needs to understand all the players. *Front. Synaptic Neurosci.* 9, 2.

Milad, M. R., Wright, C. I., Orr, S. P., Pitman, R. K., Quirk, G. J., Rauch, S. L., 2007. Recall of fear extinction in humans activates the ventromedial prefrontal cortex and hippocampus in concert. *Biol. Psychiatry.* 62, 446-454.

Moser, E. I., Krobert, K. A., Moser, M. B., Morris, R. G., 1998. Impaired spatial learning after saturation of long-term potentiation. *Science.* 281, 2038-2042.

Mugnaini, E., Walberg, F., Hauglie-Hanssen, E., 1967. Observations on the fine structure of the lateral vestibular nucleus (Deiters' nucleus) in the cat. *Exp. Brain Res.* 4, 146-186.

Muller, D., Toni, N., Buchs, P. A., 2000. Spine changes associated with long-term potentiation. *Hippocampus* 10, 596-604.

Myers, K. M., Ressler, K. J., Davis, M., 2006. Different mechanisms of fear extinction dependent on length of time since fear acquisition. *Learn. Mem.* 13, 216-223.

Nabavi, S., Fox, R., Proulx, Ch. D., Lin, J. Y., Tsien, R. Y., Malinow, R., 2014. Engineering a memory with LTD and LTP. *Nature* 511, 348-352.

Napieralski, J. A., Banks, R. J., Chesselet, M. F., 1998. Motor and somatosensory deficits following uni- and bilateral lesions of the cortex induced by aspiration or thermocoagulation in the adult rat. *Exp. Neurol.* 154, 80-88.

Nimchinsky, E. A., Sabatini, B. L., Svoboda, K., 2002. Structure and function of dendritic spines. *Annu. Rev. Physiol.* 64, 313-353.

Nishijima, H., Ueno, T., Funamizu, Y., Ueno, S., Tomiyama, M., 2018. Levodopa treatment and dendritic spine pathology. *Mov. Disord.* 33, 877-888.

Nishiyama, J., 2019. Plasticity of dendritic spines: Molecular function and dysfunction in neurodevelopmental disorders. *Psychiatry Clin. Neurosci.* 73, 541-550.

Niswender C. M., Conn P. J , 2010. Metabotropic glutamate receptors: physiology, pharmacology, and disease *Annu. Rev. Pharmacol. Toxicol.* 50, 295-322.

Nowak, L., Bregestovski, P., Ascher, P., Herbet, A., Prochiantz, A., 1984. Magnesium gates glutamate-activated channels in mouse central neurons. *Nature.* 307, 462-465.

Nudo, R. J., Plautz, E. J., Frost, S. B., 2001. Role of adaptive plasticity in recovery of function after damage to motor cortex. *Muscle Nerve* 24, 1000-1019.

Nufer, T. M., Wu, B. J., Boyce, Z., Steffensen, S. C., Edwards, J. G., 2023. Ethanol blocks a novel form of iLTD, but not iLTP of inhibitory inputs to VTA GABA neurons. *Neuropsychopharmacology* doi: 10.1038/s41386-023-01554-y.

O'Connell, C., O'Malley, A., Regan, C. M., 1997. Transient, learning-induced ultrastructural change in spatially-clustered dentate granule cells of the adult rat hippocampus. *Neuroscience.* 76, 55-62.

Olvera-Cortes, M. E., Cervantes, M., Gonzalez-Burgos, I., 2002. Place-learning, but not cue-learning training, modifies the hippocampal theta rhythm in rats. *Brain Res. Bull.* 58, 261-270.

Olvera-Cortes, M. E., Garcia-Alcantar, I., Gutierrez-Guzman, B., Hernandez-Perez, J. J., Lopez-Vazquez, M. A., Cervantes, M., 2012. Differential learning-related changes in theta activity during place learning in young and old rats. *Behav. Brain Res.* 226, 555-562.

Olvera-Cortes, M. E., Guevara, M. A., Gonzalez-Burgos, I., 2004. Increase of the hippocampal theta activity in the Morris water maze reflects learning rather than motor activity. *Brain Res. Bull.* 62, 379-384.

Olvera-Cortés, M. E., Hernández-Pérez, J. J., López-Vázquez, M. A., 2015. Fundamentos electrofisiológicos de la transmisión sináptica relacionada con el procesamiento de información mnémica [Electrophysiological foundations of synaptic transmission related to memory information processing], in: González-Burgos, I. (Ed.), *Psicobiología de la Memoria. Un Enfoque Interdisciplinario [Psychobiology of Memory. An Interdisciplinary Approach].* Bios-Médica, Guadalajara, pp. 51-78.

Olvera-Cortés, E., López-Vázquez, M. A., Beas-Zárate, C., González-Burgos, I., 2005. Neonatal exposure to monosodium glutamate disrupts place learning ability in adult rats. *Pharmacol. Biochem. Behav.* 82, 247-251.

Oudega, M., Perez, M. A., 2012. Corticospinal reorganization after spinal cord injury. *J. Physiol.* 590, 3647-3663.

Pérez-Vega, M. I., Feria-Velasco, A., González-Burgos, I., 2000. Prefrontocortical serotonin depletion results in plastic changes of prefrontocortical pyramidal neurons, underlying a greater efficiency of short-term memory. *Brain Research Bulletin.* 53, 291-300.

Pfaff, D. W., 1980. *Estrogens and Brain Function.* Springer-Verlag, New York.

Portilla-de Buen, E., Salgado-Ceballos, H., González-Tapia, D., Leal-Cortés, C., Mondragón-Lozano, R., Sánchez-Torres, S., Álvarez-Mejía, L., Fabela-Sánchez, O., Martínez-Torres, N. I., González-Ramírez, M. M., Vázquez-Hernández, N., González-Burgos, I., 2017. Spinogenesis and plastic changes in the dendritic spines of spinal cord motoneurons after traumatic injury in rats. *Arch. Med. Res.* 48, 609-615.

Ramón y Cajal, S., 1909. *Histologie du Systeme nerveux de l'homme et des vertébrés [Histology of the Nervous System of Man and Vertebrates].* Maloine, París.

Ramos, A. A., Hamdan, A. C., Machado, L., 2020. A meta-analysis on verbal working memory in children and adolescents with ADHD. *Clin. Neuropsychol.* 34, 873-898.

Reza-Zaldivar, E. E., Hernández-Sápiens, M. A., Minjarez, B., Gómez-Pinedo, U., Sánchez-González, V. J. Márquez-Aguirre, A. L., Canales-Aguirre, A. A., 2020. Dendritic spine and synaptic plasticity in Alzheimer's disease: A focus on microRNA. *Front. Cell. Dev. Biol.* 8, 255.

Roberson, E. D., Defazio, R. A., Barnes, C. A., Alexander, G. E., Bizon, J. L., Bowers, D., Foster, Th. C., Glisky, E. L., Levin, B. E., Ryan, L., Wright, C. B., Geldmacher, D. S., 2012. Challenges and opportunities for characterizing cognitive aging across species. *Front. Aging Neurosci.* 4, 6.

Romero-Guerrero, Ch., Guevara, M. A., Hernández-González, M., González-Burgos, I., 2022. Pentylenetetrazol-induced seizures are followed by a reduction in the multiunitary activity of hippocampal CA1 pyramidal neurons. *Epilepsy Behav.* 137, 108922.

Rose, S. P., 1995. Cell-adhesion molecules, glucocorticoids and long-term-memory formation. *Trends Neurosci.* 18, 502-506.

Sá, S. I., Madeira, M. D., 2005. Estrogen modulates the sexually dimorphic synaptic connectivity of the ventromedial nucleus. *J. Comp. Neurol.* 484, 68-79.

Sala, C., Segal, M., 2014. Dendritic spines: the locus of structural and functional plasticity. *Physiol. Rev.* 94, 141-188

Salinska, E., Wojciech, D., Lazerewicz, J. W., 2005. The role of excitotoxicity in neurodegeneration. *Folia Neuropathol.* 43, 322-339.

Scoville, W. B., Milner, B., 1957. Loss of recent memory after bilateral hippocampal lesions. *J. Neurol. Neurosurg. Psychiatry.* 20, 11-21.

Segal, M., Korkotian, E., Murphy, D. D., 2000. Dendritic spine formation and pruning: common cellular mechanisms? *Trends Neurosci.* 23, 53-57.

Selkoe, D. J., 1992. Aging brain, aging mind. *Sci. Am.* 267, 134-142.

Silva, A. J., 2003. Molecular and cellular cognitive studies of the role of synaptic plasticity in memory. *J. Neurobiol.* 54, 224-237.

Soria-Fregozo, C., Pérez-Vega, M. I., González-Burgos, I., Feria-Velasco, A., Beas-Zárate, C., 2008. Prefrontal serotonergic denervation induces increase in the density of 5-HT_{2A} receptors in adult rat prefrontal cortex. *Neurochem. Res.* 33, 2350-2357.

Spacek, J., Harris K. M., 1997 Three-dimensional organization of smooth endoplasmic reticulum in hippocampal CA1 dendrites and dendritic spines of the immature and mature rat. *J. Neurosci.* 17, 190-203.

Spencer, G. E., Klumperman, J., Syed, N. I., 1998. Neurotransmitters and neurodevelopment. Role of dopamine in neurite outgrowth, target selection and specific synapse formation. *Perspect. Dev. Neurobiol.* 5, 451–467.

Squire, L. R., Kandel, E. R., 1999. *Memory: from Mind to Molecules.* WH Freeman and Company, New York.

Tan, J.-L., Zhang, F., Liu, H.-J., Li, Z., 2018. Hyperbaric oxygen ameliorated the lesion scope and nerve function in acute spinal cord injury patients: A retrospective study. *Clin. Biochem.* 53, 1-7.

Thompson, R. F., 1990. Neural mechanisms of classical conditioning in mammals. *Philos. Trans. R. Soc. Lond. B. Biol. Sci.* 329, 161-170.

Turrigiano, G. G., Nelson, S. B., 2004. Homeostatic plasticity in the developing nervous system. *Nat. Rev. Neurosci.* 5, 97-107.

Vázquez-Hernández, N., González-Tapia, D. C., Martínez-Torres, Nestor I., González-Tapia, D., González-Burgos, I., 2017. Different patterns of motor activity induce differential plastic changes in pyramidal neurons in the motor cortex of rats: A Golgi study. *Neurosci. Lett.* 657, 27-31.

Vázquez-Hernández, N., Martínez-Torres, N. I., González-Burgos, I., 2021. Plastic changes to dendritic spines in the cerebellar and prefrontal cortices undelie the decline in motor coordination and working memory during successful aging. *Behav. Brain Res.* 400, 113014.

Vázquez-Hernández, N., Martínez-Torres, N., Tejeda-Martínez, A., Flores-Soto, M., Salgado-Ceballos, H., González-Burgos, I., 2022. Rehabilitation on a treadmill induces plastic changes in the dendritic spines of spinal motoneurons associated with improved execution after a pharmacological injury to the motor cortex in rats. *J. Chem. Neuroanat.* 125, 102159.

Velázquez-Zamora, D. A., García-Segura, L. M., González-Burgos, I., 2012b. Effects of selective estrogen receptor modulators on allocentric working memory performance and on dendritic spines in medial prefrontal cortex pyramidal neurons of ovariectomized rats. *Horm. Behav.* 61(4):512-517. 2012.

Velázquez-Zamora, D. A., González-Ramírez, M. M., Beas-Zárate, C., González-Burgos, I., 2011. Egocentric working memory impairment and dendritic spine plastic changes in prefrontal neurons after NMDA receptor blockade, in rats. *Brain Res.* 1402, 101-108.

Velázquez-Zamora, D. A., González-Tapia, D., González-Ramírez, M. M., Flores-Soto, M. E., Vázquez-Valls, E., Cervantes, M., González-Burgos, I., 2012a. Plastic changes in dendritic spines of hippocampal CA1 pyramidal neurons from ovariectomized rats after estradiol treatment. *Brain Res.* 1470, 1-10.

Velázquez-Zamora, D. A., Martínez-Degollado, M., González-Burgos, I., 2011. Morphological development of dendritic spines on rat cerebellar Purkinje cells. *Int. J. Dev. Neurosci.* 29, 515-520.

Vinogradova, O. S., 1995. Expression, control, and probable functional significance of the neuronal theta-rhythm. *Prog. Neurobiol.* 45, 523-583.

Volfovsky, N., Parnas, H., Segal, M., Korkotian, E., 1999. Geometry of dendritic spines affects calcium dynamics in hippocampal neurons: theory and experiments. *J. Neurophysiol.* 81, 450-462.

Walker, C. K., Herskowitz, J. H., Dendritic spines: Mediators of cognitive resilience in aging and alzheimer's disease. *Neuroscientist* 27, 487-505.

Warren, S. G., Juraska, J. M., 1997. Spatial and nonspatial learning across the rat estrous cycle. *Behav. Neurosci.* 111, 259-266.

Welniarz, Q., Dusart, I., Roze, E., 2017. The corticospinal tract: Evolution, development, and human disorders. *Dev. Neurobiol.* 77, 810-829.

Westrum, L. E., 1970. Observations on initial segments of axons in the prepyriform cortex of the rat. *J. Comp. Neurol.* 139, 337-356.

Wickens J., 1988. Electrically coupled but chemically isolated synapses: dendritic spines and calcium in a rule for synaptic modification. *Prog. Neurobiol.* 31, 507-528.

Woolley, C. S., Gould, E., Frankfurt, M., McEwen, B. S., 1990. Naturally occurring fluctuation in dendritic spine density on adult hippocampal pyramidal neurons. *J. Neurosci.* 10, 4035-4039.

Yan, J., Zhang, Y., Jia, Z., Taverna, F. A., McDonald, R. J., Muller, R. U., Roder, J. C., 2002. Place-cell impairment in glutamate receptor 2 mutant mice. *J. Neurosci.* 22, RC204.

Yang, Y., Wang, X., Frerking, M., Zhou, Q., 2008. Spine expansion and stabilization associated with long-term potentiation. *J. Neurosci.* 28, 5740-5751.

Yu, X., Zuo, Y., 2011. Spine plasticity in the motor cortex. *Curr. Opin. Neurobiol.* 21, 169-174.

Chapter 6

Estrogen-Mediated Signaling in Dendritic Spines that Promote Memory Formation

Victoria Luine[1,*]
and Maya Frankfurt[2]
[1]Department of Psychology, Hunter College of CUNY, New York, USA
[2]Hofstra Northwell School of Nursing and Physician Assistant Studies, Hofstra University, Hempstead, New York, USA

Introduction

Although the presence of dendritic spines was initially demonstrated using silver stains in the late 1800's, their functional importance remained controversial for more than a century. Methods such as Golgi impregnation and methylene blue gave different results concerning their presence, but finally both Golgi and Ramon y Cajal, winners of the Noble Prize in 1906, recognized the existence of dendritic spines. However, whether they were functionally relevant was still debated (Golgi, 1903), and it was not until the 1980s using electron microscopic techniques, combined with other methods, that the importance of dendritic spines in synaptic transmission was firmly established (Frotscher and Leranth, 1986). Still, the significance and function(s) of spines continued to be debated. DeFelipe (2017) provides an excellent history of these controversies as well as images from these early neuroanatomists. In this chapter, we expand the discussion on dendritic spines and provide evidence that estradiol enhances memory by increasing the density of dendritic spines located on pyramidal neurons in the hippocampus and prefrontal cortex. Underlying molecular mechanisms for spinogenesis and memory enhancements are also detailed.

In: Dendritic Spines
Editor: Ignacio González-Burgos
ISBN: 979-8-89113-080-7

Learning and Memory Measurements

Cognition, in its most basic context, is the ability to learn, retain and recall information. In humans, it also includes a multidimensional set of intellectual functions such as judgment and evaluation. Since we are researching rodents, estrogenic effects on cognition are applied in more basic contexts, primarily to learning and memory. However, it should be noted that effects of estradiol have been demonstrated in a number of cognitive domains in humans, and these are not the focus of this review (Hampson et al., 2015). Nonetheless, hormonal enhancements on cognition in rodents need to be congruent with existing theories on mechanisms for learning and memory. Current mechanisms posit that subjects first acquire or learn information (learning phase) and then the new information requires a consolidation period within specific brains areas, and finally, when appropriate or useful, the stored information is retrieved/recalled for use (memory) (McGaugh, 1989). This time-dependent process is depicted in Figure 1, and researchers have used this framework for developing the post-training protocol for assessing memory. In it, drugs or hormones are given after subjects acquire information such as the location of a platform in a water maze or after subjects have explored objects in recognition memory tasks, the training trial/T1 Trial. This paradigm is based on the premise that following training, learning, or sampling trials, the newly acquired information requires consolidation and storage and that drugs or hormones could influence these processes during the immediate period following training. For example, rats receiving "memory enhancing drugs" like amphetamine or noradrenergic agonists during this period show better retention of what they learned in retention trials (McGaugh, 1989). An important additional finding is that effectiveness of post-training treatments is time-limited, i.e., only treatments given with 1-3 h following post training/sampling trials are effective while treatments given later are ineffective. Thus, a window for memory consolidation exists within which drugs can enhance (or impair) the process.

Figure 1 shows results of this paradigm using DES, a potent synthetic estrogen, in object recognition tests. When DES is given immediately before the training trial/T1, it enhances retention in the T2 trial given 4 h later, i.e., the treated ovariectomized (OVX) rats spend more time exploring at the new than the old object. However, it is unclear whether DES affected learning or memory because it was given before both the learning and the retention trial. When DES is given immediately following the training trial/T1, subjects spend more time exploring the new object indicating that DES enhances the consolidation of memory. However, if DES treatment is delayed, given 2 h post T1, treated subjects no longer show enhanced memory consolidation because they spend the same amount of time exploring the old and new objects. This window during which memory consolidation occurs and when drugs are effective can be longer or shorter than 2 h depending on a number of factors including the difficulty of the task, sex of subject and dose of drug or hormone administered (See Luine, 2015 for further discussion). Another advantage of post training test protocols is that consolidation enhancements following post-trial injections rule out the possibility that treatments derive from effects on learning because they are given after the learning trial (Packard, 1998). Neurochemical processes underlying consolidation of memory and its enhancement are currently being unraveled (see section 6).

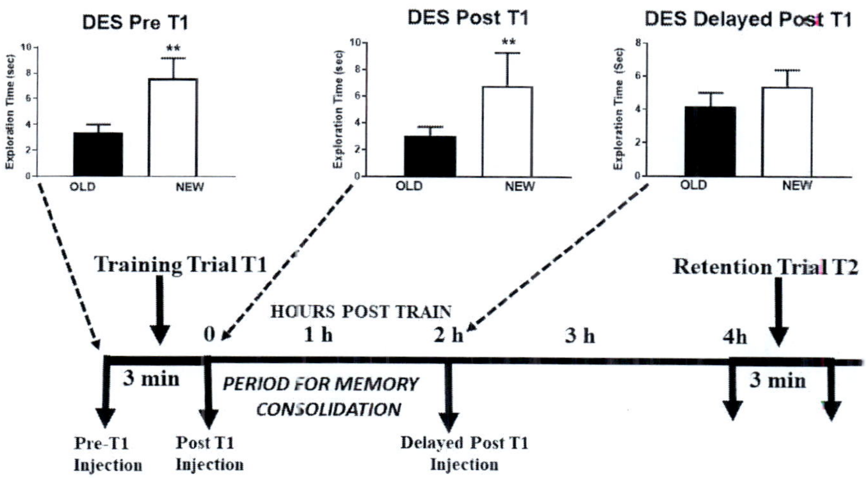

Figure 1. Timeline showing that memory is consolidated during the two hours after training trial (T1) in object recognition task. Left graph: DES, 15 μg SC, was given 30 min before Training Trial, T1, and object recognition was enhanced 4 h later, Retention Trial, T2. ** $p < 0.01$. Center graph: DES, 15 μg SC, was given immediately post T1 trial, and object recognition was enhanced 4 h later, Retention Trial, T2. ** $p < 0.01$. Right graph: DES, 15 μg SC, was given 2 h post T1 trial, and object recognition was not enhanced 4 h later, Retention Trial, T2. Thus, treatments given 2 h after T1 do not affect memory consolidation. Data from Luine et al., 2003.

Assessing whether hormones or drugs affect learning processes (training/learning trial, T1) is more difficult because non-mnemonic variables can affect performance on cognitive tasks. Such variables include affective state (arousal, anxiety, mood, motivation), regulatory or homeostatic processes such as thirst and hunger, sensory-perception processing such as vision and olfaction and motor influences such as activity levels (Luine, 2008). These variables can alter performance of tasks without affecting the mechanisms of learning.

Estrogens Enhance Cognitive Processes

Estrogens enhance cognition in rodents, non-human primates, and humans. Early studies showed that chronic estrogen treatments (days to weeks), given to OVX rodents enhance performance of learning and memory tasks (Luine, 2008). More recent studies show that acute administration of estrogens (minutes to an hour) enhance consolidation of memory (Luine et al., 2018b). These effects occur via different mechanisms and involve different estrogen receptors, both nuclear receptors (nERs) and membrane receptors (mERs). Long term effects are mediated via classical estrogen receptors, nERs, located in the cytoplasm which bind estradiol and then the complex translocates to the cell nucleus where it acts as a nuclear transcription factor by binding to estrogen response elements (ERE) on DNA and stimulating transcription of specific genes whose proteins then alter neural functions including memory. In the brain, increased expression of proteins related to neurotransmitters such as acetylcholine and monoamines and synaptic function are prominent effects of estrogen (Luine, 2014). Here, we review the rapid effects of estrogens on enhancements of memory consolidation. Rapid effects occur through mechanisms that differ from the chronic effects and rely on changes in signaling pathways and dendritic spine plasticity through binding to mERs (Frick et al., 2015; Luine et al., 2018b).

Estrogens Enhance Memory Consolidation

With increasing evidence in the late 1990's for the existence of estrogen receptors on membranes of cell bodies, axons, spines, presynaptic terminals and near post-synaptic neurotransmitter receptors (Blaustein, 1992; Milner et al., 2001), we and others reasoned that hormones might rapidly enhance neural functions and began to explore acute estrogenic effects on cognition. To investigate this question, we adopted recognition memory tests which allow a post training approach, discussed above. Unlike previous maze tasks which require several training trails over several days, these tasks involve a single training trial where rats explore two identical objects in an open field (sample or training trial, T1). The ability to remember a training object is tested by substituting a new or novel object for one of the familiar objects in the recognition trial (memory trial, T2). Since rats have a natural predilection for novelty, more time spent exploring the new object indicates memory for the identity of the now old, training object. Location or place memory can also be tested using the same T1 trial, but in the recognition T2 trial, one of the two objects is moved to a new location in the field. Further information and details for conducting these tests can be found in Ennaceur and Souza Silva (Ennaceur, 2018).

In a series of experiments, we were the first to investigate rapid effects of estrogens on recognition memory (Luine et al., 2003). As discussed above and in Figure 1, when DES is given 30 min before a T1 trial or immediately after a T1 trial for object recognition, memory is enhanced since rats spent significantly more time, approximately twice as long, exploring the new object than the old object in the T2 recognition trial given 4 h later. However, when DES is given 2h post T1 and retention tested 4 h later, no memory enhancements were found which is consistent with the memory consolidation theory. Estradiol also enhances object placement memory when given immediately following T1 (Frye et al., 2007; Inagaki et al., 2012; Inagaki et al., 2010; Jacome et al., 2010; Luine et al., 2003). The dose-response curves for these estrogen effects are inverted U curves, consistent with membrane mediation, and lower doses of estradiol require administration within 45 min after T1 to enhance consolidation (Inagaki et al., 2010). In addition, effective doses for rapid effects of estrogens are lower than for chronic effects (Inagaki et al., 2012). Thus, these experiments provided initial evidence that estrogens rapidly enhance memory consolidation.

Experiments in the Frick laboratory with mice confirmed and expanded these findings. Using post training injections of cyclodextrin encapsulated estradiol which is rapidly metabolized, they showed enhanced memory consolidation in object recognition and placement in OVX C57BL/6 mice (Fernandez et al., 2008; Frick et al., 2010; Gresack and Frick, 2004, 2006; Gresack et al., 2007a, b; Lewis et al., 2008). The rapid metabolism of this encapsulated estrogen provides further evidence that the hormone acts rapidly and does not need to be present at the time of testing. Using another learning task, inhibitory avoidance, Rhodes and Frye (2004) found that estradiol immediately, but not 1,2, or 3 h post-training increased crossover latencies as compared to vehicle treatment in mice. Post training effects of estrogens have been also demonstrated in Morris water maze tests (Packard, 1998). The estrogen agonists, DPN and PPT, have also been tested, and enhance recognition memory tasks when given post training in both mice and rats (See Table 1).

Table 1. Summary of studies showing estrogens enhance memory consolidation and/or dendritic spine density when injection is given immediately post training and measurements made minutes to hours later. Treatments were given to ovariectomized or castrated rats except * which were mice. Treatments given immediately following training session and memory was tested at time indicated in brackets, for example (40 min = 40 min after training session). Bolded entries highlight that memory and spine effects were measured in the same subjects. SC = subcutaneous. CA1 = pyramidal neurons in hippocampus. mPFC = medial prefrontal cortex. OR = object recognition. OP = object placement. IA = Inhibitory Avoidance. MWM = Morris Water Maze. SR = Social Recognition. DES = diethystilbestrol (synthetic estrogen). DPN = diarylpropionitrile (ERβ agonist). PPT = propyl-pyrazole-triol (ERα agonist). GP1 = GPER agonist

Treatment	Memory Test	Spine Change		Reference
Females	Effect	CA1	mPFC	
S.C. Injection				
E_2 (2-4 h)	↑OR, ↑OP	↑	↑	Inagaki et al., 2010; 2012; Luine et al., 2003; Jacome et al., 2010; Frye et al., 2007
E_2 (4 h)	↑ IA	?	?	Rhodes and Frey, 2004
DES (4 h)	↑ OR, ↑ OP	?	?	Luine et al., 2003
E_2 (40 min)*	↑ OR, = OP	↑	?	Phan et al., 2012; 2015
E_2 (24 h)	↑ MWM	?	?	Packard & Teather (1997)
DPN (40 min)*	= OR, = OP = OR, ↑ OP	↓ =	? ?	Phan et al., 2011; Phan et al., 2015
DPN (4 h)	↑ OR, ↑ OP = OP	?	?	Jacome et al., 2010; Frye et al., 2007
PPT (40 min)*	↑ OR, ↑ OP	↑	?	Phan et al., 2011; 2015
PPT (4 h)	= OR, = OP ↑ OP	?	?	Jacome et al., 2010; Frye et al., 2007
G-1 (40 min)	↑ OR, ↑ OP	↑	?	Gabor et al., 2015

Treatment	Memory Test Effect	Spine Change		Reference
Females		CA1	mPFC	
S.C. Injection				
Hippocampal Injection				
E_2 (15 min)	↑ SR	↑	?	Sheppard et al., 2021
E_2 (48 h)*	↑ OR	?	?	Fernandez et al., 2008; Fortress et al., 2013; 2014
E_2 (48h)	↑ OR	?	?	Fan et al., 2010
E_2 (48 h)*	↑ OR, ↑ OP	?	?	Boulware et al., 2013
E_2 (48 h)*	?	↑	↑	Tuscher et al., 2016
DPN (48 h)*	↑ OR, ↑ OP	?	?	Boulware et al., 2013
PPT (48 h)*	↑ OR, ↑ OP	?	?	Boulware et al., 2013
E_2 (24 h)	↑ MWM	?	?	Packard and Teather, 1997
G-1 (40 min)*	?	↑	?	Kim et al., 2019
G-1 (48 h)*	↑ OR, ↑ OP	?	?	Kim et al., 2019
Males				
SC Injection				
E_2 (2 h)	↑ OP	↑	?	Jacome et al., 2016
E_2 (2 h)	?	↑	?	Avila et al., 2017
Hippocampal Injection				
E_2 (24 h)	↑ MWM	↑	?	Packard et al., 1996

Since the hippocampus is a critical mediator of learning and memory, Frick and colleagues tested whether direct placement of estrogens intra-hippocampally would be sufficient to enhance recognition memory. A number of experiments show significant enhancements by estradiol in young and middle-aged mice in this laboratory (Boulware et al., 2013; Fan et al., 2010; Fernandez et al., 2008; Fortress et al., 2013; Fortress et al., 2014; Kim et al., 2016; Pereira et al., 2014; Zhao et al., 2012; Zhao et al., 2010). Moreover, interactions between the hippocampus and prefrontal cortex in enhancing recognition memory have been shown recently (Tuscher et al., 2019). Memory in other tasks such as social recognition is also rapidly enhanced by estradiol (Ervin et al., 2013). Thus, a substantial literature now supports our initial results showing that acute administration of estrogens enhances memory in female rodents.

Effects of estrogen on memory are not limited to females since several studies have demonstrated enhancements in males. Post training injection of estradiol to castrate male rats enhances object placement (Jacome et al., 2016). Direct injection into the hippocampus of gonadally intact or castrate mice enhances object or place recognition (Koss et al., 2018) and Morris Water

Maze in male rats (Packard et al., 1996). As is the case with females (though far fewer experiments have been completed), gonadectomy and chronic treatments of males with estrogens also enhances performance on some memory tasks (Aubele et al., 2008; Daniel et al., 2003; Frick et al., 2015; Gibbs, 2005). Thus, as shown in Table 1, consistent data from both rats and mice and in both sexes across multiple laboratories demonstrates the memory-enhancing effects of estradiol. The next sections discuss possible mechanisms for these changes including increases in density of dendritic spines and alterations in cell signaling cascades.

Estrogens Increase Spinogenesis

In 2023, it is well accepted that spines are highly plastic structures and that they are key mediators of cognitive function (Bailey et al., 2015; Russo et al., 2010; Segal, 2017; Segal et al., 2010). Here, we move a step further in the evolution of spine function and provide evidence that the gonadal hormone, estradiol, exerts its cognitive enhancing properties through actions on dendritic spines present primarily in the hippocampal formation and the prefrontal cortex. The majority of the work has been on estradiol, and that the mechanism involves dendritic spines has been almost as torturous as the history of spines themselves (Luine and Frankfurt, 2020). It was approximately fifty years ago that estrogens were shown to cross the blood brain barrier and regulate reproductive behaviors (Pfaff, 1968). Prior to this discovery, gonadal hormones were thought not to influence neural functions directly but rather to alter other chemicals which then acted in brain to regulate behaviors. In 2023, it is recognized that multiple receptors for estrogens are present throughout the brain and located in cell nuclei, axons, cell membranes of neurons, dendritic spines, and pre- and post-synaptic terminals (Almey et al., 2015). Moreover, gonadal hormones impact a wide variety of behaviors, besides sexual behaviors, including mood and affective disorders as well as the highest order of neural function, cognition. Our focus is to understand the rapid effects of estrogens in increasing dendritic spine density and enhancing the consolidation of memory, and we describe changes in multiple cell signaling pathways in the hippocampus and prefrontal cortex which underlie these effects. Delineating the mechanisms whereby estrogens enhance cognitive function may provide significant advances for treatment of dementias as well as age and neuro-degenerative disease related memory loss.

Dendritic remodeling is a well-accepted mechanism contributing to the process of learning and memory. For example, in conditioning paradigms, acquisition of new memories is associated with increases in dendritic spine density in the CA1 region of the hippocampus in both female (Beltrán-Campos et al., 2011) and male rats (Jedlicka et al., 2008; Leuner et al., 2003). In addition, existing spines undergo structural alterations that result in long term potentiation (LTP) (Bourne and Harris, 2008; Morgado-Bernal, 2011). Overall, memories appear to require alterations in dendritic spines and long term potentiation (LTP) (Matsuzaki et al., 2011). In relation to possible estrogen effects on the hippocampus, receptors for estradiol and changes in spine density in rats during the estrous cycle were reported in the 1990s (Woolley et al., 1990; Woolley and McEwen, 1993). Further, chronic treatments with estrogens increase dendritic spine density in the CA1 area of the rat hippocampus [See Table 1; (Gould et al., 1990; Li et al., 2004; Luine and Frankfurt, 2012)]. Some studies also indicate that the medial prefrontal cortex (mPFC) also contributes to memory, and estrogens given chronically increase dendritic spine density in this area (Table 1) (Khan et al., 2013).

Based on our observation that estrogens could rapidly, within a few hours, enhance recognition memory (Inagaki et al., 2010; Luine et al., 2003), we assessed acute effects of estrogen on dendritic spine density on pyramidal cells in the CA1 region of the hippocampus and the mPFC. In OVX female rats and mice, estradiol increased dendritic spine density on basal dendrites in CA1 by 29% (Inagaki et al., 2012) and in apical and basal dendrites in the mPFC by 16-27% within 30 - 40 min (Inagaki et al., 2012; Phan et al., 2012b). In addition, MacLusky et al., (MacLusky et al., 2005) reported a dose dependent increase in the postsynaptic density (PSD, a synaptic marker) within 30 minutes when estradiol was given to OVX rats. Agonists for both estrogen receptors, ERα and ERβ, also alter spine density in CA1. In OVX mice, PPT, an ERα agonist increased dendritic spine density in CA1 within one hour, whereas DPN, an ERβ agonist, decreased spine density (Phan et al., 2011).

More recently, in collaboration with the Frick laboratory, we found that direct intra-hippocampal infusion of estradiol increased spine density on apical and basal dendrites of CA1 pyramidal cells by approximately 17 - 30% within 30 minutes and density remained elevated 2h post injection (Tuscher et al., 2016). No changes in spine density were observed in either the dentate gyrus or the ventromedial nucleus of the hypothalamus following intra-hippocampal injection. However, in the mPC, dendritic spine density increased 21% after 2 h, but not after 30 min, showing that increased spine density in CA1 precedes the increase in the mPFC. Thus, estradiol mediated activity in the dorsal hippocampus may drive mPFC spinogenesis. Hippocampal infusion of G-1, an agonist for the G-protein-couple estrogen receptor (GPER), also increased CA1 spine density within 40 min (Kim et al., 2019).

A similar pattern of estrogenic effects on spine density as in OVX females was recently found in castrated (CAS) males after sc injection of estradiol. Estradiol was associated with a 31% increase in dendritic spine density within 30 minutes on the apical branches of CA1 pyramidal neurons (Jacome et al., 2016), but it had no effect in the dentate gyrus. Increased spine density in CA1 was maintained 2h after injection at the time of memory testing (Jacome et al., 2016). Application of the ERα agonist, PPT, or estradiol (Murakami et al., 2006) to hippocampal slices from male rats increased dendritic spine density within 2 hours, but the ERβ agonist, DPN, did not. Thus, acute estradiol exerts similar effects in both sexes, rapidly increasing spine density in brain areas important for memory function. We have also reported that testosterone administration to CAS male rats rapidly increases dendritic spines in a pattern similar to estradiol. Within 30 min, spines on the apical dendrites of CA1 pyramidal cells increased by 21% and remained elevated at 2 h (Jacome et al., 2016). A summary of rapid, in vivo, effects of estradiol on dendritic spine density can be found in Table 1.

Most studies have assessed hormone-dependent changes in total spines; however, spines are present as distinct types as first described by Ramon y Cajal (see Introduction). Spine subtype classification varies somewhat among researchers, but dendritic spines generally consist of a protrusion and may have a bulbous termination (in the case of mushroom spines). One may also distinguish between stubby spines and thin spines with a smaller head, which lack any terminal enlargement and have extremely thin filopodia. Thin spines are precursors to mature mushroom spines (Bourne and Harris, 2008; Urbanska et al., 2012; von Bohlen Und Halbach, 2009). In CA1, mushroom spines are mainly excitatory (glutamatergic) and contain post synaptic density protein (PSD95).

In an attempt to more fully understand and describe estrogenic effects on spines, we applied simultaneous Golgi-Cox Laser Scanning Confocal Microscopy and immunohistochemistry (GC-IHC) to the brains of CAS male rats that received estradiol 30 and 120 min before sacrifice (Avila et al., 2017).

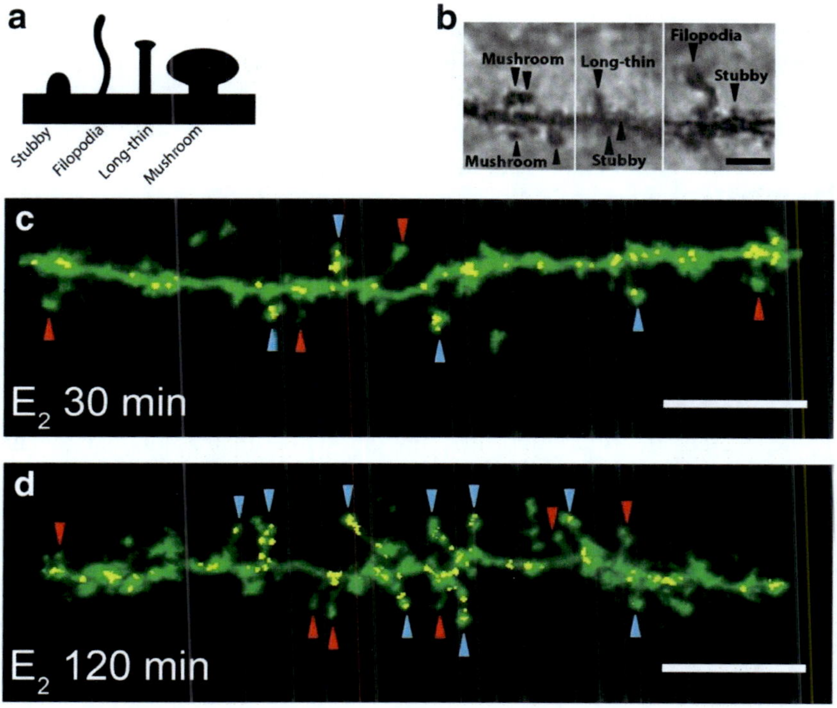

Figure 2. Effects of estradiol in CA1 using photomicrography of simultaneous Golgi-Cox scanning confocal microscopy and immunohistochemistry in hippocampus. Hippocampal CA1 spines captured at 100x magnification. Estradiol increases co-localization of GluA2 and PSD95 within Mushroom spines. Representative dendrites from E30 min (A) and E120 min (B). Yellow voxels indicate co-localized GluA2 and PSD95 within Mushroom spines. All carats indicate mushroom spines. Red carats indicate mushroom spines that do not contain co-localized GluA2 and PSD95. Blue carats indicate mushroom spines that contain co-localized GluA2 and PSD95. Scale bar: 5μm. Figure reprinted from Avila et al., 2017 by permission.

Four spine types (stubby, filopodial, long thin and mushroom) were identified by Golgi impregnation (Figure 2 a,b), and localization of PSD95 and GluA2 receptors in the spines was evaluated by IHC. GluA2 is an AMPA receptor which is located within the synaptic membrane and is important for memory consolidation (Migues et al., 2010), and some evidence suggests that mERs can functionally couple to GluA2 receptors in hippocampus (Boulware et al., 2005). Stubby and filopodial spines did not change after estradiol, but the density of long-thin spines decreased 63% at 30 min and the density of mushroom spines increased 60% at 120 min (Avila et al., 2017). Further, the density of mushroom spines, which colocalized with GluA2/PSD-95, increased approximately three-fold between 30 and 120 min following estradiol treatment (Avila et al., 2017). Figure 2 c,d is a photomicrograph of simultaneous Golgi-cox scanning confocal microscopy and IHC showing GluA2/PSD-95 containing CA1 spines with and without estradiol treatment. These results indicate a time course for a shift in the density of various spine subtypes in CA1 following estradiol and suggest that estradiol increases spine density by increasing the rate of maturation of extant spines through the transformation of long-thin spines into mature, synapse containing spines, i.e., GluA2/PSD-95 mushroom spines.

Consistent with our results, Sellers et al., 2015, found similar, but more rapid, changes in cultured cortical neurons where increased synaptogenesis and PSD-95 levels were found 30 min following estradiol. Also consistent with these results and the role of mushroom spines in memory, increased mushroom spine density in CA1 produced by water maze training is prevented by ovariectomy in rats (Beltrán-Campos et al., 2011) and using Golgi techniques alone, Li et al., 2004 reported increased density of mushroom spines in CA1 of OVX mice following 5 days of estradiol treatment. Thus, estrogens increase the density of dendritic spines at doses and time intervals consistent with its enhancements in memory consolidation. The intra-cellular mechanisms elicited by estradiol are discussed in the next section and provide evidence that the relationship between increased spine density by estradiol and enhancements in memory maybe causal.

Estrogen Increases Signaling Mechanisms

Learning and memory is primarily mediated by neuronal plasticity which includes strengthening of existing synapses, modulation of intrinsic excitability, adult neurogenesis, and synaptogenesis (Giese and Mizuno, 2013). An important focus of researchers for understanding estrogenic enhancements in cognitive function has been on the latter mechanism and unlike the slow progress of research on the role of dendritic spines in the brain and of estrogens in promoting cognitive function, determining the underlying mechanisms for estrogenic effects on spinogenesis has been rapid and prodigious. Information that non-nuclear, membrane receptors for estrogens are present in the CNS (see Almey et al., 2015 for review) provided the foundation for this new focus of research, and estrogen binding to these receptors were found to activate a number of behaviors including reproduction and motivation (Johnson et al., 2022). Early work centered on kinases because both mERs and kinases are present in dendritic spines which suggested there might be efficient coupling between postsynaptic events and these proteins (Hojo et al., 2008; Milner et al., 2005; Mukai et al., 2007). Thus, the serine/threonine kinases, including ERK, MAPK, PKA and PKC were initially investigated. In CA1, an area critically important area for memory, as well as in the dentate gyrus (DG) and CA3 regions of the hippocampus, estradiol and a variety of estrogen agonists were shown to increase phosphorylation of these kinases and their target proteins (Frick et al., 2010; Hojo et al., 2015). Figure 3 is a schematic of the signaling cascades which are activated by estrogen binding to membrane receptors, and while the receptors and phosphorylated intermediates are different, the final commonality is activation of protein synthesis which leads to increases in the number of dendritic spines and in the levels of proteins which sub-serve synaptic function. In CA1 specifically, estradiol bound to mER and NMDAR or GLUR activates ERK/MAPK signaling (Boulware et al., 2013; Woolley et al., 1997) which in turn activates phosphorylation of the transcription factor cAMP response element-binding protein (CREB) on DNA in the nucleus (Boulware et al., 2005) and through epigenetic processes alters transcription of proteins. Changes induced include a decrease in transcription of cofilin, which allows for increased polymerization of actin used for assembly of spines and increases in transcription of actin, and RhoA, which are necessary for filopodial extension and spine formation, and hence, a resultant increase in the density of mature dendritic spines (Kramár et al., 2013).

Estrogen also signals through Ca^{+}-dependent P13K → Akt → mTOR to activate protein synthesis (Akama and McEwen, 2003). This pathway appears to be activated by the interaction/binding of estrogens to membrane ERα and β which are associated with glutamate receptors, mGluR (Gross and Mermelstein, 2020). Enhanced protein synthesis via this pathway occurs outside of action in the cell nucleus. Estrogen or membrane bound estrogen may also signal enhanced spinogenesis through the brain derived neurotrophic factor (BDNF) membrane receptor, TrkB, via phosphorylation of P13K and Akt and activation of mTOR which in turn enhances protein synthesis (Srivastava et al., 2013). This pathway appears to be indirectly activated by an estrogen dependent increase in BDNF levels which in turn increases binding to TrkB and activation of the pathway (Luine and Frankfurt, 2013).

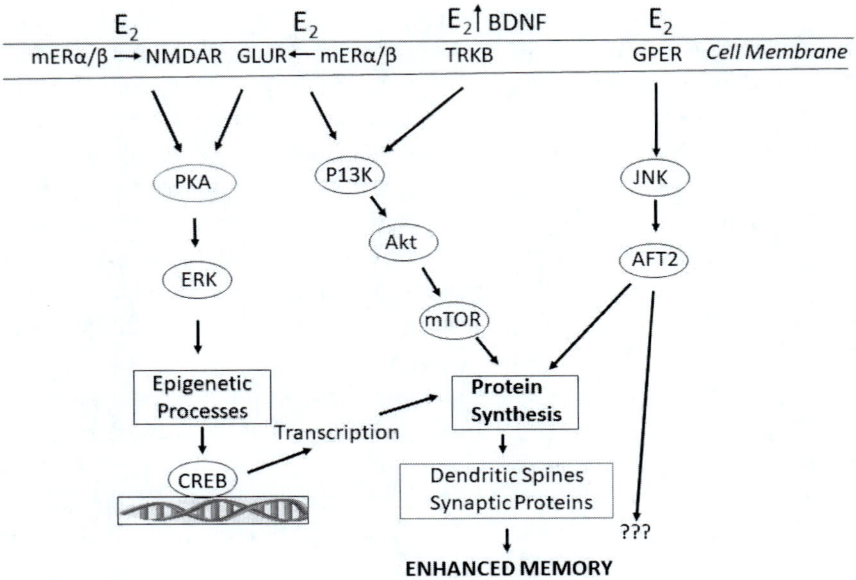

Figure 3. Schematic diagram depicting membrane receptor mediated mechanisms for rapid effects of estrogens on memory consolidation in rodents. Circulating E_2 can bind to various mERs which interact with other membrane receptors to increase several phosphorylation pathways which ultimately result in increased protein synthesis. These actions result in structural and neurochemical changes at existing synapses as well as synaptogenesis that enhance memory formation, consolidation, storage and retrieval. Binding to NMDA or mGLuR receptors promotes phosphorylation of the intermediates, PKA and ERK, which through epigenetic processes facilitate phosphorylation of the transcription factor cAMP-response-element-binding protein (CREB) located on DNA in the cell nucleus and increased protein synthesis. Activation of GLUR signals through P13K and increases protein synthesis in the cytoplasm. E_2 indirectly activates the TRKB pathway by increasing levels of its ligand, BDNF. Binding of E_2 to the GPER receptor activates JNK and AFT2 which in turn increase protein synthesis, but it is unclear whether E_2 (or another ligand) enhances memory. Abbreviations: ERK, extracellular regulated kinase; PKA, protein kinase A; P13K/AKT, phosphatidylinositol3-kinase/Akt pathway; JNK, c-Jun N-terminal kinase; ATF2, transcription factor 2.

Another phosphorylation pathway, linked to the G protein-coupled ER1 receptor (GPER), has also recently been linked to estrogen action. Formerly known as the GPR 30 orphan receptor (Funakoshi et al., 2006), GPER is abundant in the hippocampus and prefrontal cortex (Hutson et al., 2019). Treatment of rodents with a GPER agonist, G-1 or a selective antagonist, G-15, enhances or impairs, respectively, several forms of memory and spinogenesis (Gabor et al., 2015; Kim et al., 2016). Interestingly, GPER does did not activate ERK, P13K or Akt but appears to act through the c-Jun N-terminal kinase (JNK) as well as the downstream transcription factor ATF2 (Kim et al., 2016). In contrast, other studies indicate interaction through the ERK network (Hart et al., 2014; Kumar et al., 2015). However, because estradiol itself does not increase phosphorylation of JNK, and JNK inhibition does not block estrogen effects on memory (Kim et al., 2016) estradiol and G-1 may act independently. Thus, further research is required to understand the possible role of GPR 30 receptor signaling in mnemonic actions of estradiol.

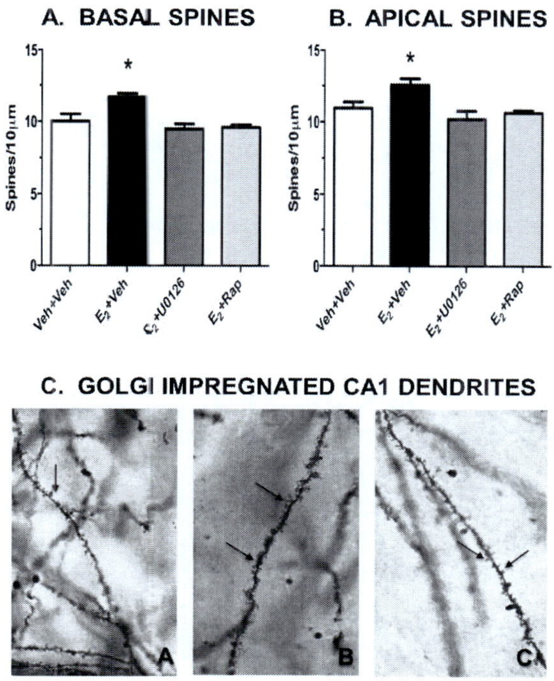

Figure 4. CA1 apical and basal spine density induced by ICV E_2 is blocked by ERK or mTOR inhibition. Two hours post ICV E_2 infusion, basal (A) and apical (B) spine density was significantly increased in CA1 relative to vehicle. These effects were blocked by DH infusion of the ERK activation inhibitor U0126 or the mTOR activation inhibitor rapamycin. Bars represent the mean ± SEM, $*p < 0.05$ relative to all other groups. C. Photomicrographs of Golgi-impregnated secondary basal dendrites of CA1 pyramidal neurons (panel A = vehicle, panel B = E_2+vehicle, panel C = E_2+U0126). Arrows denote spines. Under oil, 63x. Data from Tuscher et al., 2016.

Involvement of the above pathways in memory consolidation and the effects of estrogen were discovered by the measurement of increased kinase levels following acute estrogen administration. Experiments were usually conducted by *in vitro* measurements in hippocampal or prefrontal cortex slices. For examples, see studies by Hojo et al., (2015) and Hasegawa et al., (2015) in hippocampal slices, an *in vivo* study by Xing et al., (2018) in rats and a review by Kato et al., (2020).

Effects of estradiol on memory consolidation and spine density have also been assessed *in vivo* following peripheral or intra-hippocampal administration of drugs blocking specific signaling pathways. In male and female rodents, hippocampal dependent learning activates ERK and mTOR, and the antagonist of mTOR, rapamycin, or the ERK antagonist, U0126 prevent consolidation of recognition memories (Fernandez et al., 2008; Fortress et al., 2013). Furthermore, dorsal hippocampal infusion of these antagonists prevents estradiol from enhancing object recognition and object placement memory consolidation in young OVX mice (Fortress et al., 2013). However, in this study it was unclear whether these antagonists would also inhibit spinogenesis under the same conditions where memory was inhibited; such a result could provide direct evidence that increases in spine density in the hippocampus underlie estrogenic enhancements in memory consolidation. Figure 4 provides the details and highlights the results of one such experiment done in a collaboration with the Frick laboratory. When estradiol was infused bilaterally into the dorsal hippocampus, basal and apical spine density in CA1 were increased 30 min and 2 h later., but when estrogen infusion was immediately preceded by infusions of rapamycin (mTOR inhibitor) or U0126 (ERK inhibitor), the increased spinogenesis was blocked (Figure 4 A,B). Figure 4C shows Golgi impregnated CA1 basal dendrites following the treatments. Thus, activation of either of these pathways was necessary for estrogen induced increases in CA1 dendritic spine density and for its previously shown enhancements of recognition memory consolidation (Fernandez et al., 2008; Fortress et al., 2013). Moreover, this experiment provides further evidence that estrogen-linked enhanced spinogenesis is fundamental to enhanced memory consolidation.

Changes in an array of synaptic proteins have also been documented *in vivo* and *in vitro* following acute estradiol administration and a few examples are discussed here. As demonstrated by monoamine levels or their metabolites 30 min after estradiol injection to OVX rats, noradrenergic and dopaminergic activities are increased by approximately 100 and 35%, respectively, in the mPFC and noradrenergic activity is decreased approximately 50% in CA1 and

35% in the DG. These results are consistent with monoaminergic systems having a long history in mediating cognitive function (McGaugh, 1989). Rapid increases in the post synaptic density protein, PSD95, (Akama and McEwen, 2003; Avila et al., 2017; MacLusky et al., 2005) and BDNF (Luine and Frankfurt, 2013) by estrogens were discussed above. Several AMPA and $GABA_A$ receptor levels are also rapidly increased by estrogens in brain regions subserving memory function (reviewed in Barabas, et al., 2018).

Conclusions and Future Directions

Substantial evidence shows that estrogens are potent enhancers of memory consolidation in female and male rodents. Effects are rapid, occurring within minutes, and involve activation of membrane receptors and promotion of signaling pathways leading to increases in dendritic spine density in the hippocampus and prefrontal cortex. However, important challenges remain such as determining whether these changes are physiologically relevant. If so, are some pathways more important than others? (See Luine and Frankfurt, 2012 for further discussion).

Understanding mechanisms whereby estrogens enhance learning and memory has enormous potential consequences for human health and well-being since this knowledge may provide the foundation for developing effective treatments for cognitive impairments. First, no safe and effective long-term treatments are currently available for menopausal/estrogen related cognitive losses. According to the US census, more than 5,000 women/day enter menopause, a major population desirous and deserving of better treatments. Furthermore, treatments for age or neurodegenerative disease-related memory loss are also lacking, and these problems present enormous societal challenges. It has also been speculated that activation of estrogen signaling mechanisms may facilitate genomic responses, and thus, nuclear crosstalk could serve as a "priming" mechanism for long term changes associated with classic receptor mechanisms which are known to contribute to cognition. Taken together, understanding of the mechanisms underlying rapid estrogen enhancements of memory and spinogenesis provide a rich potential for basic and clinical research.

References

Akama, K. T., McEwen, B. S., 2003. Estrogen stimulates postsynaptic density-95 rapid protein synthesis via the Akt/protein kinase B pathway. *J. Neurosci.* 23, 2333-2339.

Almey, A., Milner, T. A., Brake, W. G., 2015. Estrogen receptors in the central nervous system and their implication for dopamine-dependent cognition in females. *Horm. Behav.* 74, 125-138.

Aubele, T., Kaufman, R., Montalmant, F., Kritzer, M. F., 2008. Effects of gonadectomy and hormone replacement on a spontaneous novel object recognition task in adult male rats. *Horm. Behav.* 54, 244-252.

Avila, J. A., Alliger, A. A., Carvajal, B., Zanca, R. M., Serrano, P. A., Luine, V. N., 2017. Estradiol rapidly increases GluA2-mushroom spines and decreases GluA2-filopodia spines in hippocampus CA1. *Hippocampus* 27, 1224-1229.

Bailey, C. H., Kandel, E. R., Harris, K. M., 2015. Structural Components of Synaptic Plasticity and Memory Consolidation. *Cold Spring Harb Perspect Biol* 7, a021758.

Barabas, K., Godo, S., Lengyel, F., Ernszt, D., Pal, J., Abraham, I. M., 2018. Rapid non-classical effects of steroids on the membrane receptor dynamics and downstream signaling in neurons. *Horm. Behav.* 104, 183-191.

Beltrán-Campos, V., Prado-Alcalá, R. A., León-Jacinto, U., Aguilar-Vázquez, A., Quirarte, G. L., Ramírez-Amaya, V., Díaz-Cintra, S., 2011. Increase of mushroom spine density in CA1 apical dendrites produced by water maze training is prevented by ovariectomy. *Brain Res.* 1369, 119-130.

Blaustein, J. D., 1992. Cytoplasmic estrogen receptors in rat brain: immunocytochemical evidence using three antibodies with distinct epitopes. *Endocrinology* 131, 1336-1342.

Boulware, M. I., Heisler, J. D., Frick, K. M., 2013. The memory-enhancing effects of hippocampal estrogen receptor activation involve metabotropic glutamate receptor signaling. *J. Neurosci.* 33, 15184-15194.

Boulware, M. I., Weick, J. P., Becklund, B. R., Kuo, S. P., Groth, R. D., Mermelstein, P. G., 2005. Estradiol activates group I and II metabotropic glutamate receptor signaling, leading to opposing influences on cAMP response element-binding protein. *J. Neurosci.* 25, 5066-5078.

Bourne, J. N., Harris, K. M., 2008. Balancing structure and function at hippocampal dendritic spines. *Annu. Rev. Neurosci.* 31, 47-67.

Daniel, J. M., Winsauer, P. J., Moerschbaecher, J. M., 2003. Castration in rats impairs performance during acquisition of a working memory task and exacerbates deficits in working memory produced by scopolamine and mecamylamine. *Psychopharmacology (Berl)* 170, 294-300.

DeFelipe, J., 2017. *Cajal's Neuronal Forest.* Oxford University Press. New York.

Ennaceur, A., Souza-Silva, M., 2018. *Handbook of Research on Object Novelty Recognition.* Elsevier/Academic Press, United Kingdom.

Ervin, K. S., Phan, A., Gabor, C. S., Choleris, E., 2013. Rapid oestrogenic regulation of social and nonsocial learning. *J. Neuroendocrinol.* 25, 1116-1132.

Fan, L., Zhao, Z., Orr, P. T., Chambers, C. H., Lewis, M. C., Frick, K. M., 2010. Estradiol-induced object memory consolidation in middle-aged female mice requires dorsal hippocampal extracellular signal-regulated kinase and phosphatidylinositol 3-kinase activation. *J. Neurosci.* 30, 4390-4400.

Fernandez, S. M., Lewis, M. C., Pechenino, A. S., Harburger, L. L., Orr, P. T., Gresack, J. E., Schafe, G. E., Frick, K. M., 2008. Estradiol-induced enhancement of object memory consolidation involves hippocampal extracellular signal-regulated kinase activation and membrane-bound estrogen receptors. *J. Neurosci.* 28, 8660-8667.

Fortress, A. M., Fan, L., Orr, P. T., Zhao, Z., Frick, K. M., 2013. Estradiol-induced object recognition memory consolidation is dependent on activation of mTOR signaling in the dorsal hippocampus. *Learn. Mem.* 20, 147-155.

Fortress, A. M., Kim, J., Poole, R. L., Gould, T. J., Frick, K. M., 2014. 17beta-Estradiol regulates histone alterations associated with memory consolidation and increases Bdnf promoter acetylation in middle-aged female mice. *Learn. Mem.* 21, 457-467.

Frick, K. M., Fernandez, S. M., Harburger, L. L., 2010. A new approach to understanding the molecular mechanisms through which estrogens affect cognition. *Biochim. Biophys. Acta* 1800, 1045-1055.

Frick, K. M., Kim, J., Tuscher, J. J., Fortress, A. M., 2015. Sex steroid hormones matter for learning and memory: Estrogenic regulation of hippocampal function in male and female rodents. *Learn. Mem.* 22, 472-493.

Frotscher, M., Leranth, C., 1986. The cholinergic innervation of the rat fascia dentata: identification of target structures on granule cells by combining choline acetyltransferase immunocytochemistry and Golgi impregnation. *J. Comp. Neurol.* 243, 58-70.

Frye, C. A., Duffy, C. K., Walf, A. A., 2007. Estrogens and progestins enhance spatial learning of intact and ovariectomized rats in the object placement task. *Neurobiol. Learn. Mem.* 88, 208-216.

Funakoshi, T., Yanai, A., Shinoda, K., Kawano, M. M., Mizukami, Y., 2006. G protein-coupled receptor 30 is an estrogen receptor in the plasma membrane. *Biochem. Biophys. Res. Commun.* 346, 904-910.

Gabor, C., Lymer, J., Phan, A., Choleris, E., 2015. Rapid effects of the G-protein coupled oestrogen receptor (GPER) on learning and dorsal hippocampus dendritic spines in female mice. *Physiol. Behav.* 149, 53-60.

Gibbs, R. B., 2005. Testosterone and estradiol produce different effects on cognitive performance in male rats. *Horm. Behav.* 48, 268-277.

Giese, K. P., Mizuno, K., 2013. The roles of protein kinases in learning and memory. *Learn. Mem.* 20, 540-552.

Golgi, C., 1903. Sulla fina anatomia degli organi centrali del sistema nervosa [On the fine anatomy of the central organs of the nervous system]. Rivista sperimentale di Freniatria 1882–1883. *Istologia Normale* [Normal Histology] (1870–1883) Milan: Ulrico Hoepli, Milan.

Gould, E., Woolley, C. S., Frankfurt, M., McEwen, B. S., 1990. Gonadal steroids regulate dendritic spine density in hippocampal pyramidal cells in adulthood. *J. Neurosci.* 10, 1286-1291.

Gresack, J. E., Frick, K. M., 2004. Environmental enrichment reduces the mnemonic and neural benefits of estrogen. *Neuroscience* 128, 459-471.

Gresack, J. E., Frick, K. M., 2006. Effects of continuous and intermittent estrogen treatments on memory in aging female mice. *Brain Res.* 1115, 135-147.

Gresack, J. E., Kerr, K. M., Frick, K. M., 2007a. Life-long environmental enrichment differentially affects the mnemonic response to estrogen in young, middle-aged, and aged female mice. *Neurobiol. Learn. Mem.* 88, 393-408.

Gresack, J. E., Kerr, K. M., Frick, K. M., 2007b. Short-term environmental enrichment decreases the mnemonic response to estrogen in young, but not aged, female mice. *Brain Res.* 1160, 91-101.

Gross, K. S., Mermelstein, P. G., 2020. Estrogen receptor signaling through metabotropic glutamate receptors. *Vitam. Horm.* 114, 211-232.

Hajszan, T., MacLusky, N. J., Leranth, C., 2005. Short-term treatment with the antidepressant fluoxetine triggers pyramidal dendritic spine synapse formation in rat hippocampus. *Eur. J. Neurosci.* 21, 1299-1303.

Hampson, E., Phillips, S. D., Duff-Canning, S. J., Evans, K. L., Merrill, M., Pinsonneault, J. K., Sadée, W., Soares, C. N., Steiner, M., 2015. Working memory in pregnant women: Relation to estrogen and antepartum depression. *Horm. Behav.* 74, 218-227.

Hart, D., Nilges, M., Pollard, K., Lynn, T., Patsos, O., Shiel, C., Clark, S. M., Vasudevan, N., 2014. Activation of the G-protein coupled receptor 30 (GPR30) has different effects on anxiety in male and female mice. *Steroids* 81, 49-56.

Hasegawa, Y., Hojo, Y., Kojima, H., Ikeda, M., Hotta, K., Sato, R., Ooishi, Y., Yoshiya, M., Chung, B. C., Yamazaki, T., Kawato, S., 2015. Estradiol rapidly modulates synaptic plasticity of hippocampal neurons: Involvement of kinase networks. *Brain Res.* 1621, 147-161.

Hojo, Y., Munetomo, A., Mukai, H., Ikeda, M., Sato, R., Hatanaka, Y., Murakami, G., Komatsuzaki, Y., Kimoto, T., Kawato, S., 2015. Estradiol rapidly modulates spinogenesis in hippocampal dentate gyrus: Involvement of kinase networks. *Horm. Behav.* 74, 149-156.

Hojo, Y., Murakami, G., Mukai, H., Higo, S., Hatanaka, Y., Ogiue-Ikeda, M., Ishii, H., Kimoto, T., Kawato, S., 2008. Estrogen synthesis in the brain--role in synaptic plasticity and memory. *Mol. Cell. Endocrinol.* 290, 31-43.

Hutson, D. D., Gurrala, R., Ogola, B. O., Zimmerman, M. A., Mostany, R., Satou, R., Lindsey, S. H., 2019. Estrogen receptor profiles across tissues from male and female Rattus norvegicus. *Biol. Sex. Differ.* 10, 4.

Inagaki, T., Frankfurt, M., Luine, V., 2012. Estrogen-induced memory enhancements are blocked by acute bisphenol A in adult female rats: role of dendritic spines. *Endocrinology* 153, 3357-3367.

Inagaki, T., Gautreaux, C., Luine, V., 2010. Acute estrogen treatment facilitates recognition memory consolidation and alters monoamine levels in memory-related brain areas. *Horm. Behav.* 58, 415-426.

Jacome, L. F., Barateli, K., Buitrago, D., Lema, F., Frankfurt, M., Luine, V. N., 2016. Gonadal hormones rapidly enhance spatial memory and increase hippocampal spine density in male rats. *Endocrinology* 157, 1357-1362.

Jacome, L. F., Gautreaux, C., Inagaki, T., Mohan, G., Alves, S., Lubbers, L. S., Luine, V., 2010. Estradiol and ERbeta agonists enhance recognition memory, and DPN, an ERbeta agonist, alters brain monoamines. *Neurobiol. Learn. Mem.* 94, 488-498.

Jedlicka, P., Vlachos, A., Schwarzacher, S. W., Deller, T., 2008. A role for the spine apparatus in LTP and spatial learning. *Behav. Brain Res.* 192, 12-19.

Johnson, C. S., Micevych, P. E., Mermelstein, P. G., 2022. Membrane estrogen signaling in female reproduction and motivation. Front. *Endocrinol.* (Lausanne) 13, 1009379.

Kato, A., Murakami, G., Hojo Y., Horie, S, Kawato, S., 2020. *Rapid effects of estradiol on dendritic spines and syncptic plasticity in the male and female hippocampus.* Oxford University Press, New York.

Khan, M. M., Dhandapani, K. M., Zhang, Q. G., Brann, D. W., 2013. Estrogen regulation of spine density and excitatory synapses in rat prefrontal and somatosensory cerebral cortex. *Steroids* 78, 614-623.

Kim, J., Schalk, J. C., Koss, W. A., Gremminger, R. L., Taxier, L. R., Gross, K. S., Frick, K. M., 2019. Dorsal hippocampal actin polymerization is necessary for activation of G-protein-coupled estrogen receptor (GPER) to increase CA1 dendritic spine density and enhance memory consolidation. *J. Neurosci.* 39, 9598-9610.

Kim, J., Szinte, J. S., Boulware, M. I., Frick, K. M., 2016. 17beta-estradiol and agonism of G-protein-coupled estrogen receptor enhance hippocampal memory via different cell-signaling mechanisms. *J. Neurosci.* 36, 3309-3321.

Koss, W. A., Haertel, J. M., Philippi, S. M., Frick, K. M., 2018. Sex differences in the rapid cell signaling mechanisms underlying the memory-enhancing effects of 17beta-estradiol. *eNeuro* 5.

Kramár, E. A., Babayan, A. H., Gall, C. M., Lynch, G., 2013. Estrogen promotes learning-related plasticity by modifying the synaptic cytoskeleton. *Neuroscience* 239, 3-16.

Kumar, A., Bean, L. A., Rani, A., Jackson, T., Foster, T. C., 2015. Contribution of estrogen receptor subtypes, ERalpha, ERbeta, and GPER1 in rapid estradiol-mediated enhancement of hippocampal synaptic transmission in mice. *Hippocampus* 25, 1556-1566.

Leuner, B., Falduto, J., Shors, T. J., 2003. Associative memory formation increases the observation of dendritic spines in the hippocampus. *J. Neurosci.* 23, 659-665.

Lewis, M. C., Kerr, K. M., Orr, P. T., Frick, K. M., 2008. Estradiol-induced enhancement of object memory consolidation involves NMDA receptors and protein kinase A in the dorsal hippocampus of female C57BL/6 mice. *Behav. Neurosci.* 122, 716-721.

Li, C., Brake, W. G., Romeo, R. D., Dunlop, J. C., Gordon, M., Buzescu, R., Magarinos, A. M., Allen, P. B., Greengard, P., Luine, V., McEwen, B. S., 2004. Estrogen alters hippocampal dendritic spine shape and enhances synaptic protein immunoreactivity and spatial memory in female mice. *Proc. Natl. Acad. Sci. USA* 101, 2185-2190.

Luine, V. N., 2008. Sex steroids and cognitive function. *J Neuroendocrinol* 20, 866-872.

Luine, V. N., 2014. Estradiol and cognitive function: past, present and future. *Horm. Behav.* 66, 602-618.

Luine, V. N., Frankfurt, M., 2012. Estrogens facilitate memory processing through membrane mediated mechanisms and alterations in spine density. Front. *Neuroendocrinol.* 33, 388-402.

Luine, V., Bowman, R., Serrano, P., 2018a. Sex differences in cognitive responses to stress in rodents, in: Silva, A. E. a. M. A. d. S. (Ed.), *Handbook of Research on Object Novelty Recognition.* Elsevier/Academic Press, New York, pp. 531-540.

Luine, V., Dohanich, G., 2008a. Sex differences in cognitive function in rodents, in: J. B. Becker, K. B., N. Geary, E. Hampson, J. P. Herman and E. A. Young (Eds.), *Sex Differences in the Brain: From Genes to Behavior.* Oxford University Press, New York, pp. 227-252.

Luine, V., Frankfurt, M., 2013. Interactions between estradiol, BDNF and dendritic spines in promoting memory. *Neuroscience* 239, 34-45.

Luine, V., Frankfurt, M., 2020. Estrogenic regulation of memory: The first 50 years. *Horm. Behav.* 121, 104711.

Luine, V., Jacome, L. F., Maclusky, N. J., 2003. Rapid enhancement of visual and place memory by estrogens in rats. *Endocrinology* 144, 2836-2844.

Luine, V., Serrano, P., Frankfurt, M., 2018b. Rapid effects on memory consolidation and spine morphology by estradiol in female and male rodents. *Horm. Behav.* 104, 111-118.

MacLusky, N. J., Luine, V. N., Hajszan, T., Leranth, C., 2005. The 17alpha and 17beta isomers of estradiol both induce rapid spine synapse formation in the CA1 hippocampal subfield of ovariectomized female rats. *Endocrinology* 146, 287-293.

Matsuzaki, M., Ellis-Davies, G. C., Kanemoto, Y., Kasai, H., 2011. Simultaneous two-photon activation of presynaptic cells and calcium imaging in postsynaptic dendritic spines. *Neural Syst. Circuits* 1, 2.

McGaugh, J. L., 1989. Dissociating learning and performance: drug and hormone enhancement of memory storage. *Brain Res. Bull.* 23, 339-345.

Migues, P. V., Hardt, O., Wu, D. C., Gamache, K., Sacktor, T. C., Wang, Y. T., Nader, K., 2010. PKMzeta maintains memories by regulating GluR2-dependent AMPA receptor trafficking. *Nat. Neurosci.* 13, 630-634.

Milner, T. A., Ayoola, K., Drake, C. T., Herrick, S. P., Tabori, N. E., McEwen, B. S., Warrier, S., Alves, S. E., 2005. Ultrastructural localization of estrogen receptor beta immunoreactivity in the rat hippocampal formation. *J. Comp. Neurol.* 491, 81-95.

Milner, T. A., McEwen, B. S., Hayashi, S., Li, C. J., Reagan, L. P., Alves, S. E., 2001. Ultrastructural evidence that hippocampal alpha estrogen receptors are located at extranuclear sites. *J. Comp. Neurol.* 429, 355-371.

Morgado-Bernal, I., 2011. Learning and memory consolidation: linking molecular and behavioral data. *Neuroscience* 176, 12-19.

Mukai, H., Tsurugizawa, T., Murakami, G., Kominami, S., Ishii, H., Ogiue-Ikeda, M., Takata, N., Tanabe, N., Furukawa, A., Hojo, Y., Ooishi, Y., Morrison, J. H., Janssen, W. G., Rose, J. A., Chambon, P., Kato, S., Izumi, S., Yamazaki, T., Kimoto, T., Kawato, S., 2007. Rapid modulation of long-term depression and spinogenesis via synaptic estrogen receptors in hippocampal principal neurons. *J. Neurochem.* 100, 950-967.

Murakami, G., Tsurugizawa, T., Hatanaka, Y., Komatsuzaki, Y., Tanabe, N., Mukai, H., Hojo, Y., Kominami, S., Yamazaki, T., Kimoto, T., Kawato, S., 2006. Comparison between basal and apical dendritic spines in estrogen-induced rapid spinogenesis of CA1 principal neurons in the adult hippocampus. *Biochem. Biophys. Res. Commun.* 351, 553-558.

Packard, M. G., 1998. Posttraining estrogen and memory modulation. *Horm. Behav.* 34, 126-139.

Packard, M. G., Kohlmaier, J. R., Alexander, G. M., 1996. Posttraining intrahippocampal estradiol injections enhance spatial memory in male rats: interaction with cholinergic systems. *Behav. Neurosci.* 110, 626-632.

Packard, M. G., Teather, L. A., 1997. Intra-hippocampal estradiol infusion enhances memory in ovariectomized rats. *Neuroreport* 8, 3009-3013.

Pereira, L. M., Bastos, C. P., de Souza, J. M., Ribeiro, F. M., Pereira, G. S., 2014. Estradiol enhances object recognition memory in Swiss female mice by activating hippocampal estrogen receptor alpha. *Neurobiol. Learn. Mem.* 114, 1-9.

Pfaff, D. W., 1968. Uptake of 3H-estradiol by the female rat brain. An autoradiographic study. *Endocrinology* 82, 1149-1155.

Phan, A., Gabor, C. S., Favaro, K. J., Kaschack, S., Armstrong, J. N., MacLusky, N. J., Choleris, E., 2012a. Low doses of 17beta-estradiol rapidly improve learning and increase hippocampal dendritic spines. *Neuropsychopharmacology* 37, 2299-2309.

Phan, A., Gabor, C. S., Favaro, K. J., Kaschack, S., Armstrong, J. N., MacLusky, N. J., Choleris, E., 2012b. Low doses of 17β-estradiol rapidly improve learning and increase hippocampal dendritic spines. *Neuropsychopharmacology* 37, 2299-2309.

Phan, A., Lancaster, K. E., Armstrong, J. N., MacLusky, N. J., Choleris, E., 2011. Rapid effects of estrogen receptor alpha and beta selective agonists on learning and dendritic spines in female mice. *Endocrinology* 152, 1492-1502.

Phan, A., Suschkov, S., Molinaro, L., Reynolds, K., Lymer, J. M., Bailey, C. D., Kow, L. M., MacLusky, N. J., Pfaff, D. W., Choleris, E., 2015. Rapid increases in immature synapses parallel estrogen-induced hippocampal learning enhancements. *Proc. Natl. Acad. Sci. USA* 112, 16018-16023.

Rhodes, M. E., Frye, C. A., 2004. Estrogen has mnemonic-enhancing effects in the inhibitory avoidance task. *Pharmacol. Biochem. Behav.* 78, 551-558.

Russo, S. J., Dietz, D. M., Dumitriu, D., Morrison, J. H., Malenka, R. C., Nestler, E. J., 2010. The addicted synapse: mechanisms of synaptic and structural plasticity in nucleus accumbens. *Trends Neurosci.* 33, 267-276.

Segal, M., 2017. Dendritic spines: Morphological building blocks of memory. *Neurobiol. Learn. Mem.* 138, 3-9.

Segal, M., Vlachos, A., Korkotian, E., 2010. The spine apparatus, synaptopodin, and dendritic spine plasticity. *Neuroscientist* 16, 125-131.

Sellers, K. J., Erli, F., Raval, P., Watson, I. A., Chen, D., Srivastava, D. P., 2015. Rapid modulation of synaptogenesis and spinogenesis by 17beta-estradiol in primary cortical neurons. *Front. Cell. Neurosci.* 9, 137.

Srivastava, D. P., Woolfrey, K. M., Penzes, P., 2013. Insights into rapid modulation of neuroplasticity by brain estrogens. *Pharmacol. Rev.* 65, 1318-1350.

Tuscher, J. J., Szinte, J. S., Starrett, J. R., Krentzel, A. A., Fortress, A. M., Remage-Healey, L., Frick, K. M., 2016. Inhibition of local estrogen synthesis in the hippocampus impairs hippocampal memory consolidation in ovariectomized female mice. *Horm. Behav.* 83, 60-67.

Tuscher, J. J., Taxier, L. R., Schalk, J. C., Haertel, J. M., Frick, K. M., 2019. Chemogenetic suppression of medial prefrontal-dorsal hippocampal interactions prevents estrogenic enhancement of memory consolidation in female mice. *eNeuro* 6.

Urbanska, M., Swiech, L., Jaworski, J., 2012. Developmental plasticity of the dendritic compartment: focus on the cytoskeleton. *Adv. Exp. Med. Biol.* 970, 265-284.

von Bohlen Und Halbach, O., 2009. Structure and function of dendritic spines within the hippocampus. *Ann. Anat.* 191, 518-531.

Woolley, C. S., Gould, E., Frankfurt, M., McEwen, B. S., 1990. Naturally occurring fluctuation in dendritic spine density on adult hippocampal pyramidal neurons. *J. Neurosci.* 10, 4035-4039.

Woolley, C. S., McEwen, B. S., 1993. Roles of estradiol and progesterone in regulation of hippocampal dendritic spine density during the estrous cycle in the rat. *J. Comp. Neurol.* 336, 293-306.

Woolley, C. S., Weiland, N. G., McEwen, B. S., Schwartzkroin, P. A., 1997. Estradiol increases the sensitivity of hippocampal CA1 pyramidal cells to NMDA receptor-mediated synaptic input: correlation with dendritic spine density. *J. Neurosci.* 17, 1848-1859.

Xing, F. Z., Zhao, Y. G., Zhang, Y. Y., He, L., Zhao, J. K., Liu, M. Y., Liu, Y., Zhang, J. Q., 2018. Nuclear and membrane estrogen receptor antagonists induce similar mTORC2 activation-reversible changes in synaptic protein expression and actin polymerization in the mouse hippocampus. *CNS Neurosci. The.r* 24, 495-507.

Zhao, Z., Fan, L., Fortress, A. M., Boulware, M. I., Frick, K. M., 2012. Hippocampal histone acetylation regulates object recognition and the estradiol-induced enhancement of object recognition. *J. Neurosci.* 32, 2344-2351.

Zhao, Z., Fan, L., Frick, K. M., 2010. Epigenetic alterations regulate estradiol-induced enhancement of memory consolidation. *Proc. Natl. Acad. Sci. USA* 107, 5605-5610.

Chapter 7

The Role of Astrocytes and Microglia in Synaptic Plasticity and Its Implications in Health

Isabel Barón-Mendoza*
Juan F. Duarte-Campos
and Aliesha González-Arenas
Departamento de Medicina Genómica y Toxicología Ambiental,
Instituto de Investigaciones Biomédicas, Universidad Nacional Autónoma de México.
Ciudad de México, México

Introduction

Synaptic plasticity is the change that synapses suffer in their structure and activity; this phenomenon modifies the strength or efficacy of synaptic transmission at preexisting synapses and plays a central role in the capacity of the brain to incorporate transitory experiences into persistent memories. Synaptic plasticity plays a critical role in the early development of neural circuitry, and evidence suggests that impairments in synaptic plasticity mechanisms contribute to several prominent neuropsychiatric disorders (Citri and Malenka, 2008). Revealing the detailed molecular mechanisms underlying synaptic plasticity is crucial for understanding the neural basis of many aspects of brain function. In this regard, astrocytes and microglia have the faculty to shape synapses in many ways, including dendritic spine modifications, synapse formation, and maturation.

* Corresponding Author's Email:

In: Dendritic Spines
Editor: Ignacio González-Burgos
ISBN: 979-8-89113-080-7

This chapter will focus on the role of astrocytes and microglia in synaptic plasticity and what happens when this synapse modulation is not typical in different neuropsychiatric, neurodevelopmental, and aging disorders.

Functions of Astrocytes in the Central Nervous System (CNS)

Astrocytes are the most prominent glial population in the central nervous system (CNS). Hence, there is a heterogenous diversity of astroglial cells depending on the different molecular and morphological features exhibited according to their function and location (Zhou et al., 2019). However, the classical and broader classification groups astrocytes into protoplasmic and fibrous. Protoplasmic astrocytes are found in gray matter and display end-feet-shaped refined branching processes that contact blood vessels. On the other hand, white matter-fibrous astrocytes have long-unbranched processes whose end feet interact with Ranvier nodes (Sofroniew and Vinters, 2010; Chaboub and Deneen, 2013).

The widely well-documented activities of astrocytes include the blood-brain barrier (BBB) formation -through junctions with endothelial cells, maintenance of extracellular ion and water balance, reuptake and metabolizing of glutamate, and supply of energy substrates to neurons. In addition, astrocytes are significantly involved in synaptic transmission modulation since their perisynaptic astrocytic processes (PAPs) envelope the post and pre-synaptic terminals to form the "tripartite synapse" (Sofroniew and Vinters, 2010; Chaboub and Deneen, 2013). Astrocytic processes enwrap about 60-90% of all synapses in the CNS, but coverage degree differs between brain regions: astrocytes envelop 60-90% of synapses in the hippocampus and cerebellum, while in the neocortex PAPs cover 30-55% of synapses (Augusto-Oliveira et al., 2020).

Moreover, during development, in several species, including rodents and humans, astrocytes are generated following neurogenesis from neural cell progenitors (NCPs) (~Embrionary day (E) 18 to Postnatal day (P) 0 in the mouse cortex); then, astrocytes expand in number (~E19-P23) and outgrowth their processes during the first postnatal weeks (P2-P30), but they keep remodeling their morphology in the adult brain. Astrocyte maturation occurs at late fetal stages (from ~P21-P30, and they continue showing mature morphology into adulthood), close to the period of synapse formation and refinement, suggesting a role of astrocytes in synaptogenesis (Farhy-Tselnicker and Allen, 2018; Perez-Catalan et al., 2021).

The first evidence of astrocyte influence on synaptogenesis came from mouse radial glial cells (RGCs) cultures. Astrocytes promote the formation of synapses in RGCs *in vitro*, leading to spontaneous activity, which suggests a crucial role of astroglia in neuronal connectivity development (Ullian et al., 2004). Furthermore, *in vivo* experiments in the rodent visual cortex during the first postnatal week show coincident raising expression of synapse-associated neuronal proteins, such as Synapsin I, Postsynaptic Density Protein 95 (PSD-95), Homer, Vesicular Glutamate Transporter 1-3 (Vglut1-3) and Glutamate Ionotropic Receptor AMPA Type Subunit 1-2 (GluA1-2), and the astrocytic proteins Thrombospondin, hevin, secreted protein acidic and rich in cysteine (SPARC), and Glypican 4/6 (Farhy-Tselnicker and Allen, 2018).

Therefore, the essential function of astrocytes not only includes neurotransmitter, ion, and water homeostasis. In addition, astrocytes are also key players in synapse formation, elimination, and plasticity depending on the soluble factors released from them or/and the direct modulation of the synapse structure by astrocytic terminals (Tan and Eroglu, 2021). Thus, the following sections will review the diverse molecular processes underlying astrocytic modulation of neural activity.

Astrocyte-Secreted Molecules Regulate Synapse Formation and Maturation

The regulation of synaptogenesis by astrocytes is associated with the multiple factors they release to neurons and synaptic structures, including proteins, lipids, and small molecules, which modulate synapses through different mechanisms (Fossati et al., 2020).

Thrombospondin (TSP1-4) is a family of extracellular matrix proteins secreted by immature astrocytes. TSPs were found to be one candidate for promoting the formation of excitatory synapses in RGCs cultures when binding to the gabapentin receptor (or the calcium channel subunit α2δ-1) (Tan et al., 2021). Likewise, a study performed in cultures of mouse cortical neurons showed that the treatment with TSP2 induced excitatory synapse formation *in vitro*. In contrast, in the postsynaptic α2δ-1 receptor knockout (KO) mice, the exposition to TSP2 did not affect synapse number. Furthermore, it was also revealed that the TSP-α2δ-1 receptor interaction induced the growth and stabilization of dendritic spines by activating the cytoskeleton remodeling protein Rac1 (Risher et al., 2018) (Figure 1).

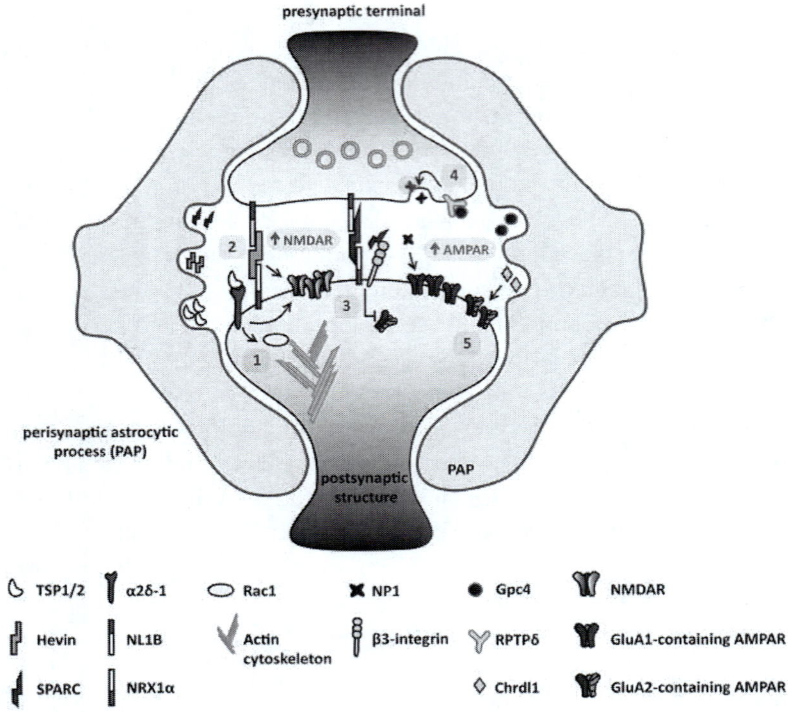

Figure 1. Astrocyte-secreted factors regulate synapse formation and maturation. Representative scheme of the mechanisms modulated by TSP1/2, hevin, NP1, and Chrdl1 astrocytic factors involved in synaptogenesis and synapse maturation. *Astrocytic factors regulating excitatory synaptogenesis:* 1) TSP (Thrombospondin 1/2) binds to the postsynaptic α2δ-1 receptor to promote the formation of NMDAR-containing (AMPAR-lacking)- silent synapses and triggers the activation of the cytoskeleton remodeling protein Rac1 to allow growth and stabilization of dendritic spines. 2) Hevin enhances the establishment of trans-synaptic complexes by bridging the presynaptic NRX1α (neurexin-1 alpha) and the postsynaptic NL1B (neuroligin-1B) proteins and recruiting NMDAR subunits into the synaptic cleft for boosting excitatory synapse formation 3) SPARC antagonizes hevin-induced synaptogenesis by either competing for NRX1α and NL1B binding sites or by getting attached to β3-integrin to downregulate the GluA2-containing AMPARs insertion into the postsynaptic terminal. *Astrocytic factors regulating synapse maturation:* 4) Gpc4 (Glypican 4) binds to the RPTPδ receptor in the neuronal presynaptic terminals to stimulate the release of the NP1 (neuronal pentraxin 1), which promotes synapse function through the recruitment of GluA1-containing AMPARs on the postsynaptic surface. 5) Chrdl1 (Chordin-like 1) increases the insertion of calcium-impermeable GluA2-containing AMPARs into the postsynaptic surface, an event necessary for synapse maturation and stabilization.

Moreover, the synaptogenic effect of TSP1 also might depend on the cell adhesion protein neuroligin 1 (NL1); when NL1 expression is knockdown using lentiviral-mediated RNA interference, the TSP1-induced synaptogenesis is blocked in early-stage cultured hippocampal neurons but is not on mature cells (Xu et al., 2010). According to the previous, secreted TSP1 and TSP2 induce synapses with a well-defined structure composed of an active presynaptic terminal but a silent postsynaptic area lacking functional AMPA receptors (AMPARs) in RGCs (Figure 1), which exhibit small synaptic events determined by recording spontaneous miniature events (mEPSC). This denotes the specific role of astrocytic TSPs in the formation of immature synapses (Christopherson et al., 2005).

On the other hand, the secreted protein acidic and rich in cysteine (SPARC) family, which comprises hevin (SPARC-like1) and SPARC, are astrocyte-secreted extracellular matrix proteins required not only for forming synapses but also for their maturation (Kucukdereli et al., 2011). Hevin expression is low in the first postnatal week in the rodent visual cortex, peaks in the second-third week, and is highly expressed at the synaptic cleft in the adult brain (Kucukdereli et al., 2011; Farhy-Tselnicker and Allen, 2018). During development, thalamic and intracortical projections compete to innervate visual cortex dendritic spines; thus, the circuit refinement and maturation depend on the release of hevin from astrocytes to stabilize Vesicular Glutamate Transporter 2-positive (VGlut2+) thalamocortical synapses. This mechanism of circuit refinement in hevin-KO mice is impaired and provokes a severe deficit in thalamocortical VGlut2+ synapse density, which can be recovered by hevin administration (Risher et al., 2014).

In addition, astrocyte-released hevin promotes the establishment of trans-synaptic complexes formed by cell adhesion molecules: the pre-synaptic neurexin-1 alpha (NRX1α) and the postsynaptic neuroligin-1B (NL1B) proteins, which are critical for thalamocortical synaptogenesis. A study found that hevin transgenic constructs added to RGCs cultures can bridge endogenous NRX1α with NL1B, which do not directly interact with each other. Thereby, hevin promotes the pre and postsynaptic specialization assembly and the recruitment of NMDA receptors (NMDARs) subunits NR1 and NR2B into the synaptic cleft of thalamocortical connections (Singh et al., 2016) (Figure 1).

More recently, it was proposed that hevin specifically boosts excitatory synaptogenesis and enhances mainly NMDAR-mediated synaptic transmission in mixed neuronal and astroglial cultures from the neonatal mouse cortex independently of neuroligin and neurexin interactions, suggesting that hevin might impact plasticity through more than one cellular mechanism (Gan and Südhof, 2020).

Conversely, the astrocytic-derived SPARC protein is considered a negative regulator of synaptic development since *in vitro* experiments in RGCs cultures revealed that SPARC exposition antagonizes hevin-induced synaptogenesis (Kucukdereli et al., 2011). Another report found that SPARC-KO mice display a higher number of AMPARs on the postsynaptic membrane of hippocampal neurons. The GluR2 subunits of AMPAR are stabilized at synapses by β3-integrin protein, whose levels in the neuronal surface are also reduced when SPARC is depleted. In consistence with this, in SPARC-deficient neuronal cultures, the exogenous SPARC peptide can attach to endogenous β3-integrin to downregulate the AMPAR insertion on the postsynaptic terminal (Figure 1). In this way, SPARC restricts the strength of maturing synapses to make them able to show synaptic plasticity (Jones et al., 2011).

Further, both hevin and SPARC protein structure displays the follistatin-like (FS) domain to interact with neuroligin and neurexin, so SPARC might be competing with hevin for cell adhesion protein-binding sites to block the establishment of trans-synaptic complexes that promote synapse formation (Fan et al., 2021) (Figure 1). Thus, the SPARC effect through distinct mechanisms is essential for fine-tuning the synapse maturation process (Jones et al., 2011; Fan et al., 2021).

Besides hevin and SPARC influence on neuroligin-neurexin complexes, astrocytic surface expresses neuroligin proteins (NLs). An evaluation performed in the mouse cortex revealed that astrocytic neuroligin 2 (NL2) participates in the formation and function of excitatory synapses. The analysis showed that the loss of astrocytic NL2 reduced synapse density in the surrounding cortical neurons *in vivo* and impaired excitatory synaptic transmission based on whole-cell patch-clamp recordings. Thus, it was hypothesized that astrocytic NL2 might attach to neuronal neurexin to induce synapse formation, but this mechanism remains uncorroborated (Stogsdill et al., 2017).

The astrocytic synaptic actions are not limited to the postsynaptic structure. Interestingly, astrocytes can also indirectly modulate synapse formation by regulating the function of presynaptic terminals during CNS development. For example, a study found that astroglial cells induce the release of the neuronal pentraxin 1 (NP1) protein from the presynaptic terminal to the postsynaptic surface, triggering the postsynaptic clustering of AMPARs in both cultured RGCs and *in vivo* in the mouse visual system. The mechanism involved in NP1 releasing relies on the secretion of glypican 4 (Gpc4) from astrocytes, which binds to the RPTPδ (Type 2a receptor protein tyrosine phosphatase δ) receptor in the neuronal presynaptic terminals. The authors demonstrated that when either RPTPδ or Gpc4 expression is downregulated, the axons from the thalamus arriving at the mouse superior colliculus and visual cortex retained NP1 within their presynaptic terminals. Hence, the secretion of NP1 from the presynaptic to the postsynaptic membrane depends on astrocyte-mediated Gpc4-RPTPδ signaling to induce synaptic recruitment of the GluA1 AMPAR subunit and promote synapse formation and function (Farhy-Tselnicker et al., 2017) (Figure 1).

Furthermore, a well-documented synapse maturation characteristic that can be regulated by astroglia cells is the switch from GluA1 to GluA2-subunit-containing AMPARs. GluA2 turns AMPARs into calcium-impermeable channels, so decreasing calcium entry into the postsynaptic synapses contributes to synapse maturation and stabilization (Blanco-Suarez et al., 2018). Chordin-like 1 (Chrdl1) protein, a BMP (Bone morphogenetic protein) pathway antagonist, is released from astrocytes to promote the establishment of mature GluA2-containing synapses in RGC cultures (Baldwin and Eroglu, 2018) (Figure 1). As expected, in vivo experiments in Chrdl1-KO mice showed a significant decrease in GluA2 levels at thalamocortical synapses, suggesting a prevalence of immature synapses. Moreover, it was also observed that Chrdl1 deficiency enhanced synaptic remodeling in response to abnormal visual stimulation in the postnatal period and adulthood. Therefore, astrocytic Chrdl1 may control the maturation time of synapses according to neuronal plasticity requirements (Blanco-Suarez et al., 2018).

In brief, astrocytes are key regulators of neuronal plasticity, playing multiple roles as both temporal promoters and repressors of synaptogenesis and synapse maturation.

The Role of Astrocytes in Synaptic Plasticity Modulation

Synaptic plasticity is involved in the modulation of synaptic strength, reorganization, and refinement of neural circuits, in response to experience, and it is essential for learning and memory (Malenka and Bear, 2004). Long-term potentiation (LTP) and long-term depression (LTD) are two forms of synaptic plasticity in which astrocytes play different roles.

The participation of astrocytes in synaptic plasticity was demonstrated in pioneer studies in which cultured hippocampal neurons only displayed

NMDAR-dependent LTP if grown in direct contact with a layer of astrocytes. This study proposed that LTP induction was associated with extracellular D-serine, a co-agonist of synaptic NMDARs derived from astrocytes, considered a key gliotransmitter (Yang et al., 2003). Nevertheless, recent studies revealed that neurons mainly produce D-serine, and the blocked synthesis of D-serine in astrocytes does not impair LTP. Despite this, astrocytes might support D-serine synaptic functions through other mechanisms, such as the *de novo* production of D-serine precursor, L-serine (Ivanov and Mothet, 2019).

Regarding the previous, astrocytes can also promote and modulate plasticity through several secreted factors. For instance, astrocytes release other plasticity-regulator gliotransmitters, such as glutamate, GABA (γ-aminobutyric acid), and ATP (Sancho et al., 2021).

Astrocyte-derived glutamate binds to ionotropic and metabotropic glutamate neuronal receptors. During the induction of short-term plasticity, glutamate secreted by astrocytes interacts with subtype 1 and 5 group I-metabotropic glutamate receptors (I mGluR1, 5) on presynaptic terminals to regulate transient potentiation of Schaffer collateral synapses in CA1 pyramidal neurons (Perea and Araque, 2007). Moreover, astrocytes can evoke calcium-dependent glutamate release, which activates NMDA postsynaptic receptors and potentiates the subsequent excitatory postsynaptic potentials (EPSPs) in neurons (Lee et al., 2007).

In addition, ATP is also co-released with glutamate from astrocytes. Experiments performed in mouse neocortical pyramidal neurons showed that excitatory synapses express purinergic P2X receptors that bind ATP. The authors observed that activation of P2X receptors triggered by both neuron and astrocyte-derived ATP was involved in the downregulation of NMDAR expression at the postsynaptic surface, thereby modulating synaptic strength (Lalo et al., 2016).

Another mechanism associated with the astrocytic modulation of LTP is calcium signaling. In astrocytes, the calcium dynamic is triggered by astrocytic G-protein-coupled receptors (GPCRs) activation in response to the binding of synaptic transmitters and factors derived from neuronal spillover. *In vivo* experiments in the mouse hippocampus demonstrated that astrocytic calcium signaling evocates the subsequent activation of inositol triphosphate receptors (IP3R2), which promotes the synthesis and release of the brain-derived neurotrophic factor (BDNF) from astrocytes. In turn, astrocytic calcium signaling-derived BDNF leads to the induction of late-phase LTP (L-LTP), a crucial process for remote memory establishment (Liu et al., 2022).

Regarding the molecules that evocate calcium concentration rise in astrocytes, neuronal-derived glutamate has been widely reported due to the activation of astrocytic Gq-coupled metabotropic glutamate receptors (particularly mGluR5). However, there is contradictory evidence about how astrocytic glutamate receptors trigger calcium dynamics, including changes in the mGluRs subtype, their expression in the young and adult brain, and the absence of an effect on synaptic plasticity in response to mGluR-specific activation (Bazargani and Attwell, 2016).

On the other hand, calcium signaling in astrocytes can also be mediated by GABAergic interneurons to modify neuronal activity. For example, in the hippocampus, potentiation of synaptic transmission is enhanced through the activation of astrocytic $GABA_B$ receptors, which trigger an intracellular calcium elevation followed by glutamate release from astrocytes. Astrocyte-derived glutamate subsequently binds to metabotropic glutamate receptors at presynaptic terminals, thereby transforming inhibitory into excitatory signals that transiently enhance synaptic transmission (Perea et al., 2016). Conversely, glutamate released from astrocytes induced by interneurons can also long-lastingly improve inhibitory synaptic transmission through the activation of kainate receptors at inhibitory terminals (Kang et al., 1998). Therefore, gliotransmitters might display multiple effects accordingly to the neural circuit requirements for activity modulation (Perea et al., 2016).

Moreover, other neurotransmitters have been proposed to modulate LTP or LTD by regulating calcium in astrocytes, such as noradrenaline, acetylcholine, and endocannabinoids, via astrocytic adrenergic, muscarinic, and cannabinoid receptors, respectively (Navarrete et al., 2012; Gómez-Gonzalo et al., 2015; Pankratov and Lalo, 2015; Liu et al., 2021). In this context, neurons can establish different modulation loops with astrocytes through neurotransmitter release stimulated by local neuronal activity.

Beyond the role of (glio/neuro)-transmitter molecules in synaptic plasticity, astrocyte-derived lipids have also been found to regulate synapse function. Astrocytes synthesize and release fatty acids and cholesterol, which form complexes with lipoproteins that neurons can take up. *In vivo* experiments in the mouse hippocampus have revealed that reducing lipid biogenesis in astrocytes due to the downregulation of sterol regulatory element binding proteins (SREBPs) results in a lower number of presynaptic vesicles, suggesting decreased neurotransmitter release. Additionally, reduced lipid availability in astrocytes leads to a decrease in LTP amplitude. However, these experiments also showed that the postsynaptic structure remains intact, indicating that astrocytic lipid supply affects synaptic plasticity establishment by principally regulating presynaptic terminal development (van Deijk et al., 2017).

Another astrocytic metabolic substrate essential for LTP is lactate. Astrocytes store glycogen, which can be converted to lactate. Then, lactate is released from astrocytes via monocarboxylate transporters (MCTs), so neurons can take it up. A report showed that in the rat hippocampus, blocking the expression of astrocytic MCT4 and MCT1 transporters led to the impairment of LTP maintenance and long-term memory formation, which was rescued by the administration of L-lactate (Suzuki et al., 2011). Therefore, astrocyte-to-neuron transport of different metabolic products appears to be a key mechanism for maintaining the metabolic demand during synaptic plasticity.

Interestingly, structural interactions established by astrocytes, such as the close spatial relation between the perisynaptic astrocytic processes (PAPs) and synapses, are also implicated in synaptic plasticity modulation since PAPs become proximate to activated synapses when LTP is induced. In the somatosensory cortex of adult mice, it was observed that PAPs motility is transiently increased, following LTP induction, to enwrap tightly enlarged potentiated spines, which leads to enhanced astrocytic coverage and spine stability (Bernardinelli et al., 2014).

Hence, the previously described findings suggest that astrocytes orchestrate a complex mechanism comprising biochemical and structural processes to achieve synaptic transmission regulation in the developing and adult brain.

Regulation of Synapse Refinement and Pruning by Astrocytes

Besides the relevant role of astrocytes in synaptogenesis, astroglia cells contribute to synapse refinement and elimination to ensure proper circuit formation (Chung et al., 2015). Neurons display an excess of synapses in the development period that must be eliminated to achieve precise neural connectivity. In adulthood, neural circuit shaping in response to experience and memory requires synapse pruning (Lee and Chung, 2021; Luo and Gao, 2021).

Regardless of the synaptogenic effect of astrocytic-secreted factors, they can also act as a "molecular brake" necessary to preserve a permissive response range in neurons to show synaptic plasticity. Such as we previously described, the SPARC protein limits the level of postsynaptic AMPA receptors at the surface of maturing synapses; however, this event is essential for neurons to induce LTP (Jones et al., 2011).

Likewise, the restrictive functions of astrocytes at synapses are implicated in the maturation process for closing the windows of high plasticity throughout brain development. For instance, presynaptic spike timing-dependent long-term depression (t-LTD) is a form of plasticity displayed by hippocampal CA3-CA1 synapses during the third postnatal week, which is subsequently lost in the fourth postnatal week and replaced by timing-dependent long-term potentiation (t-LTP). The t-LTD-to-t-LTP shift is proposed to underlie the shaping and refinement of neural connections into mature adult circuits. In this context, the induction of presynaptic t-LTP relies on the release of adenosine and glutamate by astrocytes (Pérez-Rodríguez et al., 2019; Falcón-Moya et al., 2020).

On the other hand, the classic intercellular interactions displayed by astrocytes also participate in the refinement of neuronal synaptic communication. Astroglia cells establish gap junctions to generate cellular communicating networks through hemichannels composed of connexin 30 and 43 (Cx30, Cx43) proteins. In this manner, astrocytes carry out conventional functions, such as glutamate clearance (Charvériat et al., 2017). However, connexins can also regulate the structure of astroglial processes by interacting with actin and tubulin cytoskeleton proteins or modulating astrocytic cell-adhesion contacts with the extracellular matrix (Pannasch et al., 2014). A study demonstrated that the Cx30-deficient mice displayed increased astrocytic processes close to the synaptic cleft and enhanced glutamate clearance in the hippocampus. Further, the authors observed that the morphological changes promoted by Cx30 to restrict the synaptic location of astrocytic processes are critical for regulating synaptic glutamate concentration and the induction of LTP associated with contextual memory performance (Pannasch et al., 2014).

Moreover, the bidirectional communication between astrocytes and neurons takes part in the accurate regulation of synaptic plasticity. One of the identified neuron-glia interplay mechanisms implicates bidirectional signaling induced by cell-to-cell contact through the Eph receptor and their ephrin ligands, which are expressed on the surface of astrocytes and neurons in the adult mouse hippocampus (Murai et al., 2003). EphrinA3 in astrocytes binds to EphA4 neuronal receptors on dendritic spines to induce spine retraction. In turn, neuron EphA4/ephrinA3-induced signaling reduces the expression of glutamate transporters in the astrocytic surface to regulate glutamate levels, leading to the modulation of LTP induction (Filosa et al., 2009; Tan and

Eroglu, 2021). Besides, it is known that Ephrin-B-EphB protein complexes are established between the synaptic membranes of neurons to enhance synapse formation and maintenance. In contrast, when astroglial ephrin-B1 attaches to neuronal EphB receptors, it interferes with the Ephrin-B-EphB interaction between the synaptic surfaces, restricting the formation of new spines in the adult hippocampus during learning (Koeppen et al., 2018; Nguyen et al., 2020) (Figure 2).

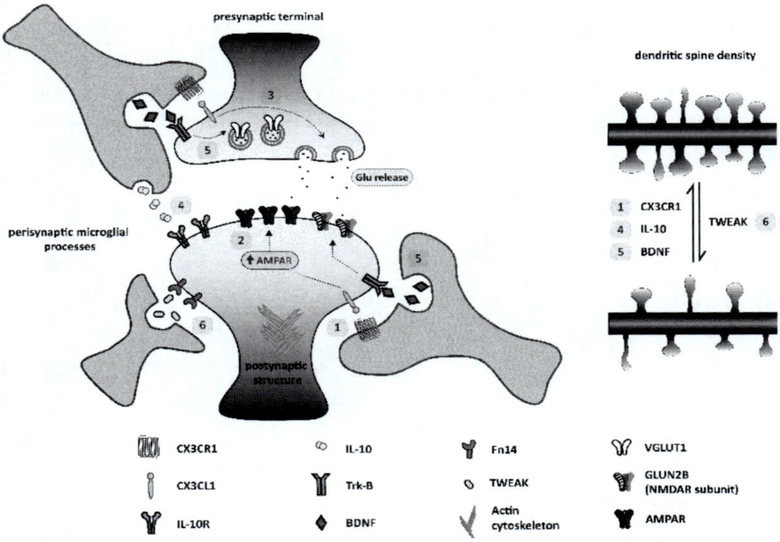

Figure 2. Microglial influence in synapse formation. Representative scheme depicting the molecular pathways regulated by microglia that are involved in synapse formation and maturation. 1) Microglial CX3CR1 (C-X3-C motif chemokine receptor 1) interaction with its ligand CX3CL1 (also known as fractalkine) on neurons is necessary for the normal acquisition of dendritic spine density during early postnatal development, 2) CX3CR1-CX3CL1 regulates hippocampal AMPAR/NMDAR ratio and connectivity by promoting AMPAR insertion into the postsynaptic membrane, 3) besides enhancing presynaptic glutamate release. 4) Microglia-derived IL-10 promotes dendritic spines and synapse formation by binding and activating its receptor IL-10R. 5) Microglial-derived BDNF (Brain-Derived Neurotrophic Factor) is relevant for spine formation. BDNF produced by microglia acts by binding and activating its receptor TrkB and it upregulates the expression of presynaptic VGLUT1 and postsynaptic GLUN2B. 6) Expression of Fn14 (Fibroblast Growth Factor-Inducible 14 receptor) promotes the formation of bulbous spines with a larger number of synapses. There, microglial-specifically-expressed TWEAK (TNF-Associated Weak Inducer of Apoptosis) that binds to Fn14 and antagonizes its spine-formation effect, showing a mechanism by which microglia can regulate the establishment of dendritic spines in an experience-dependent fashion.

In particular, it is known that some of the signaling mechanisms associated with the participation of astrocytes in synapse elimination result in the triggering of their phagocytic ability. Phagocytosis is the process by which astrocytes identify, engulf, and digest particles such as synapses and apoptotic cells, among other cell debris (Luo and Gao, 2021). Clearing synapses helps maintain homeostasis in both brain development and the normal functioning of the mature central nervous system (Lee and Chung, 2021).

The most studied mechanisms for astrocyte-mediated synapse clearance are the MEGF10 (Multiple Epidermal Growth Factor-Like Domains Protein 10) and MERTK (MER Proto-Oncogene, Tyrosine Kinase) phagocytic pathways. MEGF10 and MERTK receptors recognize "eat me" signals to initiate the engulfment process (Lee and Chung, 2021; Luo and Gao, 2021).

The MEGF10 receptor binds with high affinity to C1q (complement component C1q), the initiating protein of the classical complement cascade, which has been proposed to tag unnecessary synapses for triggering their elimination by astrocytic phagocytosis (Stevens et al., 2007; Iram et al., 2016) (Figure 2). On the other hand, MERTK receptor signaling crosstalk with αvβ5 integrin pathway, leading to Rac1 activation, which controls the rearrangement of the actin cytoskeleton underlying phagocytosis (Wu et al., 2005; Chung et al., 2013) (Figure 2). Even though many of the specific signals targeting synapses for elimination remain unclear, some studies have suggested candidates, such as phosphatidylserine, which is recognized either by MEGF10 via C1q or MERTK via Gas6 and ProS (Protein S) protein ligands (Konishi et al., 2022).

The role of MEGF10 and MERTK has been shown to regulate synapse engulfment during mouse visual system development when retinogeniculate segregation is induced by retinal activity. An excess of synapses is formed from RGCs axonal inputs to neurons in the dorsal lateral geniculate nucleus (dLGN) and subsequently removed through glial phagocytosis dependent on spontaneous retinal wave activity (Stevens et al., 2007; Chung et al., 2013). In this case, MEGF10 and MERTK-regulated astrocytic phagocytosis preferentially target weaker-activated synapses (Chung et al., 2013).

Furthermore, *in vivo* experiments in the adult mouse hippocampus found that astrocytes can phagocyte excitatory and inhibitory synapses; nevertheless, only the engulfment of excitatory synapses was increased in response to environmental enrichment-induced neuronal activity. Moreover, the phagocytosis of the excitatory synapses was demonstrated to rely on MEGF10 function; as expected, the depletion of astrocytic MEGF10 expression alters LTP induction and impairs hippocampus-associated memory formation.

According to the previous evidence, MEGF10-mediated astrocytic phagocytosis may participate in the renewal of synaptic memory traces in the adult hippocampus (Lee et al., 2021).

Astrocyte-Microglia Interplay in the Regulation of Neuronal Activity

In addition to the role of astrocytes as phagocytic cells, they also can modulate microglia, the principal phagocytes in CNS, to promote microglial-mediated synapse engulfment. For instance, astrocytes attached to synapses secrete the interleukin-1 family cytokine interleukin (IL)-33 in the mouse thalamus and spinal cord to induce excitatory synapse pruning via microglial IL1RL1 receptor signaling (Figure 2). Interestingly, IL-33 is highly expressed in windows of synaptic maturation, such as in the dLGN during the eye-opening period. Conversely, the absence of IL-33 results in elevated spontaneous firing in the thalamus of young adult mice because of an excessive number of excitatory synapses. Thus, astrocytes preserve neural activity homeostasis through microglia-synaptic pruning (Vainchtein et al., 2018).

The previous evidence exposes the relevance of the astrocyte effect on microglia during the remodeling of neuronal networks. Still, a critical bidirectional interplay from microglia to astrocytes is involved in synaptic plasticity modulation. Recently, it was shown that microglia cells are essential for maintaining the homeostasis of astrocyte syncytium, which is required for adequate synaptic transmission in the mouse hippocampus, by regulating the expression of the gap junction connexin protein. In agreement, microglia depletion severely disrupts astrocyte coupling, leading to reduced strength of synaptic transmission at baseline state and after LTP induction (Du et al., 2022).

Another study on acute hippocampal slices found that lipopolysaccharide (LPS)-stimulated microglia can also increase spontaneous excitatory postsynaptic currents when releasing small amounts of ATP that bind to the purinergic receptor P2Y1 (P2Y1R) in astrocytes. In turn, astrocytes secrete glutamate, which activates neuronal metabotropic glutamate receptors to modulate the activity of AMPA receptor-containing synapses. These data suggest astrocyte-microglia communication might control neuronal activity even under inflammatory conditions (Pascual et al., 2012).

Therefore, beyond astrocyte modulation, the role of the microglia cells in synaptic plasticity comprises several processes associated with microglia-neuron interactions. Hence, the mechanisms underlying the microglia-dependent formation and elimination of synapses will be discussed hereunder.

Microglia Functions in the Central Nervous System (CNS)

Microglia are the most abundant pool of phagocytic mononuclear cells in the central nervous system (CNS), constituting up to 10% of the total cells (Colonna and Butovsky, 2017). Microglia emerge during early development – embryonic day (E) 8 in mice– from erythromyeloid progenitors in the yolk sac, and colonize the brain by E9.5 (Ginhoux et al., 2010; Smolders et al., 2019). In humans, microglial cells can be detected in the developing CNS as early as the third gestational week (Hutchins et al., 1990), with an increasing cell influx by 4.5-5.5 gestational weeks, that peaks at week 16 (Menassa and Gomez-Nicola, 2018; Rezaie et al., 2005; Verney et al., 2010). Microglial cell numbers are maintained in the brain through coupled proliferation and apoptosis rates during lifespan (Askew et al., 2017) without any significant contribution from peripheral cells under physiological conditions (Ajami et al., 2007).

Microglia undergo a cell number expansion after their entry into the mouse brain, where the number of cells increases during the first two postnatal weeks and shows a reduction starting at the third postnatal week, reaching a stable cell density by the sixth week of life (Alliot et al., 1999; Kim et al., 2015; Nikodemova et al., 2015). Microglial proliferation has been documented in the postnatal rat brain, which peaks on P9 (Dalmau et al., 2003), although such phenomenon hasn't been shown to occur as extensively in the mouse brain (Eyo et al., 2016; Eyo and Wu, 2019).

Apoptosis is believed to be the main mechanism underlying microglial cell number loss in the postnatal brain, albeit it has been poorly documented, as apoptotic microglia was rarely detected in the rat brain between E18-P9 (Dalmau et al., 2003), and microglia displays comparable densities between wild-type (WT) and *Bax*-deficient mice (Eyo et al., 2016). However, microglial apoptosis shows a boost starting on P21 in the mouse brain (Nikodemova et al., 2015), indicating a complex pattern of proliferation and apoptosis that should be carefully assessed in a timely- and region-specific manner.

Microglial cells perform a variety of functions in the developing and adult brain, which range from promoting axonal outgrowth and fasciculation (Pont-Lezica et al., 2014; Squarzoni et al., 2014), removal of apoptotic cell bodies (Wakselman et al., 2008; Fourgeaud et al., 2016), to supporting adult hippocampal neurogenesis (Diaz-Aparicio et al., 2019; Sierra et al., 2010).

Microglia are able to regulate brain homeostatic processes such as neurotransmission. For example, activation of microglia in the spinal cord through its receptor CX3CR1 induces the production of IL-1β which can in turn stimulate and activate NDMA receptors on the post-synapse that later releases eicosanoid messengers that ultimately enhance presynaptic neurotransmitter release (Clark et al., 2015), indicating that microglia can promote synaptic strengthening under physiological conditions.

The early arrival of microglia into the brain during development and in posnatal stages poses them as relevant contributors to synaptic development and refinement since they can promote synaptic formation and removal through various molecular processes (Blagburn-Blanco et al., 2022; Reemst et al., 2016), which will be described in the following sections.

Cell-to-Cell Contact Mediated Synapse Formation by Microglia

Microglia are highly motile cells that sense their environment through their multiple ramified cell processes. Using in vivo two-photon imaging of fluorescently labeled microglia and neurons, Wake et al. (2009) demonstrated that microglia make transient (~5 min) contact with synapses at a frequency of about once per hour (Wake et al., 2009). Furthermore, such contacts are activity-dependent, and they become reduced in frequency when neuronal activity is lowered. Hence, multiple studies have aimed to explore if microglial direct contact with synapses can have an impact on synaptic stability or plasticity.

For instance, microglial cell-to-cell contact drives synaptic formation in the developing cortex. Using *in vivo* multi-photon imaging, a study demonstrated that microglial contact with pyramidal neurons in the somatosensory cortex between P8 and P10 induces filopodia formation at a higher rate than in those dendrites without microglial contact (Miyamoto et al., 2016). Furthermore, authors showed that microglial contact with neurons elicits a local Ca^{2+} elevation in dendrites, which promotes actin accumulation and filopodia formation, producing more stable filopodia than in those dendrites contacted by microglia in which Ca^{2+} elevation is absent, suggesting that microglial contact accompanied by local Ca^{2+} increase are relevant for spine formation.

Another study described a similar phenomenon using organotypic cultures where they observed that microglia does indeed contact dendritic spines and promotes the formation of filopodia protruding from preexistent spine heads (spine head filopodia) in ~39% of spines contacted by microglia (Weinhard et al., 2018). However, they observed that spine heads are a preferred contact site for microglia over dendritic shafts, and they also described the formation of transient dendritic spines that appeared upon microglial contact and subsequently disappeared (~13% of total contacted microglia-contacted spines). Likewise, there is evidence that microglial contact with dendritic spines in the visual cortex promotes transient changes on dendritic spine sizes, inducing an enlargement in 62% of contacted spines and a shrinkage in 32% of them, which accounts for a significant increase in average spine sizes in the presence of putative microglial contact, versus in the absence of it (Tremblay et al., 2010) suggesting altogether that a highly dynamic interaction between microglia and neurons contributes to spine formation through cell contact.

In the olfactory bulb, early postnatal depletion of microglial population using PLX5622, an inhibitor of the colony-stimulating factor 1 receptor (CSF-1R), reduces the dendritic spine density in adult-born granular cells (Reshef et al., 2017) suggesting that microglia are necessary for promoting spine formation in these neurons during early development. Moreover, depletion of microglial cells is linked to a higher proportion of longer-lasting dendritic spines and a lower proportion of newly formed spines, which points toward an altered pattern of dendritic spine turnover in the olfactory bulb of microglia-deficient animals.

C-X3-C motif chemokine receptor 1 (CX3CR1) a membrane-bound receptor that is expressed on microglia and macrophages in the CNS, binds its ligand CX3CL1 (also known as fractalkine) which is expressed in a membrane-bound form over neurons or in its soluble form after proteolytic cleavage, and it's been proposed as a main signaling axis between microglia and neurons under physiological conditions (Arnoux and Audinat, 2015). In that same study, Reshef et al. (2017) showed that microglial-neuron interactions through the CX3CR1-CX3CL1 axis might be relevant for spine formation. Using *Cx3cr1* knockout (KO) mice, authors described that adult-born granular cells display a decreased dendritic spine density, with a higher proportion of small-sized spines and lower proportion of large spines (Reshef et al., 2017). Since a correlation between shape and maturity of dendritic spines has been suggested (Kasai et al., 2003) this could indicate that microglia-neuron contact mediated by CX3CR1 is relevant, not only for the establishment of dendritic spines on adult-born neurons, but also for spine maturation in the olfactory bulb (Figure 3).

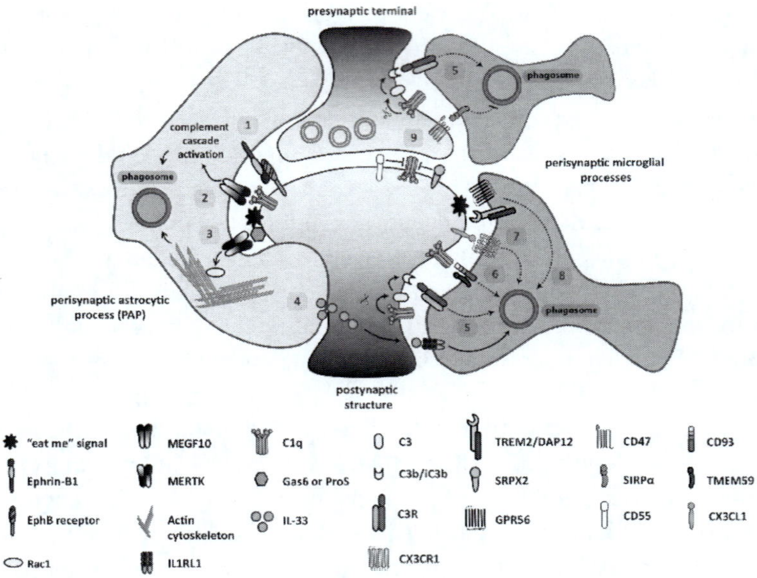

Figure 3. Role of astrocytes and microglia in synapse refinement and elimination. Representative scheme of the processes modulated by astrocytes and microglia for synapse refinement and pruning. *Astrocytic mechanisms*: 1) The binding of the astrocytic ephrin-B1 ligand to EphB neuronal receptors interferes with establishing the Ephrin-B-EphB complexes between neuronal synaptic surfaces required for synapse formation and maintenance, thus inducing spine retraction. "Eat me" signals at synapses can be recognized by astrocytes through 2) the MEGF10 (Multiple Epidermal Growth Factor-Like Domains Protein 10) receptor, which binds to C1q (complement component C1q) for triggering complement cascade activation-dependent synapse phagocytosis, or 3) the astrocytic MERTK (MER Proto-Oncogene, Tyrosine Kinase) receptor via Gas6 and ProS ligands, to induce Rac1 activation and the subsequent the rearrangement of the actin cytoskeleton underlying phagocytosis. 4) Astrocytes can indirectly promote excitatory synaptic pruning by releasing IL-33 (Interleukin-33) to activate microglial IL1RL1 receptor signaling, thereby inducing microglial synapse engulfment. *Microglial mechanisms:* 5) The general model for synapse removal mediated by the complement proteins comprises the initial binding of C1q, which can subsequently promote the binding and cleavage of C3 that opsonizes synapses to be recognized by microglia through its receptor

C3R (CD11b/CD18) and later phagocytosed. Such process occurs in a synapse activity-dependent manner. 5) Tmem59 (Transmembrane Protein 59) mediates synaptic pruning by interacting with and stabilizing CD93, a C1q receptor, that facilitates spine engulfment. 7) CX3CR1 (C-X3-C motif chemokine receptor 1)-CX3CL1 signaling has been shown to be necessary for synaptic pruning since it regulates mice lacking this signaling axis display supernumerary synapses. 8) Receptors like TREM2 and GPR56 have been shown to be relevant for dendritic spine remotion. TREM2 regulates the content of synaptic proteins along with dendritic spine density, albeit the "eat-me" signal that mediates such effect under physiological conditions is unknown. GPR56 is a microglial receptor that was shown to drive synapse removal by binding to exposed PS (phosphatidylserine), which is an "eat-me signal". Interestingly a microglia-specific splicing isoform of this receptor is necessary for PS recognition. 9) Neuronal CD47 is recognized by microglial SIRPα (Signal Regulator Protein Alpha), and it functions as a "don't eat me" signal by inhibiting phagocytic processes. Other molecules like CD55 and SRPX2 (Sushi Repeat Protein X-linked 2) have been shown to directly bind C1q and inhibit the complement cascade reaction that drives synapse engulfment, although the exact cellular pathways that operate in the context of synapse modification are not fully understood.

Furthermore, the relation between the CX3CR1-CX3CL1 axis and spine maturation has been functionally explored using acute hippocampal slices from *Cx3cr1* KO mice. (Basilico et al., 2019). *Cx3cr1* KO mice showed a reduced AMPA/NMDA ratio at 6-7 weeks of age, along with a defective CA3-CA1 connectivity which suggests that synapses fail to mature in these animals (Figure 3). Further, CA1 pyramidal neurons from *Cx3cr1* KO mice displayed a reduced probability of presynaptic release, while postsynaptic function was comparable to that of WT mice, suggesting that changes in synaptic transmission in *Cx3cr1* KO are due to altered presynaptic mechanisms (Figure 3). Interestingly, microglia from *Cx3cr1* KO mice display morphological alterations and impaired motility, suggesting that a reduction in the CX3CR1-CX3CL1 axis-mediated contact between microglia and neurons in the hippocampus may result in altered synaptic refinement and maturation, which can have a further behavioral impact, as it had been previously suggested (Zhan et al., 2014).

Synaptic Formation Mediated by Soluble Factors

Besides the very wide arrange of cell-surface molecules that microglia express to sense their environment, they're notable producers of soluble factors (e.g., cytokines, growth factors) that can contribute to synaptic plasticity.

In primary cultured hippocampal neurons, the addition of primary microglia on day in vitro (DIV) 7 without direct contact induced an increase in the number of dendritic spines, along with excitatory and inhibitory synapses (Lim et al., 2013). The authors identified IL-10 as the main promoter of synapse formation under such conditions. Furthermore, when recombinant IL-10 was applied to primary neuron cultures, synapse formation was comparably enhanced, while adding an anti-IL-10 neutralizing antibody hindered the formation of new synapses (Figure 3). Similarly, when the expression of IL-10 receptor (IL-10R) was silenced in primarily cultured neurons prior to microglia exposure, synaptic formation was equally inhibited. Interestingly, the authors also found that IL-1β, a proinflammatory cytokine that is readily produced after tissue damage or pathogen exposure (Voet et al., 2019), antagonizes the formation of new synapses even in the presence of IL-10, suggesting that neuroinflammatory events can promptly downregulate synaptic formation during critical developmental stages (Lim et al., 2013).

BDNF is a 14 kDa soluble neuropeptide that is produced in the CNS by neurons and glial cells, and it participates in synaptic plasticity events by binding with high affinity to the Tropomyosin receptor kinase B (TrkB) receptor and upregulating a number of cellular processes that can ultimately promote structural and functional plasticity of synapses (C. S. Wang et al., 2022). Interestingly, microglia are an important cellular source of BDNF and can therefore regulate synaptic formation, as it was shown by Parkhurst et al. (2013). Here the authors observed that by suppressing microglial-specific BDNF expression, animals showed a reduced content of postsynaptic glutamate receptor (NMDA) receptor subunit epsilon-2 (GluN2B) and the presynaptic glutamate transporter 1 (VGlut1) in whole brain isolated synaptosomes, as compared with control mice (Parkhurst et al., 2013) (Figure 3). Additionally, mice with deficient expression of microglial BDNF showed a decreased motor-learning-induced spine formation upon rotarod training. Animals lacking microglial BDNF also showed a reduced freezing fear response to the conditioned auditory cue in a fear conditioning paradigm, further demonstrating that microglia participate in synaptic formation through BDNF signaling.

Recently, it was shown that synaptic refinement in the retinogeniculate system is regulated by a non-phagocytic mechanism that involves the microglial production of the cytokine TNF-associated weak inducer of apoptosis (TWEAK) which binds its receptor fibroblast growth factor-inducible (Fn14) on the surface of thalamic relay neurons (Cheadle et al., 2020). Mechanistically, Fn14 promotes the formation of bulbous – larger and wider – dendritic spines, while TWEAK antagonizes this effect by binding its receptor, thus reducing the number of bulbous dendritic spines. Interestingly, both the microglial expression of TWEAK and neuronal expression of Fn14 are upregulated upon sensory stimulation between P20 and P27. Therefore, sensory deprivation during this critical time window results in a decreased density of bulbous dendritic spines, since Fn14 fails to be upregulated by sensory experience in late dark-reared (LDR) mice (Cheadle et al., 2020) (Figure 3).

Microglia-Mediated Synaptic Elimination during Development

Microglia are largely known to perform synaptic elimination through phagocytic mechanisms that target synaptic contacts with low activity, while preserving highly active synaptic elements. Indeed, it has been shown that dendritic spines in mouse primary visual cortex that get putatively contacted by microglia are more frequently eliminated over a two-day period after contact, versus those spines without microglia contact (Tremblay et al., 2010). Interestingly, only small spines were lost after microglia contact, while no larger-sized spines were lost, further reinforcing the notion that less active synapses are prone to elimination by microglial contact (Tremblay et al., 2010).

The complement system has been well characterized to participate in a process known as synaptic pruning, in which microglia recognizes opsonized synaptic elements and removes them by phagocytic cellular mechanisms (Figure 2). Interestingly, this process occurs in a neural activity dependent manner, since synapses with lower activity are more likely to be removed (Soteros and Sia, 2022).

The implication of the complement system in synapse elimination was initially described at the retinogeniculate synapse of the visual system, in which synaptic refinement occurs in a timely-specific manner during early development (Stevens et al., 2007). The initiating protein of the classic cascade of the complement, C1q, is expressed by P5 in different CNS regions such as the retinal inner plexiform layer, the brain cortex and the dorsolateral geniculate nucleus (dLGN) of the thalamus, however its expression is developmentally downregulated (Stevens et al., 2007). There, authors demonstrated that C1q colocalizes with synapse-specific proteins like SV2 and PSD-95; while C3, a pivotal protein in the complement cascade, displays a distribution pattern like that of C1q. When C1q expression is constitutively eliminated, animals display an increased density of dendritic spines and, most importantly, they fail to undergo synaptic refinement and eye-specific segregation in the dLGN (Stevens et al., 2007). A later study found that *C1q* KO mice also show increased bouton density and enhanced connectivity of excitatory synapses in cortical layer V, indicating that C1q is also necessary for mediating synapse refinement in this region (Chu et al., 2010) (Figure 2).

Evidence linking the complement system and microglial-mediated synaptic pruning under physiological conditions was later described by Schafer et al., (2012). At early developmental stages when ocular dominance is stablished and synaptic segregation occurs in the dLGN of the thalamus, microglia phagocytose inputs from retinal ganglion cells (RGC) which are labeled for degradation as they colocalize with the lysosomal marker CD68 in phagocytic cells (Schafer et al., 2012). Microglial elimination of presynaptic elements is dependent on neural activity as the authors showed that upon tetrodotoxin or forskolin administration, synaptic engulfment by microglia increases or decreases, respectively. Furthermore, genetic ablation of either complement protein C3 or its receptor CR3 (CD11b/CD18) results in a decreased engulfment of synaptic inputs by microglia de in the dLGN.

Moreover, it was shown that the complement system might be relevant for synaptic pruning mediated by the microglia in the hippocampus. In a contextual fear conditioning paradigm, mice showed a naturally decreased freezing response to the aversive context 35 days after training vs 5 days after training, thus, memory retention decays over time. When microglia were depleted pharmacologically through dietary PLX3397 administration, freezing response did not decrease by day 35 post-training (C. Wang et al., 2020). Authors found colocalizing signal of proteins C1q, PSD-95 and CD68 within GFP-expressing microglia, suggesting the complement system and a phagocytic mechanism might be involved in forgetting the aversive context. To test such hypothesis, authors induced the expression of decay-accelerating factor (DAF, or CD55), an inhibitor of both classical and alternative complement pathways, in engram-forming neurons. Likewise, freezing response to the aversive context did not decay 35 days after training in animals with neuronal overexpression of CD55 (C. Wang et al., 2020) (Figure 2), indicating that microglial-driven synaptic elimination is dependent on complement elements and it's intimately involved in functional properties of the hippocampus, such as memory forming and forgetting.

More recently, additional molecular players have been identified to participate in complement-mediated synaptic elimination. For instance, Tmem-59, a single-transmembrane domain surface protein, was shown to interact and stabilize CD93, a receptor to C1q. When *Tmem-59* was specifically deleted from microglia, there was an increased proportion of both immature- and mature-like dendritic spines in the hippocampus and brain cortex of male mice on P19-22, which was persistent into adulthood (Meng et al., 2022) (Figure 2). Such structural changes were accompanied by an increased excitatory synaptic transmission in CA1 pyramidal neurons, further confirming that the complement system is key in hippocampal synaptic remodeling. However, the exact relation of Tmem-59 with CD93 and the likely intracellular pathways that it regulates are not fully understood.

Therefore, the proposed mechanism for complement-mediated elimination of synapses involves initial tagging of synapses by C1q and C3, which can be further cleaved and activated to be recognized by their specific receptors on microglia surface to trigger their subsequent phagocytosis and elimination (Figure 2). The mechanistic views on complement-mediated synaptic elimination have been reviewed extensively elsewhere (Borucki et al., 2020; Presumey et al., 2017; Soteros and Sia, 2022).

More recent studies have identified the role of microglia in pruning synapses from other CNS nervous system such as the laver V of the prefrontal cortex (PFC) in mice during adolescence, where microglial engulfment significantly increases on P39 (Mallya et al., 2019). In the same fashion, microglial-mediated synaptic pruning occurs at the medial nucleus of the trapezoid body (MNTB) in the auditory brainstem circuit. Pharmacological depletion of microglia in this CNS area leads to increased number of polyinnervated MNTB neurons on P13, after hearing onset, consistent with defective synaptic pruning (Milinkeviciute et al., 2019), suggesting that microglial synaptic pruning during development is a widely distributed process in the CNS, although future research should seek to elucidate particular aspects and molecular mechanisms that govern region-specific pruning.

Besides the complement proteins, several other molecules have been described to participate in synapse remodeling across different brain regions. *Cx3cr1* KO mice display a transient reduction in microglia density at the CA1 hippocampal subregion on P8, which normalizes by P40 (Paolicelli et al., 2011). In the same way, *Cx3cr1* KO mice displayed an increased PSD-95+ immunoreactivity with an increased dendritic spine density in the same region, along with a significant increase in amplitude and frequency of miniature excitatory postsynaptic current (mEPSC), which altogether suggests the presence excessive and immature synapses in *Cx3cr1* KO mice when microglia are transiently depleted in the hippocampus (Paolicelli et al., 2011;

(Figure 2). In fact, recent work showed that microglia display a differential role in modulating LTP establishment between dorsal hippocampus (DH) and ventral hippocampus (VH). When microglia were pharmacologically depleted using PLX5622 or minocycline, it resulted in a reduced LTP amplitude in the DH, while it increased in the VH (De Felice et al., 2022). Interestingly, *Cx3cr1* KO mice recapitulated what was described in the pharmacological depletion model, further underlining the importance of CX3CR1-CX3CL1 axis for hippocampal connectivity.

The triggering receptor expressed on myeloid cells 2 (TREM2) is a membrane-bound, innate immune receptor that is solely expressed by microglia in the brain parenchyma. While it doesn't have intrinsic enzymatic activity, it associates with its co-receptors DAP12/10 for intracellular signaling (Qin et al., 2021). Interestingly, *Trem2* deficiency in mice causes pronounced changes in hippocampal microglia activation and density, resulting in a phenotype with overall hindered phagocytosis (Filipello et al., 2018). *Trem2* KO animals displayed and increased hippocampal content of synaptic proteins PSD-95, SHANK2 and SV2A, which is accompanied by increased dendritic spines density and an augmented neuronal excitatory activity at CA1 on P18-P20. Furthermore, authors demonstrated that TREM2 expression on microglia is required to perform engulfment of synaptic elements (Figure 2). It is of note that *Trem2* KO animals displayed markedly increased grooming time and reduced social behaviors in a three-chamber test on P90, as compared to WT animals, pinpointing the relevance of receptor TREM2 for synaptic refinement which can have long-lasting impacts on behavioral aspects. A later work showed that microglial TREM2 is necessary for limiting astrocytic engulfment of synapses in the cortex and hippocampus during early development (Jay et al., 2019), indicating that multiple cellular pathways and cell types may be involved in synaptic pruning downstream of TREM2.

Additionally, locally exposed phosphatidylserine (PS) has been shown to be an "eat me" signal that tags synapses for elimination by microglia in the dLGN and the hippocampus (Scott-Hewitt et al., 2020). Another work showed that a microglia-specific isoform of the receptor GPR56 is necessary for synaptic developmental refinement in the dLGN and the hippocampus, as it directly binds PS and promotes spine elimination (Figure 2) (Li et al., 2020). Conversely, genetic elimination of GPR56 results in an increased density of synapses in both the hippocampus and the dLGN. Interestingly, microglial MERTK can recognize exposed PS over inhibitory post-synapsis and subsequently eliminating them in the cortex, the hippocampus and in the inferior colliculus (Park et al., 2021), however, it is unknown if this receptor serves to equally recognize excitatory synapses.

Finally, some mechanisms that prevent synapses from being removed by microglia have been described. For instance, CD47 is a transmembrane protein of immunoglobulin superfamily that is expressed ubiquitously across tissues, and it binds its receptor, signal regulator protein alpha (SIRPα), that is expressed on phagocytic cells to readily inhibit phagocytosis of self-elements (Okazawa et al., 2005). Since CD47 expression had been identified on synapses, it was hypothesized that it might serve as a "don't eat me" signal for preventing synaptic stripping in that context.

A study showed that CD47 localizes to dLGN synapses during pruning stages (P5) and it's recognized by microglial SIRPα. CD47 KO mice showed increased synaptic engulfment in the dLGN and fewer VGlut2+ retinogeniculate synapses (Figure 2). Remarkably, such changes persist well into adulthood, where dLGN relay neurons receive significantly less retinal inputs as compared to WT mice (Lehrman et al., 2018). Authors also showed that CD47 preferentially localizes to highly active synapses, which made microglia from *Cd47* KO animals less capable of recognizing and targeting weaker synapses, suggesting an additional mechanism by which stronger synapses are protected from pruning.

Sushi repeat protein X-linked 2 (SRPX2) is a secreted protein expressed by neurons that was shown to directly bind C1q and inhibit complement activation, resulting in synapse protection (Figure 2). Upon SRPX2 deletion, retinogeniculate synapses were reduced in number in the dLGN on P10, which was accompanied by a reduction in the number of functional inputs to dLGN, consistent with a phenotype of enhanced synaptic pruning (Cong et al., 2020). Interestingly, authors also showed that SRPX2 acts by inhibiting complement activation in layer IV but not in layers II/III, of the somatosensory cortex of adult animals, indicating that its effect might be especially regulated.

Typical synapse refinement also involves the elimination of inhibitory synapses in the brain as well. Following a pharmacological depletion of microglial cells using the drug PLX5622, the number of synapses from parvalbumin (PV)+ cells on excitatory neurons was increased in the somatosensory cortex of microglia-depleted mice as compared with intact animals by P15 (Favuzzi et al., 2021). The authors also described that a subset of GABA-responsive microglia engage in contact preferentially with inhibitory synapses (~60% inhibitory synapses for GABA-receptive microglia vs ~14% inhibitory synapses for GABA-non receptive microglia). Interestingly, microglial-specific elimination of $GABA_{B1}R$ reduces microglial contacts with PV boutons and it alters the expression of genes involved with synapse remodeling, suggesting a complex crosstalk between neuronal and microglial subpopulations that is key for tuning cortical brain connectivity.

Astrocytes and Microglia-Associated Synaptic Dysfunction in Neuropsychiatric, Neurodevelopmental, and Aging Disorders

As we previously discussed, under healthy conditions, astrocytes and microglia functions are essential for controlling the synaptic formation, remodeling, and elimination in both the developmental and adult brain. Hence, different synaptic impairment events throughout life have been associated with astrocytic and microglial dysfunction.

The synaptic alterations found in some developmental disorders, such as the Fragile X syndrome and autism spectrum disorder (ASD), are correlated with astrocytic abnormalities, which might underlie an important part of their etiology (Molofsky et al., 2012).

Fragile X syndrome (FXS) is an inherited intellectual disability caused by an alteration in the X-linked *FMR1* gene. FXS patients display a broad spectrum of phenotypes, such as an overabundance of immature spines that might be attributed to astrocytic and microglial atypical activities (Hodges et al., 2017). For example, a study showed that astrocyte specific *Fmr1* KO mice display an accumulation of immature spines in the motor cortex during adolescence (Hodges et al., 2017). Furthermore, the abnormal synaptic development in the FXS condition has been related to changes in astrocytic factors release; hevin and SPARC levels in the cortex and hippocampus of *Fmr1* KO mice are dysregulated in specific postnatal periods when their expression must be accurately controlled to establish proper synapse formation and refinement (Wallingford et al., 2017). Another study found that astrocytes show a decreased formation of tripartite synapses at the CA1 hippocampal region of adult male *Fmr1* KO mice when analyzed through three-dimensional electron microscopy (Jawaid et al., 2018).

Interestingly, that same study analyzed synaptic pruning mediated by microglia in the hippocampus of both WT and *Fmr1* KO on P21, when developmental pruning peaks, finding that the latter display reduced pruning by microglia on that day (Jawaid et al., 2018), suggesting that impaired synaptic pruning by microglia can be related to the overall increased dendritic spine density observed in these animals.

It is of note that primary cultures of postnatal microglia from *Fmr1* KO mice show no evident baseline differences in cytokine production or phagocytic activity. However, microglia from *Fmr1* KO mice display an enhanced expression of proinflammatory genes *Il6*, *Tnfa*, *Il1b* and *Nos2* up to 6 hours after exposure to an activating stimulus with LPS, when compared to that of WT mice (Parrott et al., 2021). Additionally, microglia from *Fmr1* KO mice show an increased phagocytic activity of extracellular material *in vitro* after receiving the same stimulus with LPS (Parrott et al., 2021), indicating that *Fmr1* deficiency in microglia can be related to altered inflammatory responses and changes in its phagocytic activity.

Similarly, systemic administration of LPS to adult *Fmr1* KO mice elicited an enhanced acute expression of hippocampal IL-1β and IL-6, compared to that of WT mice, while no differences in glial protein expression (IBA-1, GFAP) were observed (Hodges et al., 2020); however, the relation of such changes in cytokine expression with microglia activation and synaptic structure *in vivo* remains to be determined.

Autism spectrum disorder (ASD) is another neurodevelopmental condition displaying astrocyte and microglia abnormalities. ASD is characterized by impairments in social interaction and communication, and the manifestation of restricted and repetitive patterns of behavior (Zeidán-Chuliá et al., 2014; Petrelli et al., 2016).

Among the ASD-related brain differences is a reduced number of cortical astrocytes, which might be implicated in the synaptic function abnormalities associated with atypical brain connectivity displayed by individuals with autism (Zeidán-Chuliá et al., 2014; Vakilzadeh et al., 2022), along with an increased density of microglial cells in the cerebral cortex (Morgan et al., 2010; Tetreault et al., 2012) that display an abnormal spatial distribution in relation to neuronal somas (Morgan et al., 2012). Histological analysis of postmortem brain samples have revealed that microglia from individuals with autism display a reactive-related morphology (Lee et al., 2017), with an increased expression proinflammatory and glial markers (Crawford et al., 2015; Edmonson et al., 2014; Vargas et al., 2005).

Regarding this, *in vitro* experiments using hiPSCs (Human induced pluripotent stem cell) derived-neural cell cultures have demonstrated that astrocytes obtained from subjects with ASD decrease the synaptic events in co-cultures with neurons from both neurotypical individuals and with autism. These synaptic anomalies might be associated with a higher production of reactive oxygen species (ROS) and the pro-inflammatory interleukin-6 (IL-6) by ASD-derived astrocytes, which interferes with synaptogenesis. In contrast, when neurons from ASD condition are co-cultured with neurotypical astrocytes, the number of synapses is enhanced, highlighting the relevance of astroglial signals to neurons for proper synapse formation and synaptic transmission (Russo et al., 2018).

Moreover, similar to FXS, an excessive number of dendritic spines was found in the prefrontal cortex of subjects with autism, which might be linked to a defective synaptic elimination process (Hutsler and Zhang, 2010). In hiPSCs-derived astrocytes from individuals with ASD, there are reduced mRNA and protein levels of the complement system component C4 that could be associated with inefficient synaptic pruning (Mansur et al., 2021).

Indeed, some experimental evidence has linked defective pruning patterns with autistic-like behaviors. For example, microglia-specific genetic ablation of *Tmem59* leads to deficiencies in synaptic pruning in the mouse prefrontal cortex and the hippocampus that are mediated by the complement proteins. Such changes are further correlated with a decreased phagocytic capacity by microglia and behavioral impairments like a reduced social interaction and increased grooming time (Meng et al., 2022). Interestingly, authors also found a decreased expression of TMEM59 in peripheral blood cells from children with autism, suggesting a likely link between complement-mediated synaptic pruning and ASD. Previous evidence had suggested that an impaired neuron-microglia signaling through the CX3CR1-CX3CL1 axis in the mouse brain might result in autistic-like behaviors (Zhan et al., 2014), highlighting the relevance of multiple microglial signaling pathways that might be involved in the emergence of ASD. Interestingly, a rare genetic variant in the *Cx3cr1* coding gene was identified in a cohort of individuals with ASD and it was correlated with increased odds for developing autism due to changes in the three-dimensional structure of its product that impact on the signaling properties of the receptor (Ishizuka et al., 2017), further highlighting the relevance of this cellular signaling pathway.

Furthermore, it was hypothesized that astrocytes and microglia reactive state might be crucial in ASD-associated synapse dysfunction. According to an analysis of the postmortem brain cortex of individuals with ASD, there is an increased number of astrocytes displaying reactive-like features, such as high GFAP (Glial fibrillary acidic protein) staining in cell body with gemistocytic appearance and poor-branched glial processes (Vakilzadeh et al., 2022). The previous data suggests that neuroinflammation is a potential mechanism involved in glia-neuron communication alterations leading to impaired synaptogenesis and synapse function.

Immune-activating challenges during early development have been associated with an increased incidence of ASD (Atladóttir et al., 2010; Croen et al., 2019; Jiang et al., 2016). Inducing a maternal immune activation (MIA) during gestational periods has indeed shown that the offspring displays increased dendritic spine density in the hippocampus which correlates with persistent behavioral changes like reduced ultrasonic vocalizations and social preference, and increased stereotypic movements (Fernández de Cossío et al., 2017; Giovanoli et al., 2016), which suggests that microglial activation during early developmental stages can have a long-lasting effect on synaptic structure and behavioral patterns due to a hypothesized switch in microglial phagocytic properties that may lead to brain connectivity impairments (Miyanishi et al., 2021).

Schizophrenia is a complex psychiatric condition determined by positive symptoms, such as psychosis, negative symptoms associated with a lack of functions, and cognitive symptoms, including difficulties in memory, attention, and executive functioning (Tarasov et al., 2020). In postmortem brain samples from patients with schizophrenia, both increased and decreased numbers of astrocytes have been found depending on the analyzed region. Also, around 70% of patients displayed astrogliosis (Tarasov et al., 2020).

Schizophrenia is characterized by a reduced dendritic spine density mainly in the prefrontal cortex of the brain (Konopaske et al., 2014). Interestingly, the human prefrontal cortex undergoes a wide synaptic remodeling during adolescence and up until early adulthood, which coincides in time with the onset of schizophrenia symptomatology (Mallya and Deutch, 2018; Petanjek et al., 2011). This has led to the hypothesis that microglial hyperactivation could be involved with a phenomenon of excessive pruning during adolescence that could be correlated with the dendritic phenotype in the brains of individuals with schizophrenia.

A study indeed found a heightened risk for developing schizophrenia that is linked to a structural diversity of alleles in the genes that encode the complement component 4 (*C4*) that is associated with an increased expression of *C4A* in the brains of schizophrenic individuals (Sekar et al., 2016), which was further confirmed in a mouse model expressing the human isoforms of *C4*, *C4A* and *C4B*, that displayed excessive loss of cortical synapses (Yilmaz et al., 2021) highlighting a role for complement proteins and schizophrenia development.

Another study found that microglia-mediated synapse elimination is enhanced when reprogrammed microglia derived from patients with schizophrenia is co-cultured *in vitro* with hiPSC-derived neuronal cells, which was further associated with the deposition of C4 complement components on synapses (Sellgren et al., 2019) suggesting a strong association of complement- and microglia-mediated elimination of synapses in this pathology.

On the other hand, the etiology of neurodegenerative conditions such as Alzheimer's disease (AD), the most common cause of dementia in older adults, includes synaptic impairments resulting from the development of amyloid-beta (Aβ) plaques and neurofibrillary tau tangles, but also due to the disrupted functioning of astrocytes and microglia (Henstridge et al., 2019).

In the AD brain, astrocytes have been proposed to display deficient clearance and elimination mechanisms that likely participate in β-amyloid accumulation. Moreover, astrocytes turn into a reactive state and release feed-forward pro-inflammatory signals to microglia and oligodendrocytes to exacerbate neurodegeneration (Preman et al., 2021). These changes in astrocyte functionality could lead to the synaptic loss underlying the progressive cognitive decline in AD patients (Hulshof et al., 2022).

The association of microglia around amyloid plaques in the brain of AD patients and animal models has led to the hypothesize that microglia may mediate synaptic loss during this neurodegenerative disease (Rajendran and Paolicelli, 2018). For instance, complement components like C1q show an increased expression during aging (Stephan et al., 2013), which suggests that a higher deposition of complement proteins to synapses might be tagging them for removal by microglia in the context of late onset AD.

A study found that C1q shows and increased content and localization to synapses in an AD mouse model. There, authors found that inhibition of either C1q, C3 or CR3 reduces synaptic loss, while observing that C1q is needed for the toxic effects of amyloid-β (Aβ) oligomers on hippocampal LTP (Hong et al., 2016). Interestingly, type I interferons have been shown to mediate microgliosis and excessive synapse removal in a mouse model of AD. Using *postmortem* brain samples from AD patients, authors found that microglia surrounding Aβ plaques overexpress a set of interferon-stimulated genes (Roy et al., 2020) further linking the neuroinflammatory states of AD with microglial activation and synaptic stripping in this pathology. More recently, membrane attack complex (MAC) was shown to participate in synaptic loss in two different AD mouse models (Carpanini et al., 2022), underpinning a larger complexity regarding the complement system activation in the context of neurodegeneration than previously expected.

Additional signaling molecules have been implicated in the pathogenesis of AD. For instance, TREM2 is a receptor that is necessary for mediating synaptic stripping under physiological conditions as previously mentioned, and loss-of-function mutations in its coding gene strongly increase risk for developing AD (Paasila et al., 2022). Since TREM2 modulates a variety of cellular functions in microglia, several pathways have been proposed to correlate alterations in this receptor and exuberant synapse removal by microglia in AD. For example, TREM2-deficient animals show a better cognitive performance compared to WT mice during aging which correlates with higher dendritic spine density and enhanced hippocampal LTP (Qu and Li, 2020). Notably, TREM2-deficiency in an AD mouse model was shown to be protective against synaptic loss at early-middle pathological stages by directly inhibiting microglial phagocytosis of synapses, while it caused an accelerated Aβ deposition at middle-late stages, alongside synaptic dysfunction (Sheng et al., 2019), indicating a complex timely-dependent role for TREM2 in AD pathogenesis.

Apolipoprotein E (APOE) has been widely associated with the onset of AD as its epsilon allele (ε)4 has been shown to be one of the strongest genetic risk factors for developing sporadic AD (Paasila et al., 2022). Interestingly, astrocytic-specific elimination of *Apoe4* can improve tauopathy-associated neurodegeneration by reducing synaptic loss and microglial phagocytosis in a mouse model of AD (Wang et al., 2021), while another study found that microglial-specific elimination of *APOE* had no overt effects on microglial shift into a diseased state or Aβ plaque load in an AD mouse model, however mice with *APOE*-deficient microglial displayed a decreased content of hippocampal PSD-95 and synaptophysin compared to that of WT mice (Henningfield et al., 2022), suggesting a putative role for APOE in synaptic maintenance.

Finally, synapse protection mechanisms seem to be deregulated in the AD brain as well, since microglial SIRPα was shown to be decreased in cortical samples from AD patients (Ding et al., 2021) highlighting the implication of multiple cellular pathways that involve a deregulated synaptic stripping by microglial in the context of AD.

Conclusion

In this chapter, we described how astrocytes and microglia collaborate in synaptic plasticity. These cells interact with neurons in two general ways to regulate synapses 1) by modifying spine dendritic density or 2) by modulating synapsis strength. Several pathways are involved in these processes; however, many intermediate steps are still unknown and should be investigated to understand how dysregulation in different cell mechanisms leads to neural disorders such as autism, schizophrenia, or AD.

References

Ajami, B., Bennett, J. L., Krieger, C., Tetzlaff, W., Rossi, F. M. V., 2007. Local self-renewal can sustain CNS microglia maintenance and function throughout adult life. *Nature Neuroscience* 10, 1538–1543.

Alliot, F., Godin, I., Pessac, B., 1999. Microglia derive from progenitors, originating from the yolk sac, and which proliferate in the brain. *Dev. Brain Res.* 117, 145–152.

Arnoux, I., Audinat, E., 2015. Fractalkine signaling and microglia functions in the developing brain. *Neural Plast.* 2015.

Askew, K., Li, K., Olmos-Alonso, A., Garcia-Moreno, F., Liang, Y., Richardson, P., Tipton, T., Chapman, M. A., Riecken, K., Beccari, S., Sierra, A., Molnár, Z., Cragg, M. S., Garaschuk, O., Perry, V. H., Gomez-Nicola, D., 2017. Coupled proliferation and apoptosis maintain the rapid turnover of microglia in the adult brain. *Cell Rep.* 18, 391–405.

Atladóttir, H. Ó., Thorsen, P., Østergaard, L., Schendel, D. E., Lemcke, S., Abdallah, M., Parner, E. T., 2010. Maternal infection requiring hospitalization during pregnancy and autism spectrum disorders. *J. Autism Dev. Disord.* 40, 1423–1430.

Augusto-Oliveira, M., Arrifano, G. P., Takeda, P. Y., Lopes-Araújo, A., Santos-Sacramento, L., Anthony, D. C., Verkhratsky, A., Crespo-Lopez, M. E., 2020. Astroglia-specific contributions to the regulation of synapses, cognition and behaviour. *Neurosci. Biobehav. Rev.* 118, 331–357.

Baldwin, K. T., Eroglu, C., 2018. Astrocytes "Chordinate" synapse maturation and plasticity. *Neuron* 100, 1010–1012.

Basilico, B., Pagani, F., Grimaldi, A., Cortese, B., Di Angelantonio, S., Weinhard, L., Gross, C., Limatola, C., Maggi, L., Ragozzino, D., 2019. Microglia shape presynaptic properties at developing glutamatergic synapses. *Glia* 67, 53–67.

Bazargani, N., Attwell, D., 2016. Astrocyte calcium signaling: the third wave. *Nat. Neurosci.* 19, 182–189.

Bernardinelli, Y., Randall, J., Janett, E., Nikonenko, I., König, S., Jones, E. V., Flores, C. E., Murai, K. K., Bochet, C. G., Holtmaat, A., 2014. Activity-dependent structural plasticity of perisynaptic astrocytic domains promotes excitatory synapse stability. *Curr. Biol.* 24, 1679–1688.

Blagburn-Blanco, S. V., Chappell, M. S., De Biase, L. M., DeNardo, L. A., 2022. Synapse-specific roles for microglia in development: New horizons in the prefrontal cortex. *Front. Mol. Neurosci.* 15, 965756.

Blanco-Suarez, E., Liu, T.-F., Kopelevich, A., Allen, N. J., 2018. Astrocyte-secreted chordin-like 1 drives synapse maturation and limits plasticity by increasing synaptic GluA2 AMPA receptors. *Neuron* 100, 1116–1132.

Borucki, D. M., Toutonji, A., Couch, C., Mallah, K., Rohrer, B., Tomlinson, S., 2020. Complement-mediated microglial phagocytosis and pathological changes in the development and degeneration of the visual system. *Front. Immunol.* 11(September), 1–16.

Carpanini, S. M., Torvell, M., Bevan, R. J., Byrne, R. A. J., Daskoulidou, N., Saito, T., Saido, T. C., Taylor, P. R., Hughes, T. R., Zelek, W. M., Morgan, B. P., 2022. Terminal complement pathway activation drives synaptic loss in Alzheimer's disease models. *Acta Neuropathol. Commun.* 10, 1–16.

Chaboub, L. S., Deneen, B., 2013. Astrocyte form and function in the developing central nervous system. *Semin. Ped. Neurol.* 20, 230–235.

Charvériat, M., Naus, C. C., Leybaert, L., Sáez, J. C., Giaume, C., 2017. Connexin-dependent neuroglial networking as a new therapeutic target. Front. *Cell. Neurosci.* 11, 174.

Cheadle, L., Rivera, S. A., Phelps, J. S., Ennis, K. A., Stevens, B., Burkly, L. C., Lee, W. C. A., Greenberg, M. E., 2020. Sensory experience engages microglia to shape neural connectivity through a non-phagocytic mechanism. *Neuron* 108, 451-468.e9.

Christopherson, K. S., Ullian, E. M., Stokes, C. C. A., Mullowney, C. E., Hell, J. W., Agah, A., Lawler, J., Mosher, D. F., Bornstein, P., Barres, B. A., 2005. Thrombospondins are astrocyte-secreted proteins that promote CNS synaptogenesis. *Cell* 120, 421–433.

Chu, Y., Jin, X., Parada, I., Pesic, A., Stevens, B., Barres, B., Prince, D. A. 2010. Enhanced synaptic connectivity and epilepsy in C1q knockout mice. *Proc. Natl. Acad. Sci* 107, 7975–7980.

Chung, W. S., Allen, N. J., Eroglu, C., 2015. Astrocytes control synapse formation, function, and elimination. *Cold Spring Harbor Persp. Biol.* 7.

Chung, W.-S., Clarke, L. E., Wang, G. X., Stafford, B. K., Sher, A., Chakraborty, C., Joung, J., Foo, L. C., Thompson, A., Chen, C. 2013. Astrocytes mediate synapse elimination through MEGF10 and MERTK pathways. *Nature* 504, 394–400.

Citri, A., Malenka, R. C. 2008. Synaptic plasticity: Multiple forms, functions, and mechanisms. *Neuropsychopharmacology* 33, 18–41.

Clark, A. K., Gruber-Schoffnegger, D., Drdla-Schutting, R., Gerhold, K. J., Malcangio, M., Sandkühler, J., 2015. Selective activation of microglia facilitates synaptic strength. *J. Neurosci.* 35, 4552–4570.

Colonna, M., Butovsky, O., 2017. Microglia function in the central nervous system during health and neurodegeneration. *Ann. Rev. Immunol.* 35, 441–468.

Cong, Q., Soteros, B. M., Wollet, M., Kim, J. H., Sia, G. M., 2020. The endogenous neuronal complement inhibitor SRPX2 protects against complement-mediated synapse elimination during development. *Nat. Neurosci.* 23, 1067–1078.

Crawford, J. D., Chandley, M. J., Szebeni, K., Szebeni, A., Waters, B., Ordway, G. A., 2015. Elevated GFAP protein in anterior cingulate cortical white matter in males with autism spectrum disorder. *Autism Res.* 8, 649–657.

Croen, L. A., Qian, Y., Ashwood, P., Zerbo, O., Schendel, D., Pinto-Martin, J., Daniele Fallin, M., Levy, S., Schieve, L. A., Yeargin-Allsopp, M., Sabourin, K. R., Ames, J. L., 2019. Infection and fever in pregnancy and autism spectrum disorders: findings from the study to explore early development. *Autism Res.* 12, 1551–1561.

Dalmau, I., Vela, J. M., González, B., Finsen, B., Castellano, B., 2003. Dynamics of microglia in the developing rat brain. *J. Comp. Neurol.* 458, 144–157.

De Felice, E., Gonçalves de Andrade, E., Golia, M. T., González Ibáñez, F., Khakpour, M., Di Castro, M. A., Garofalo, S., Di Pietro, E., Benatti, C., Brunello, N., Tascedda, F., Kaminska, B., Limatola, C., Ragozzino, D., Tremblay, M. E., Alboni, S., Maggi, L. 2022. Microglial diversity along the hippocampal longitudinal axis impacts synaptic plasticity in adult male mice under homeostatic conditions. *J. Neuroinfl.* 19, 292.

Diaz-Aparicio, I., Paris, I., Sierra-Torre, V., Plaza-Zabala, A., Rodríguez-Iglesias, N., Márquez-Ropero, M., Beccari, S., Abiega, O., Alberdi, E., Matute, C., Bernales, I., Schulz, A., Otrokocsi, L., Sperlagh, B., Happonen, K. E., Lemke, G., Maletic-Savatic, M., Valero, J., ierra, A., 2019. Microglia actively remodels adult hippocampal neurogenesis through the phagocytosis secretome. *BioRxiv.* 40, 1453–1482.

Ding, X., Wang, J., Huang, M., Chen, Z., Liu, J., Zhang, Q., Zhang, C., Xiang, Y., Zen, K., Li, L., 2021. Loss of microglial SIRPα promotes synaptic pruning in preclinical models of neurodegeneration, in: *Nature Communications* (Vol. 12, Issue 1).

Du, Y., Brennan, F. H., Popovich, P. G., Zhou, M., 2022. Microglia maintain the normal structure and function of the hippocampal astrocyte network. *Glia* 70, 1359–1379.

Edmonson, C., Ziats, M. N., Rennert, O. M., 2014. Altered glial marker expression in autistic post-mortem prefrontal cortex and cerebellum. *Mol. Autism* 5, 1–9.

Eyo, U. B., Miner, S. A., Weiner, J. A., Dailey, M. E., 2016. Developmental changes in microglial mobilization are independent of apoptosis in the neonatal mouse hippocampus. *Brain Behav. Immun.* 55, 49–59.

Eyo, U. B., Wu, L.-J., 2019. Microglia: Lifelong patrolling immune cells of the brain. *Prog. Neurobiol.* 179, 101614.

Falcón-Moya, R., Pérez-Rodríguez, M., Prius-Mengual, J., Andrade-Talavera, Y., Arroyo-García, L. E., Pérez-Artés, R., Mateos-Aparicio, P., Guerra-Gomes, S., Oliveira, J. F., Flores, G., 2020. Astrocyte-mediated switch in spike timing-dependent plasticity during hippocampal development. *Nat. Commun.* 11, 1–14.

Fan, S., Gangwar, S. P., Machius, M., Rudenko, G., 2021. Interplay between hevin, SPARC, and MDGAs: Modulators of neurexin-neuroligin transsynaptic bridges. *Structure* 29, 664–678.

Farhy-Tselnicker, I., Allen, N. J., 2018. Astrocytes, neurons, synapses: a tripartite view on cortical circuit development. *Neural Dev.* 13, 1–12.

Farhy-Tselnicker, I., van Casteren, A. C. M., Lee, A., Chang, V. T., Aricescu, A. R., Allen, N. J., 2017. Astrocyte-secreted glypican 4 regulates release of neuronal pentraxin 1 from axons to induce functional synapse formation. *Neuron* 96, 428–445.

Favuzzi, E., Huang, S., Saldi, G. A., Binan, L., Ibrahim, L. A., Fernández-Otero, M., Cao, Y., Zeine, A., Sefah, A., Zheng, K., Xu, Q., Khlestova, E., Farhi, S. L., Bonneau, R., Datta, S. R., Stevens, B., Fishell, G., 2021. GABA-receptive microglia selectively sculpt developing inhibitory circuits. *Cell* 184, 4048-4063.e32.

Fernández de Cossío, L., Guzmán, A., van der Veldt, S., Luheshi, G. N., 2017. Prenatal infection leads to ASD-like behavior and altered synaptic pruning in the mouse offspring. *Brain Behav. Immun.* 63, 88–98.

Filipello, F., Morini, R., Corradini, I., Zerbi, V., Canzi, A., Michalski, B., Erreni, M., Markicevic, M., Starvaggi-Cucuzza, C., Otero, K., Piccio, L., Cignarella, F., Perrucci,

F., Tamborini, M., Genua, M., Rajendran, L., Menna, E., Vetrano, S., Fahnestock, M., Matteoli, M., 2018. The microglial innate immune receptor TREM2 is required for synapse elimination and normal brain connectivity. *Immunity* 48, 979-991.e8.

Filosa, A., Paixão, S., Honsek, S. D., Carmona, M. A., Becker, L., Feddersen, B., Gaitanos, L., Rudhard, Y., Schoepfer, R., Klopstock, T., 2009. Neuron-glia communication via EphA4/ephrin-A3 modulates LTP through glial glutamate transport. *Nat. Neurosci.* 12, 1285–1292.

Fossati, G., Matteoli, M., Menna, E., 2020. Astrocytic factors controlling synaptogenesis: a team play. *Cells* 9, 2173.

Fourgeaud, L., Traves, P. G., Tufail, Y., Leal-Bailey, H., Lew, E. D., Burrola, P. G., Callaway, P., Zagorska, A., Rothlin, C. V., Nimmerjahn, A., Lemke, G., 2016. TAM receptors regulate multiple features of microglial physiology. *Nature* 532, 240–244.

Gan, K. J., Südhof, T. C., 2020. SPARCL1 promotes excitatory but not inhibitory synapse formation and function independent of neurexins and neuroligins. *J. Neurosci.* 40, 8088–8102.

Ginhoux, F., Greter, M., Leboeuf, M., Nandi, S., See, P., Gokhan, S., Mehler, M. F., Conway, S. J., Ng, L. G., Stanley, E. R., Samokhvalov, I. M., Merad, M., 2010. Fate mapping analysis reveals that adult microglia derive from primitive macrophages. *Science* 330, 841–845.

Giovanoli, S., Weber-Stadlbauer, U., Schedlowski, M., Meyer, U., Engler, H., 2016. Prenatal immune activation causes hippocampal synaptic deficits in the absence of overt microglia anomalies. *Brain Behav. Immun.* 55, 25–38.

Gómez-Gonzalo, M., Navarrete, M., Perea, G., Covelo, A., Martín-Fernández, M., Shigemoto, R., Luján, R., Araque, A., 2015. Endocannabinoids induce lateral long-term potentiation of transmitter release by stimulation of gliotransmission. *Cereb. Cortex* 25, 3699–3712.

Henningfield, C. M., Arreola, M. A., Soni, N., Spangenberg, E. E., Green, K. N., 2022. Microglia-specific ApoE knock-out does not alter Alzheimer's disease plaque pathogenesis or gene expression. *Glia* 70, 287–302.

Henstridge, C. M., Hyman, B. T., Spires-Jones, T. L., 2019. Beyond the neuron–cellular interactions early in Alzheimer disease pathogenesis. *Nat. Rev. Neurosci.* 20, 94–108.

Hodges, J. L., Yu, X., Gilmore, A., Bennett, H., Tjia, M., Perna, J. F., Chen, C.-C., Li, X., Lu, J., Zuo, Y., 2017. Astrocytic contributions to synaptic and learning abnormalities in a mouse model of fragile X syndrome. *Biol. Psychiatry* 82, 139–149.

Hodges, S. L., Nolan, S. O., Tomac, L. A., Muhammad, I. D. A., Binder, M. S., Taube, J. H., Lugo, J. N. 2020. Lipopolysaccharide-induced inflammation leads to acute elevations in pro-inflammatory cytokine expression in a mouse model of Fragile X syndrome. *Physiol. Behav.* 215(December 2019), 112776.

Hong, S., Beja-Glasser, V. F., Nfonoyim, B. M., Frouin, A., Li, S., Ramakrishnan, S., Merry, K. M., Shi, Q., Rosenthal, A., Barres, B. A., Lemere, C. A., Selkoe, D. J., Stevens, B., 2016. Complement and microglia mediate early synapse loss in Alzheimer mouse models. *Science* (New York, N.Y.), 352, 712–716.

Hulshof, L. A., Van Nuijs, D., Hol, E. M., Middeldorp, J., 2022. The role of astrocytes in synapse loss in Alzheimer's disease: A systematic review. *Front. Cell. Neurosci.* 16, 899251.

Hutchins, K. D., Dickson, D. W., Rashbaum, W. K., Lyman, W. D., 1990. Localization of morphologically distinct microglial populations in the developing human fetal brain: implications for ontogeny. *Dev. Brain Res.* 55, 95–102.

Hutsler, J. J., Zhang, H., 2010. Increased dendritic spine densities on cortical projection neurons in autism spectrum disorders. *Brain Res.* 1309, 83–94.

Iram, T., Ramirez-Ortiz, Z., Byrne, M. H., Coleman, U. A., Kingery, N. D., Means, T. K., Frenkel, D., El Khoury, J., 2016. Megf10 is a receptor for C1Q that mediates clearance of apoptotic cells by astrocytes. *J. Neurosci.* 36, 5185–5192.

Ishizuka, K., Fujita, Y., Kawabata, T., Kimura, H., Iwayama, Y., Inada, T., Okahisa, Y., Egawa, J., Usami, M., Kushima, I., Uno, Y., Okada, T., Ikeda, M., Aleksic, B., Mori, D., Someya, T., Yoshikawa, T., Iwata, N., Nakamura, H., Ozaki, N., 2017. Rare genetic variants in CX3CR1 and their contribution to the increased risk of schizophrenia and autism spectrum disorders. *Transl. Psych.* 7, e1184-8.

Ivanov, A. D., Mothet, J.-P., 2019. The plastic d-serine signaling pathway: Sliding from neurons to glia and vice-versa. *Neurosci. Lett.* 689, 21–25.

Jawaid, S., Kidd, G. J., Wang, J., Swetlik, C., Dutta, R., Trapp, B. D., 2018. Alterations in CA1 hippocampal synapses in a mouse model of fragile X syndrome. *Glia* 66, 789–800.

Jay, T. R., von Saucken, V. E., Muñoz, B., Codocedo, J. F., Atwood, B. K., Lamb, B. T., Landreth, G. E., 2019. TREM2 is required for microglial instruction of astrocytic synaptic engulfment in neurodevelopment. *Glia* 67, 1873–1892.

Jiang, H. Y., Xu, L. lLan, Shao, L., Xia, R. M., Yu, Z. H., Ling, Z. X., Yang, F., Deng, M., Ruan, B., 2016. Maternal infection during pregnancy and risk of autism spectrum disorders: A systematic review and meta-analysis. *Brain Behav. Immun.* 58, 165–172.

Jones, E. V., Bernardinelli, Y., Tse, Y. C., Chierzi, S., Wong, T. P., Murai, K. K., 2011. Astrocytes control glutamate receptor levels at developing synapses through SPARC–β-integrin interactions. *J. Neurosci.* 31, 4154–4165.

Kang, J., Jiang, L., Goldman, S. A., Nedergaard, M., 1998. Astrocyte-mediated potentiation of inhibitory synaptic transmission. *Nat. Neurosci.* 1, 683–692.

Kasai, H., Matsuzaki, M., Noguchi, J., Yasumatsu, N., Nakahara, H., 2003. Structure–stability–function relationships of dendritic spines. *Trends Neurosci.* 26, 360–368.

Kim, I., Mlsna, L. M., Yoon, S., Le, B., Yu, S., Xu, D., Koh, S., 2015. A postnatal peak in microglial development in the mouse hippocampus is correlated with heightened sensitivity to seizure triggers. *Brain Behav.* 5, 1–4.

Koeppen, J., Nguyen, A. Q., Nikolakopoulou, A. M., Garcia, M., Hanna, S., Woodruff, S., Figueroa, Z., Obenaus, A., Ethell, I. M., 2018. Functional consequences of synapse remodeling following astrocyte-specific regulation of Ephrin-B1 in the adult hippocampus. *J. Neurosci.* 38, 5710–5726.

Konishi, H., Koizumi, S., Kiyama, H., 2022. Phagocytic astrocytes: Emerging from the shadows of microglia. *Glia* 70, 1009–1026.

Konopaske, G. T., Lange, N., Coyle, J. T., Benes, F. M., 2014. Prefrontal cortical dendritic spine pathology in schizophrenia and bipolar disorder. *JAMA Psychiatry* 71, 1323–1331.

Kucukdereli, H., Allen, N. J., Lee, A. T., Feng, A., Ozlu, M. I., Conatser, L. M., Chakraborty, C., Workman, G., Weaver, M., Sage, E. H., 2011. Control of excitatory

CNS synaptogenesis by astrocyte-secreted proteins Hevin and SPARC. *Proc. Natl. Acad. Sci.* 108, E440–E449.

Lalo, U., Palygin, O., Verkhratsky, A., Grant, S. G. N., Pankratov, Y., 2016. ATP from synaptic terminals and astrocytes regulates NMDA receptors and synaptic plasticity through PSD-95 multi-protein complex. *Sci. Rep.* 6, 1–20.

Lee, A. S., Azmitia, E. C., Whitaker-Azmitia, P. M., 2017. Developmental microglial priming in postmortem autism spectrum disorder temporal cortex. *Brain Behav.* Immun. 62, 193–202.

Lee, C. J., Mannaioni, G., Yuan, H., Woo, D. H., Gingrich, M. B., Traynelis, S. F. 2007. Astrocytic control of synaptic NMDA receptors. *J. Physiol.* 581, 1057–1081.

Lee, J.-H., Kim, J., Noh, S., Lee, H., Lee, S. Y., Mun, J. Y., Park, H., Chung, W.-S , 2021. Astrocytes phagocytose adult hippocampal synapses for circuit homeostasis. *Nature* 590, 612–617.

Lee, S. Y., Chung, W.-S., 2021. The roles of astrocytic phagocytosis in maintaining homeostasis of brains. *J. Pharmacol. Sci.* 145, 223–227.

Lehrman, E. K., Wilton, D. K., Litvina, E. Y., Welsh, C. A., Chang, S. T., Frouin, A., Walker, A. J., Heller, M. D., Umemori, H., Chen, C., Stevens, B., 2018. CD47 protects synapses from excess microglia-mediated pruning during development. *Neuron* 100, 120-134.e6.

Li, T., Chiou, B., Gilman, C. K., Luo, R., Koshi, T., Yu, D., Oak, H. C., Giera, S., Johnson-Venkatesh, E., Muthukumar, A. K., Stevens, B., Umemori, H., Piao, X., 2020. A splicing isoform of GPR56 mediates microglial synaptic refinement via phosphatidylserine binding. *EMBO J.* 39, 1–20.

Lim, S. H., Park, E., You, B., Jung, Y., Park, A. R., Park, S. G., Lee, J. R., 2013. Neuronal synapse formation induced by microglia and interleukin 10. *PLoS ONE* 8, 1–13.

Liu, J.-H., Zhang, M., Wang, Q., Wu, D.-Y., Jie, W., Hu, N.-Y., Lan, J.-Z., Zeng, K., Li, S.-J., Li, X.-W., 2022. Distinct roles of astroglia and neurons in synaptic plasticity and memory. *Mol. Psychiatry* 27, 873–885.

Liu, X., Ying, J., Wang, X., Zheng, Q., Zhao, T., Yoon, S., Yu, W., Yang, D., Fang, Y., Hua, F., 2021. Astrocytes in neural circuits: Key factors in synaptic regulation and potential targets for neurodevelopmental disorders. *Front. Mol. Neurosci.* 14.

Luo, T., Gao, T.-M., 2021. A novel phagocytic role of astrocytes in activity-dependent elimination of mature excitatory synapses. *Neurosci. Bull.* 37, 1256–1259.

Malenka, R. C., Bear, M. F., 2004. LTP and LTD: an embarrassment of riches. *Neuron* 44, 5–21.

Mallya, A. P., Deutch, A. Y., 2018. (Micro)glia as effectors of cortical volume loss in schizophrenia. *Schizophrenia Bull.* 44, 948–957.

Mallya, A. P., Wang, H.-D., Lee, H. N. R., Deutch, A. Y., 2019. Microglial pruning of synapses in the prefrontal cortex during adolescence. *Cereb. Cortex* 29, 1634–1643.

Mansur, F., Teles e Silva, A. L., Gomes, A. K. S., Magdalon, J., de Souza, J. S., Griesi-Oliveira, K., Passos-Bueno, M. R., Sertié, A. L., 2021. Complement c4 is reduced in ipsc-derived astrocytes of autism spectrum disorder subjects. *Int. J. Mol. Sci.* 22, 7579

Menassa, D. A., Gomez-Nicola, D., 2018. Microglial dynamics during human brain development. *Front. Immunol.* 9(MAY), 1014.

Meng, J., Han, L., Zheng, N., Wang, T., Xu, H., Jiang, Y., Wang, Z., Liu, Z., Zheng, Q., Zhang, X., Luo, H., Can, D., Lu, J., Xu, H., Zhang, Y. W., 2022. Microglial Tmem59 deficiency impairs phagocytosis of synapse and leads to autism-like behaviors in mice. *J. Neurosci.* 42, 4958–4979.

Milinkeviciute, G., Henningfield, C. M., Muniak, M.A., Chokr, S. M., Carr, C., 2019. Microglia regulate pruning of specialized synapses in the auditory brainstem. 13(August), 1–19.

Miyamoto, A., Wake, H., Ishikawa, A. W., Eto, K., Shibata, K., Murakoshi, H., Koizumi, S., Moorhouse, A. J., Yoshimura, Y., Nabekura, J., 2016. Microglia contact induces synapse formation in developing somatosensory cortex. *Nat. Commun.* 7.

Miyanishi, K., Sato, A., Kihara, N., Utsunomiya, R., Tanaka, J., 2021. Synaptic elimination by microglia and disturbed higher brain functions. *Neurochem. Int.* 142(August 2020), 104901.

Molofsky, A. V., Krenick, R., Ullian, E., Tsai, H., Deneen, B., Richardson, W. D., Barres, B. A., Rowitch, D. H. 2012. Astrocytes and disease: a neurodevelopmental perspective. *Genes Dev.* 26, 891–907.

Morgan, J. T., Chana, G., Abramson, I., Semendeferi, K., Courchesne, E., Everall, I. P., 2012. Abnormal microglial-neuronal spatial organization in the dorsolateral prefrontal cortex in autism. *Brain Res.* 1456, 72–81.

Morgan, J. T., Chana, G., Pardo, C. A., Achim, C., Semendeferi, K., Buckwalter, J., Courchesne, E., Everall, I. P., 2010. Microglial activation and increased microglial density observed in the dorsolateral prefrontal cortex in autism. *Biol. Psychiatry* 68, 368–376.

Murai, K. K., Nguyen, L. N., Irie, F., Yamaguchi, Y., Pasquale, E. B., 2003. Control of hippocampal dendritic spine morphology through ephrin-A3/EphA4 signaling. *Nat. Neurosci.* 6, 153–160.

Navarrete, M., Perea, G., de Sevilla, D. F., Gómez-Gonzalo, M., Núñez, A., Martín, E. D., Araque, A., 2012. Astrocytes mediate in vivo cholinergic-induced synaptic plasticity. *PLoS Biol.* 10, e1001259.

Nguyen, A. Q., Koeppen, J., Woodruff, S., Mina, K., Figueroa, Z., Ethell, I. M., 2020. Astrocytic ephrin-B1 controls synapse formation in the hippocampus during learning and memory. *Front. Synaptic Neurosci.* 12, 10.

Nikodemova, M., Kimyon, R. S., De, I., Small, A. L., Collier, L. S., Watters, J. J., 2015. Microglial numbers attain adult levels after undergoing a rapid decrease in cell number in the third postnatal week. *J. Neuroimmunol.* 278, 280–288.

Okazawa, H., Motegi, S., Ohyama, N., Ohnishi, H., Tomizawa, T., Kaneko, Y., Oldenborg, P.-A., Ishikawa, O., Matozaki, T., 2005. Negative regulation of phagocytosis in macrophages by the CD47-SHPS-1 system. *J. Immunol.* 174, 2004–2011.

Paasila, P. J., Aramideh, J. A., Sutherland, G. T., Graeber, M. B., 2022. Synapses, microglia, and lipids in Alzheimer's disease. *Front. Neurosci.* 15(January).

Pankratov, Y., Lalo, U., 2015. Role for astroglial α1-adrenoreceptors in gliotransmission and control of synaptic plasticity in the neocortex. *Front. Cell. Neurosci.* 9, 230.

Pannasch, U., Freche, D., Dallérac, G., Ghézali, G., Escartin, C., Ezan, P., Cohen-Salmon, M., Benchenane, K., Abudara, V., Dufour, A., 2014. Connexin 30 sets synaptic strength by controlling astroglial synapse invasion. *Nat. Neurosci.* 17, 549–558.

Paolicelli, R. C., Bolasco, G., Pagani, F., Maggi, L., Scianni, M., Panzanelli, P., Giustetto, M., Ferreira, T. A., Guiducci, E., Dumas, L., Ragozzino, D., Gross, C. T., 2011. Synaptic pruning by microglia is necessary for normal brain development. *Science* 333, 1456–1458.

Park, J., Choi, Y., Jung, E., Lee, S.-H., Sohn, J.-W., Chung, W.-S. 2021., Microglial MERTK eliminates phosphatidylserine-displaying inhibitory post-synapses. *EMBO J.* 40, e107121.

Parkhurst, C. N., Yang, G., Ninan, I., Savas, J. N., Yates, J. R., Lafaille, J. J., Hempstead, B. L., Littman, D. R., Gan, W. B., 2013. Microglia promote learning-dependent synapse formation through Brain-Derived Neurotrophic Factor. *Cell* 155, 1596–1609.

Parrott, J. M., Oster, T., Lee, H. Y., 2021. Altered inflammatory response in FMRP-deficient microglia. *IScience* 24.

Pascual, O., Ben Achour, S., Rostaing, P., Triller, A., Bessis, A., 2012. Microglia activation triggers astrocyte-mediated modulation of excitatory neurotransmission. *Proc. Natl. Acad. Sci.* 109, E197–E205.

Perea, G., Araque, A., 2007. Astrocytes potentiate transmitter release at single hippocampal synapses. *Science* 317, 1083–1086.

Perea, G., Gómez, R., Mederos, S., Covelo, A., Ballesteros, J. J., Schlosser, L., Hernández-Vivanco, A., Martín-Fernández, M., Quintana, R., Rayan, A., 2016. Activity-dependent switch of GABAergic inhibition into glutamatergic excitation in astrocyte-neuron networks. *Elife* 5, e20362.

Perez-Catalan, N. A., Doe, C. Q., Ackerman, S. D., 2021. The role of astrocyte-mediated plasticity in neural circuit development and function. *Neural Dev.* 16, 1–14.

Pérez-Rodríguez, M., Arroyo-García, L. E., Prius-Mengual, J., Andrade-Talavera, Y., Armengol, J. A., Pérez-Villegas, E. M., Duque-Feria, P., Flores, G., Rodríguez-Moreno, A., 2019. Adenosine receptor-mediated developmental loss of spike timing-dependent depression in the hippocampus. *Cereb. Cortex,* 29, 3266–3281.

Petanjek, Z., Judaš, M., Šimić, G., Rašin, M. R., Uylings, H. B. M., Rakic, P., Kostović, I., 2011. Extraordinary neoteny of synaptic spines in the human prefrontal cortex. *Proc. Natl. Acad. Sci. USA* 108, 13281–13286.

Petrelli, F., Pucci, L., Bezzi, P. 2016. Astrocytes and microglia and their potential link with autism spectrum disorders. *Front. Cell. Neurosci.* 10(February), 1–8.

Pont-Lezica, L., Beumer, W., Colasse, S., Drexhage, H., Versnel, M., Bessis, A., 2014. Microglia shape corpus callosum axon tract fasciculation: functional impact of prenatal inflammation. *Eur. J. Neurosci.* 39, 1551–1557.

Preman, P., Alfonso-Triguero, M., Alberdi, E., Verkhratsky, A., Arranz, A. M., 2021. Astrocytes in Alzheimer's disease: pathological significance and molecular pathways. *Cells* 10, 540.

Presumey, J., Bialas, A. R., Carroll, M. C., 2017. Complement system in neural synapse elimination in development and disease, in: *Advances in Immunology* (1st ed , Vol. 135). Elsevier Inc. USA.

Qin, Q., Teng, Z., Liu, C., Li, Q., Yin, Y., Tang, Y., 2021. TREM2, microglia, and Alzheimer's disease. *Mech. Ageing Dev.* 195(December 2020), 111438.

Qu, W., Li, L., 2020. Loss of TREM2 confers resilience to synaptic and cognitive impairment in aged mice. *J. Neurosci.* 40, 9552–9563.

Rajendran, L., Paolicelli, R. C., 2018. Microglia-mediated synapse loss in Alzheimer's disease. *J. Neurosci.* 38, 2911–2919.

Reemst, K., Noctor, S. C., Lucassen, P. J., Hol, E. M., 2016. The indispensable roles of microglia and astrocytes during brain development. *Front. Human Neurosci.* 10(NOV2016), 1–28.

Reshef, R., Kudryavitskaya, E., Shani-Narkiss, H., Isaacson, B., Rimmerman, N., Mizrahi, A., Yirmiya, R., 2017. The role of microglia and their CX3CR1 signaling in adult neurogenesis in the olfactory bulb. *ELife* 6, 1–30.

Rezaie, P., Dean, A., Male, D., Ulfig, N., 2005. Microglia in the cerebral wall of the human telencephalon at second trimester. *Cereb. Cortex* 15, 938–949.

Risher, W. C., Kim, N., Koh, S., Choi, J. E., Mitev, P., Spence, E. F., Pilaz, L. J., Wang, D., 2018. Thrombospondin receptor α2δ-1 promotes synaptogenesis and spinogenesis via postsynaptic. *J. Cell Biol.* 217, 3747-3765.

Risher, W. C., Patel, S., Kim, I. H., Uezu, A., Bhagat, S., Wilton, D. K., Pilaz, L.-J., Alvarado, J. S., Calhan, O. Y., Silver, D. L., 2014. Astrocytes refine cortical connectivity at dendritic spines. *Elife* 3, e04047.

Roy, E. R., Wang, B., Wan, Y. W., Chiu, G., Cole, A., Yin, Z., Propson, N. E., Xu, Y., Jankowsky, J. L., Liu, Z., Lee, V. M. Y., Trojanowski, J. Q., Ginsberg, S. D., Butovsky, O., Zheng, H., Cao, W., 2020., Type I interferon response drives neuroinflammation and synapse loss in Alzheimer disease. *J. Clin. Invest.* 130, 1912–1930.

Russo, F. B., Freitas, B. C., Pignatari, G. C., Fernandes, I. R., Sebat, J., Muotri, A. R., Beltrão-braga, P. C. B., 2018., Modeling the interplay between neurons and astrocytes in autism using human induced pluripotent stem cells. *Biol. Psychiatry* 83, 569–578.

Sancho, L., Contreras, M., Allen, N. J., 2021. Glia as sculptors of synaptic plasticity. *Neurosci. Res.* 167, 17–29.

Schafer, D. P., Lehrman, E. K., Kautzman, A. G., Koyama, R., Mardinly, A. R., Yamasaki, R., Ransohoff, R. M., Greenberg, M. E., Barres, B. A., Stevens, B., 2012., Microglia sculpt postnatal neural circuits in an activity and complement-dependent manner. *Neuron* 74, 691–705.

Scott-Hewitt, N., Perrucci, F., Morini, R., Erreni, M., Mahoney, M., Witkowska, A., Carey, A., Faggiani, E., Schuetz, L. T., Mason, S., Tamborini, M., Bizzotto, M., Passoni, L., Filipello, F., Jahn, R., Stevens, B., Matteoli, M., 2020, Local externalization of phosphatidylserine mediates developmental synaptic pruning by microglia. *EMBO J.* 39, e105380.

Sekar, A., Bialas, A. R., De Rivera, H., Davis, A., Hammond, T. R., Kamitaki, N., Tooley, K., Presumey, J., Baum, M., Van Doren, V., Genovese, G., Rose, S. A., Handsaker, R. E., Daly, M. J., Carroll, M. C., Stevens, B., McCarroll, S. A., 2016. Schizophrenia risk from complex variation of complement component 4. *Nature* 530, 177–183.

Sellgren, C. M., Gracias, J., Watmuff, B., Biag, J. D., Thanos, J. M., Whittredge, P. B., Fu, T., Worringer, K., Brown, H. E., Wang, J., Kaykas, A., Karmacharya, R., Goold, C. P., Sheridan, S. D., Perlis, R. H., 2019. Increased synapse elimination by microglia in schizophrenia patient-derived models of synaptic pruning. *Nat. Neurosci.* 22, 374–385.

Sheng, L., Chen, M., Cai, K., Song, Y., Yu, D., Zhang, H., Xu, G., 2019. Microglial Trem2 induces synaptic impairment at early stage and prevents amyloidosis at late stage in APP/PS1 mice. *FASEB Journal: Official Publication of the Federation of American Societies for Experimental Biology,* 33, 10425–10442.

Sierra, A., Encinas, J. M., Deudero, J. J. P., Chancey, J. H., Enikolopov, G., Overstreet-Wadiche, L. S., Tsirka S. E., Maletic-Savatic, M., 2010. Microglia shape adult hippocampal neurogenesis through apoptosis-coupled phagocytosis. *Cell Stem Cell.* 7, 483–495.

Singh, S. K., Stogsdill, J. A., Pulimood, N. S., Dingsdale, H., Kim, Y. H., Pilaz, L.-J., Kim, I. H., Manhaes, A. C., Rodrigues Jr., W. S., Pamukcu, A., 2016. Astrocytes assemble thalamocortical synapses by bridging NRX1α and NL1 via hevin. *Cell* 164, 183–196.

Smolders, S. M. T., Kessels, S., Vangansewinkel, T., Rigo, J. M., Legendre, P., Brône, B., 2019. Microglia: Brain cells on the move. *Prog. Neurobiol.* 178, 101612.

Sofroniew, M. V., Vinters, H. V., 2010. Astrocytes: biology and pathology. *Acta Neuropathol.* 119, 7–35.

Soteros, B. M., Sia, G. M., 2022. Complement and microglia dependent synapse elimination in brain development. *WIREs Mech. Dis.* 14, 1–17.

Squarzoni, P., Oller, G., Hoeffel, G., Pont-Lezica, L., Rostaing, P., Low, D., Bessis, A., Ginhoux, F., Garel, S., 2014. Microglia modulate wiring of the embryonic forebrain. *Cell Rep.* 8, 1271–1279.

Stephan, A. H., Madison, D. V., Mateos, J. M., Fraser, D. A., Lovelett, E. A., Coutellier, L., Kim, L., Tsai, H. H., Huang, E. J., Rowitch, D. H., Berns, D. S., Tenner, A. J., Shamloo, M., Barres, B. A., 2013. A dramatic increase of C1q protein in the CNS during normal aging. *J. Neurosci.* 33, 13460–13474.

Stevens, B., Allen, N. J., Vazquez, L. E., Howell, G. R., Christopherson, K. S., Nouri, N., Micheva, K. D., Mehalow, A. K., Huberman, A. D., Stafford, B., Sher, A., Litke, A. M. M., Lambris, J. D., Smith, S. J., John, S. W. M., Barres, B. A., 2007. The classical complement cascade mediates CNS synapse elimination. *Cell* 131, 1164–1178.

Stogsdill, J. A., Ramirez, J., Liu, D., Kim, Y. H., Baldwin, K. T., Enustun, E., Ejikeme, T., Ji, R.-R., Eroglu, C., 2017. Astrocytic neuroligins control astrocyte morphogenesis and synaptogenesis. *Nature* 551, 192–197.

Suzuki, A., Stern, S. A., Bozdagi, O., Huntley, G. W., Walker, R. H., Magistretti, P. J., Alberini, C. M., 2011. Astrocyte-neuron lactate transport is required for long-term memory formation. *Cell* 144, 810–823.

Suzuki, K., Sugihara, G., Ouchi, Y., Nakamura, K., Futatsubashi, M., Takebayashi, K., Yoshihara, Y., Omata, K., Matsumoto, K., Tsuchiya, K. J., Iwata, Y., Tsujii, M., Sugiyama, T., Mori, N., 2013. Microglial activation in young adults with autism spectrum disorder. *Arch. Gen. Psychiatry* 70, 49–58.

Tan, C. X., Eroglu, C., 2021. Cell adhesion molecules regulating astrocyte–neuron interactions. *Curr. Opin. Neurobiol.* 69, 170–177.

Tan, C. X., Lane, C. J. B., Eroglu, C., 2021. Role of astrocytes in synapse formation and maturation, in: *Current Topics in Developmental Biology* (Vol. 142, pp. 371–407). Elsevier.

Tarasov, V. V., Svistunov, A. A., Chubarev, V. N., Sologova, S. S., Mukhortova, P., Levushkin, D., Somasundaram, S. G., Kirkland, C. E., Bachurin, S. O., Aliev, G.,

2020. Alterations of astrocytes in the context of schizophrenic dementia. *Front. Pharmacol.* 10, 1612.

Tetreault, N. A., Hakeem, A. Y., Jiang, S., Williams, B. A., Allman, E., Wold, B. J., Allman, J. M., 2012. Microglia in the cerebral cortex in autism. *J. Aut. Dev. Disord.* 42, 2569–2584.

Tremblay, M. Ě., Lowery, R. L., Majewska, A. K., 2010. Microglial interactions with synapses are modulated by visual experience. *PLoS Biol.* 8.

Ullian, E. M., Christopherson, K. S., Barres, B. A., 2004. Role for glia in synaptogenesis. *Glia* 47, 209–216.

Vainchtein, I. D., Chin, G., Cho, F. S., Kelley, K. W., Miller, J. G., Chien, E. C., Liddelow, S. A., Nguyen, P. T., Nakao-Inoue, H., Dorman, L. C., 2018. Astrocyte-derived interleukin-33 promotes microglial synapse engulfment and neural circuit development. *Science* 359, 1269–1273.

Vakilzadeh, G., Falcone, C., Dufour, B., Hong, T., Noctor, S. C., Martínez-Cerdeño, V., 2022. Decreased number and increased activation state of astrocytes in gray and white matter of the prefrontal cortex in autism. *Cereb. Cortex* 32, 4902–4912.

van Deijk, A. F., Camargo, N., Timmerman, J., Heistek, T., Brouwers, J. F., Mogavero, F., Mansvelder, H. D., Smit, A. B., Verheijen, M. H. G., 2017. Astrocyte lipid metabolism is critical for synapse development and function in vivo. *Glia* 65, 670–682.

Vargas, D. L., Nascimbene, C., Krishnan, C., Zimmerman, A. W., Pardo, C. A., 2005. Neuroglial activation and neuroinflammation in the brain of patients with autism. *Ann. Neurol.* 57, 67–81.

Verney, C., Monier, A., Fallet-Bianco, C., Gressens, P., 2010. Early microglial colonization of the human forebrain and possible involvement in periventricular white-matter injury of preterm infants. *J. Anat.* 217, 436–448.

Voet, S., Srinivasan, S., Lamkanfi, M., Loo, G., 2019. Inflammasomes in neuroinflammatory and neurodegenerative diseases. *EMBO Mol. Med.* 11, 1–16.

Wake, H., Moorhouse, A. J., Jinno, S., Kohsaka, S., Nabekura, J., 2009. Resting microglia directly monitor the functional state of synapses in vivo and determine the fate of ischemic terminals. *J. Neurosci.* 29, 3974–3980.

Wakselman, S., Béchade, C., Roumier, A., Bernard, D., Triller, A., Bessis, A., 2008. Developmental neuronal death in hippocampus requires the microglial CD11b integrin and DAP12 immunoreceptor. *J. Neurosci.* 28, 8138–8143.

Wallingford, J., Scott, A. L., Rodrigues, K., Doering, L. C., 2017. Altered developmental expression of the astrocyte-secreted factors hevin and SPARC in the fragile X mouse model. *Front. Mol. Neurosci.* 10, 268.

Wang, C. S., Kavalali, E. T., Monteggia, L. M., 2022. BDNF signaling in context: From synaptic regulation to psychiatric disorders. *Cell* 185, 62–76.

Wang, C., Xiong, M., Gratuze, M., Bao, X., Shi, Y., Andhey, P. S., Manis, M., Schroeder, C., Yin, Z., Madore, C., Butovsky, O., Artyomov, M., Ulrich, J. D., Holtzman, D. M., 2021. Selective removal of astrocytic APOE4 strongly protects against tau-mediated neurodegeneration and decreases synaptic phagocytosis by microglia. *Neuron* 109, 1657-1674.e7.

Wang, C., Yue, H., Hu, Z., Shen, Y., Ma, J., Li, J., Wang, X.-D., Wang, L., Sun, B., Shi, P., Wang, L., Gu, Y., 2020. Microglia mediate forgetting via complement-dependent synaptic elimination. *Science* 367, 688–694.

Weinhard, L., Di Bartolomei, G., Bolasco, G., Machado, P., Schieber, N. L., Neniskyte, U., Exiga, M., Vadisiute, A., Raggioli, A., Schertel, A., Schwab, Y., Gross, C. T., 2018. Microglia remodel synapses by presynaptic trogocytosis and spine head filopodia induction. *Nat. Commun.* 9.

Wu, Y., Singh, S., Georgescu, M.-M., Birge, R. B., 2005. A role for Mer tyrosine kinase in αvβ5 integrin-mediated phagocytosis of apoptotic cells. *J. Cell Sci.* 118, 539–553.

Xu, J., Xiao, N., Xia, J., 2010. Thrombospondin 1 accelerates synaptogenesis in hippocampal neurons through neuroligin 1. *Nat. Neurosci.* 13, 22–24.

Yang, Y., Ge, W., Chen, Y., Zhang, Z., Shen, W., Wu, C., Poo, M., Duan, S., 2003. Contribution of astrocytes to hippocampal long-term potentiation through release of D-serine. *Proc. Natl. Acad. Sci.* 100, 15194–15199.

Yilmaz, M., Yalcin, E., Presumey, J., Aw, E., Ma, M., Whelan, C. W., Stevens, B., McCarroll, S. A., Carroll, M. C., 2021. Overexpression of schizophrenia susceptibility factor human complement C4A promotes excessive synaptic loss and behavioral changes in mice. *Nat. Neurosci.* 24, 214–224.

Zeidán-Chuliá, F., Salmina, A. B., Malinovskaya, N. A., Noda, M., Verkhratsky, A., Moreira, J. C. F., 2014. The glial perspective of autism spectrum disorders. *Neurosci. Biobehav. Rev.* 38, 160–172.

Zhan, Y., Paolicelli, R. C., Sforazzini, F., Weinhard, L., Bolasco, G., Pagani, F., Vyssotski, A. L., Bifone, A., Gozzi, A., Ragozzino, D., Gross, C. T., 2014. Deficient neuron-microglia signaling results in impaired functional brain connectivity and social behavior. *Nat. Neurosci.* 17, 400–406.

Zhan, Y., Paolicelli, R. C., Sforazzini, F., Weinhard, L., Bolasco, G., Pagani, F., Vyssotski, A. L., Bifone, A., Gozzi, A., Ragozzino, D., Gross, C. T., 2014. Deficient neuron-microglia signaling results in impaired functional brain connectivity and social behavior. *Nat. Neurosci.* 17, 400–406.

Zhou, B., Zuo, Y., Jiang, R., 2019. Astrocyte morphology: Diversity, plasticity, and role in neurological diseases. *CNS Neurosci. Ther.* 25, 665–673.

Chapter 8

Pathological Plasticity of Dendritic Spines Derived from Malnutrition

Humberto Martínez-Orozco*
Zyanya Gloria Mena Casas
and Sofía Díaz-Cintra
Instituto de Neurobiología, Universidad Nacional Autónoma de México,
Campus Juriquilla-Queretaro, Querétaro, México

Introduction

Nutrition is the act by which an organism takes in food and absorbs essential nutrients to support all physiological processes and sustain life. Humans have developed new technologies to meet the demand for food and facilitate survival (Valoppi et al. al., 2021). However, malnutrition, which encompasses undernutrition (wasting, stunting, and underweight), vitamin or mineral imbalances, overweight, obesity, and diet-related non-communicable diseases, remains a pressing global health problem that requires greater efforts to address it. According to data from the UNICEF, World Health Organization (WHO), and World Bank inter-agency team, in 2020, 45.4 million children worldwide were wasted (low weight-for-height ratio) and 38.9 million overweight.

One approach to tackle malnutrition has been the research of brain areas associated with food intake regulation (Rudebeck & Izquierdo, 2022). The circuitries in these areas establish neural connections through synapses, pre-synaptic (axons) and post-synaptic elements, and dendritic spines. The latter

* Corresponding Author's Email:

In: Dendritic Spines
Editor: Ignacio González-Burgos
ISBN: 979-8-89113-080-7

are special, plastic formations that play a role in neural communication. Dendritic spines are dynamically formed, maintained, and eliminated over time; hence, their density and morphology vary depending on the life stage (Runge et al., 2020).

Poor nutrition during pregnancy (Cortés-Albornoz et al., 2021), the postnatal period (Hortensius et al., 2019), childhood (Galler et al., 2021), and even adolescence (Galler et al., 2017) can lead to alterations in brain growth. Consequently, one of the biggest concerns about malnutrition is its impact on neurodevelopment. Dendritic spine plasticity requires nutrients. Therefore, moderate, and severe malnutrition can impair brain functions by causing alterations in dendritic spine morphology and plasticity, thus disrupting synaptic transmission, brain plasticity, decision-making for health care, and cognition. Furthermore, dendritic spine alteration is a hallmark of neurodevelopmental disorders (Nishiyama, 2019). For example, evidence has implicated dendritic spine dysfunction as a central substrate of the pathogenesis of dementia (Herms and Dorostkar, 2016).

This chapter addresses relevant aspects of malnutrition and its impact on neurodevelopment. Subchapters describe distinct evidence underscoring the effects of malnutrition on dendritic spine formation and morphology in critical stages of life. Finally, this chapter will discuss how malnutrition-induced changes in dendritic spines contribute to the appearance of neurodevelopmental disorders like visual, auditory, and motor alterations, as well as neuropsychiatric illnesses (schizophrenia and Alzheimer's disease), during the postnatal stage and later stages of life.

An Overview of Malnutrition Prevalence

According to the WHO, malnutrition refers to *"deficiencies, excesses or imbalances in a person's intake of energy and/or nutrients"* (WHO, 2020), and it includes both undernutrition and overnutrition. Undernutrition is characterized by low height-for-age and micronutrient-related malnutrition, whereas overnutrition includes overweight, obesity and diet-related diseases. The causes of undernutrition have been associated with reduced dietary intake, malabsorption, increased loss of weight, and elevated energy expenditure. Undernutrition results in loss of functions in muscle, cardio-respiratory, gastrointestinal, and immune systems implicated in the appearance of psychosocial disorders such as apathy, anxiety, depression, and self-neglect (Saunders & Smith, 2010). Overnutrition leads to excessive weight gain which results from dense caloric intake as a consequence of the high availability of hypercaloric foods at home, consumption preference for sweet and fatty/fried foods and salty snacks, skipping meals, a sedentary lifestyle, stressful environments, and influenced by changes in family structure, time spent by parents with their children, economic status, accessibility to ready-to-eat/cook foods, urbanization, psychosocial stress, as well as aggressive marketing relating to nutritional and physical activity habits (Mathur & Pillai, 2019).

Humans can make healthy eating and dietary choices to nourish the body and maintain optimal health. But how does the brain control food intake? This function is strictly regulated by sensory, homeostatic, and hedonic neural circuits, which balance energy intake and expenditure (Yapici, et al., 2016). The failure to regulate food perception and appetite results in maladaptive eating behaviors and an increase in the occurrence of metabolic syndromes and eating disorders. The main structures implicated in such regulations are the limbic brain regions; for example, the prefrontal cortex, which plays an important role in foraging or food-seeking behaviors and planning to maximize food procurement based on the assessment of its availability (Rudebeck & Izquierdo, 2022). The prefrontal cortex is also involved in decision-making under free-choice conditions (Funahashi, 2017). Furthermore, several studies support the role of this region in self-control of food intake (Alonso-Alonso & Pascual-Leone, 2007).

Given that the prefrontal cortex of adults is mature, information about how certain foods can impact a person's health in the future increases their awareness and allows them to make healthier food choices (Gustafson, 2022). Children, however, exhibit reduced activity in the dorsolateral prefrontal cortex compared with adults, which is associated with making fewer healthy choices like avoiding tasty foods (van Meer et al., 2017). Therefore, prefrontal cortex maturation in childhood is crucial to develop the ability to make healthy food choices; this implies consciously determining whether the foods that are being consumed are sufficiently nutritious and balanced.

The psychological status of an individual influences the persistence of nutritional disbalances including under and overnutrition. A cross-sectional study showed that only 40% of the 562 pregnant women recruited for this study consumed enough fiber and vegetables a day, which was associated with their mental well-being (Nguyen et al., 2022). Malnutrition and depression are common during pregnancy, and they increase the risk for behavioral and cognitive problems in the offspring (Bleker et al., 2019). In this context, mental health affectations lead to reduced self-care behavior, promoting unbalanced nutrition. Importantly, if nutritious food is readily available, then making unhealthy decisions towards malnourishment could represent a problem with the brain circuitries involved in feeding and self-care modulation.

The Impact of Malnutrition on Neurodevelopment

During pregnancy, several processes like neurulation, neurogenesis, microglial entry, neuronal migration, synaptogenesis, gliogenesis, and myelination are crucial for fetal brain development (Doi et al., 2022). According to an imaging study (Jernigan et al., 2011), a brain growth spurt occurs at postnatal age that drastically changes brain morphology and promotes oligodendrocyte proliferation and myelin deposition. Early childhood is characterized by an orderly pattern of myelination in white matter regions. From late childhood (6 years old), the brain changes become subtler, and differences in the cranial vault size are little. Nonetheless, further changes in the brain architecture, like those in the gray matter volumes, take place in later life (Jernigan and Tallal, 1990; Jernigan et al., 1991; Jernigan et al., 2011). Therefore, it is important to understand that the brain is a dynamic tissue that changes throughout life, but some events, like those in neurodevelopment, must occur precisely to maintain the integrity of brain functions.

Just as any physiological process, neurodevelopment and brain maturation require nutrient supply from daily dietary intake. When nutrition is deficient, brain development is affected. Research from the fifties and sixties suggested that gestational deficiency of folic acid (vitamin B9) played a part in embryonic malformations (Nelson et al., 1952) and congenital defects in the fetus (Goetsch, 1962). Years later, the Medical Research Council Vitamin Study Research Group confirmed that folic acid oral supplementation during gestation prevented fetal neural tube defects in 72% of the cases (MRC, 1991).

Today, it is well known that pre- and postnatal deficiency of distinct key nutrients can disrupt the anatomy, chemistry, physiology, and metabolism of brain cells involved in critical processes of neurodevelopment (Cusick & Georgieff, 2016). Accordingly, deficient intake of specific macromolecules such as proteins, some minerals like iron, copper, zinc, and iodine, and different types of vitamins is associated with impairments in neurogenesis, neurite outgrowth, synaptogenesis, myelination, neurotransmitter metabolism, or electrical efficiency in the entire brain or specific regions. Malnutrition also alters brain function and structure by reducing cell numbers, cell migration, myelination, synaptogenesis, hippocampal formation, and neurotransmission during child development (Cakir et al., 2019). The study by Cakir and colleagues also showed that malnourished children exhibited low language, motor, social, self-care, and general development scores compared to controls.

On the other hand, as Barker postulated in the nineties, the *in utero* environment as a result of undernutrition during pregnancy can induce fetal adaptations known as *"genetic programming"* that predispose an individual to develop chronic diseases in later life (Barker, 1998). Maternal overnutrition with a high-fat diet (HFD) is a clear example of fetal programming with the manifestation of brain health outcomes in adulthood. Maternal HFD affects prenatal brain development of the offspring by producing morphological alterations such as reductions in the density of dendritic spines in the primary somatosensory area; fewer dendrite branches and reduced total dendritic length in the hippocampus; reduced dopaminergic neuron projection, formation, refinement, and maturation in the midbrain; and decreased myelination in the medial prefrontal cortex (mPFC) (Urbonaite et al., 2022). In humans, maternal obesity and diabetes, which can be originated from excessive intake of hypercaloric diets, increase the risk of behavioral disorders in the offspring like attention-deficit/hyperactivity disorder (Rodriguez et al., 2018) and autism spectrum disorders (Krakowiak et al., 2012). Furthermore, animal studies revealed that maternal obesity and HFD consumption induce placental dysfunction and inflammation (Sullivan et al., 2012) which could alter neurodevelopment and fetal programming, and result in predisposition to inflammation-associated psychopathologies like Alzheimer's disease (AD).

Malnutrition Effects on Dendritic Spines during Crucial Stages of Life

Gestation and Lactation

Inadequate dietary intake before and during pregnancy can produce micronutrient deficits in the mother and have consequences on the child's brain and cognitive development (Young & Ramakrishnan, 2020). When malnutrition is present during pregnancy, there is a greater impact on the neurodevelopment of the offspring, leaving lifelong sequelae. Moreover, different alterations in dendritic spine morphology and plasticity in the offspring, because of malnutrition before and during gestation, have been described in rodents subjected to starvation, food restriction, deficient intake of macronutrients like protein, vitamins, or minerals, and hypercaloric diets. However, it is important to consider that malnutrition-induced changes in dendritic spines occur not only in the offspring but also in the mother. For example, dams fed a low-casein diet before and during pregnancy and lactation had fewer spines in apical dendrites of layers III and V of the cerebral cortex compared with controls. These changes persisted until postpartum day 60 (Leuba & Rabinowicz, 1979a).

Starvation and Food Restriction

Starvation and food restriction experiments serve to study physiological responses to the limited availability or lack of food. An early study in the rat brain showed that postnatal starvation alters spines of lamina V pyramidal neurons in the cingulate cortex, causing a reduction in spine numbers and alterations in differentiation and morphology (Schönheit & Haensel, 1984). Also, a 50% restriction of daily food intake in rats from gestation day 7 until the end of lactation produces modifications in the hippocampal dendritic spines of the offspring, as well as a reduction in both apical and basal dendritic trees of CA1 pyramidal cells and decreased density of thin-type spines in apical neurons and a reduction in the type of mushroom spines in both apical and basal dendrites (Zhang et al., 2013).

Food restriction for under 30 days interferes with the development of neurons in the facial nucleus of rats, thus altering neuronal activation and integration of afferent sensory signals involved in facial expression (Torrero et al., 2003). It has also been observed that dendritic density decreases in layer II/III neurons of the anterior cingulate cortex, which could affect emotional, cognitive, and motor functions (Salas et al., 2015). This could be related to the maternal response of licking the anogenital region of the pups, which is altered by food deprivation (Salas et al., 1998). According to one study, neonatal food restriction and binaural ear occlusion in the rat resulted in a decrease in the number and extension of the distal part of motor neuron dendrites in layer V of the motor cortex (Torrero et al., 2005). According to other studies, perinatal food restriction causes a reduction in the number and extension of dendrites in the nucleus of the solitary tract, which can be reversed in a nutritional and sensory manner during lactation (Rubio et al., 2004). In this sense, the recovery of dendritic spines is a form of experience-dependent plasticity.

Low-Protein Diet

Most of the studies linking anatomo-morphological modifications of dendritic spines with prenatal and postnatal malnutrition have been performed in animals and have widely focused on the effects of low-protein diets containing 60-70% less protein than standard diets. Protein deprivation causes abnormalities in body and brain weight in early life (Figure 1) and decreased dendritic spines of pyramidal neurons in layer V of the frontal, occipital, and parietal cortices. Additionally, synaptic loss is more marked in brain regions that receive projections from non-specific multi-synaptic systems than in areas associated with specific afferent systems or those receiving callous projections (Salas, Diaz, & Nieto, 1974; Salas, 1980).

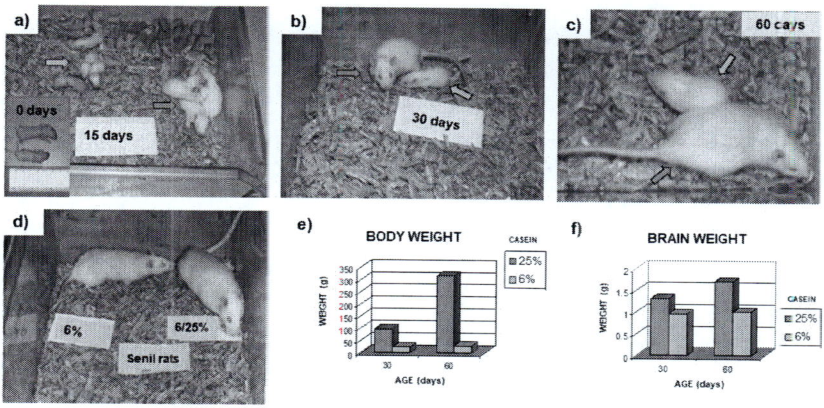

Figure 1. Effect of exposure to a low-protein diet during pregnancy and lactation in the offspring of Sprague-Dawley rats. Abnormalities in size of the newborn due to protein deficiency in the mother are evident from (a) days 0 to 15 with extended effects until (b) day 30, (c) day 60 and (d) advanced age. At days 30 and 60, pups from malnourished mothers fed with 6% casein had significantly low (e) body and (f) brain weights compared to controls fed a 25% casein diet.

In the eighties, Kemper and collaborators used the Golgi-Cox technique to evaluate dendrite morphology in the neostriatum and reticular formation of pups from mothers subjected to a low-casein diet (8% kcal) before mating and during gestation and lactation. The authors observed reduced synaptic spine density in the heavily spined dominant neuron of the neostriatum, as well as a marked reduction in the dendritic length of neurons in the reticular formation (Kemper et al., 1978). Through the rapid Golgi technique, other researchers have demonstrated that chronic pre-gestational exposure to a low-protein diet (6-8% kcal) causes a reduction in cell size, number of dendritic spines, and complexity of dendritic branching in granule cells of the fascia dentata of the hippocampus of Sprague-Dawley rat offspring (Díaz-Cintra et al., 1991) (Figures 2, 3).

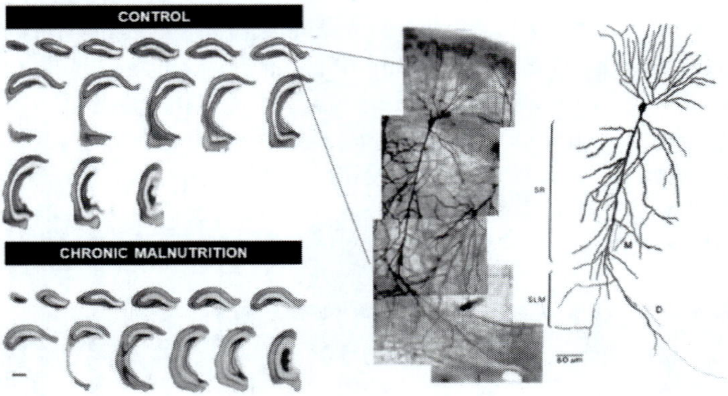

Figure 2. Pre-gestational chronic malnutrition by a low-protein diet diminishes the hippocampal volume of the offspring. Dendritic spine alterations are assessed by rapid Golgi staining to reveal the neuronal architecture, which is drawn on paper for enhanced image analysis (modified from Cintra et al., 1997).

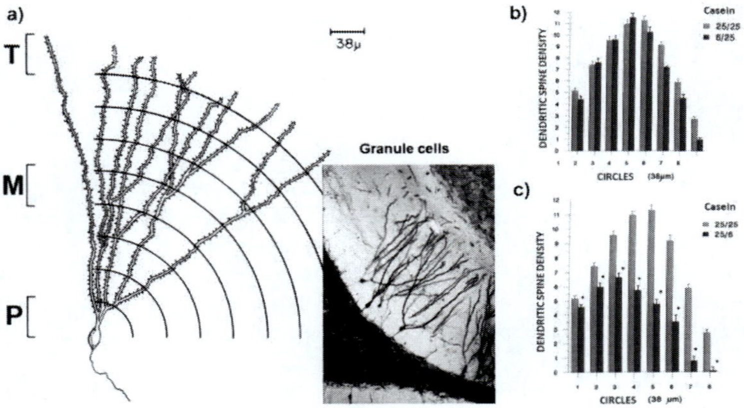

Figure 3. Effects of protein deprivation during pregnancy and lactation on the dendritic spines of granule cells in the offspring. Example of a camera lucida drawing of a granule cell neuron to show the location in concentric rings (each 38 μm apart). The panels b and c indicate dendritic spine density located in each concentric ring for rats fed with a normal casein diet during pregnancy and lactation (25/25), a low-protein diet (6%) only during gestation (6/25), and deficient protein intake during the lactation period (25/6). A reduction in dendritic spine density in the granule cells is only seen for pups from protein malnourished mothers during lactation, which highlights that deficiencies in this period are critical for granule cell morphology (modified from Cintra et al., Brain Res, 1990).

Studies by Cintra et al. have reported alterations in circadian sleep because of protein malnutrition during pregnancy (Cintra et al., 2002). Sleep control involves the participation of the aminergic brain stem nucleus raphe dorsalis, where it has been found that well-nourished animals display an increase in dendritic spines and morphology at 30 and 90 postnatal days of age, but these parameters were disrupted due to the protein deficiency in both ages (Díaz-Cintra et al., 1981). In another study, rats were prenatally and postnatally exposed to a low-protein diet to analyze dendritic spines in the *locus coeruleus* (involved in arousal). According to the findings, protein deficiency impeded normal age-related changes in dendritic spine density and morphological alterations of multipolar neurons in this region. The 6% casein diet caused selective changes in each cell type of this nucleus; thus, on this anatomical basis, these studies suggest the possibility of functional deficiencies for the *locus coeruleus* noradrenergic system when this type of malnutrition is maintained during brain development. Possible repercussions include muscle hyperactivity and poor response to environmental stimuli in a state of vigilance, as this noradrenergic nucleus is involved in the maintenance of such functions (Díaz-Cintra et al., 1984).

Further, a similar study was carried out with pups from malnourished rats due to a low-protein diet. These pups displayed a decrease in body and brain weight associated with morphological changes, as well as an altered number of spines in basal and apical dendrites of pyramidal cells in layers II/III and V of the visual cortex (Díaz-Cintra et al., 1990). In addition, other authors have shown that pre-gestational and postnatal exposure of rats to a low-protein diet (8% kcal) affects the cerebellar development of the female offspring by producing changes in Purkinje cell spines and morphology, which alters the synaptic profile (Hillman & Chen, 1981).

Altogether, this evidence exemplifies how undernutrition related to protein deficits impacts the offspring by altering dendritic spine plasticity in brain areas highly involved in sensorimotor, emotional, and cognitive functions.

Mineral Deficiency

Iron is important for oxygen delivery, electron transport, and enzymatic activity and its deficient intake during pregnancy disrupts fetal brain development (Georgieff, 2020). For example, poor iron intake causes fetal/neonatal anemia, which has severe consequences for brain function even in adulthood. Perinatal iron deficiency in rats alters apical dendritic morphology in CA1 pyramidal neurons in the offspring—an effect that persists until adulthood (Jorgenson et al., 2003). Moreover, an iron-deficient diet during pregnancy downregulates the expression of proteins critical for dendrite growth in the hippocampus of rat pups at postnatal day 7, with further alterations in peak branching, thinner third-generation branches, and decreased diameter of spine heads at postnatal day 15 (Brunette et al., 2010). Consistently with these findings in rodents, a magnetic resonance imaging (MRI) study has reported that neonates of pregnant adolescents with lower iron intake during the third semester exhibit alterations in gray matter that suggest decreased arborization in dendrites (Monk et al., 2016). Therefore, the lasting effects of iron deficiency could be due to severe alterations in dendritic spine plasticity.

Iodine is important for thyroid hormone synthesis. A previous study showed a reduction in the number of dendritic spines on the shaft of pyramidal neurons in the cerebral cortex of the offspring as a result of a severe iodine-deficient diet (Chen et al., 1986), and recent work showed that iodine deficiency during pregnancy and lactation produces in the pups a mild decrease in the density of dendritic spines in CA1 pyramidal neurons of the hippocampus (Min et al., 2017). Therefore, iodine also has a relevant role in dendritic spine plasticity during neurodevelopment.

Zinc is another mineral required for proteins and nucleic acids and participates in several processes that are valuable for normal neurodevelopment like dendrite pruning (D'Amelio et al., 2010). Studies by Dvergsten and collaborators have evaluated the effects of zinc deficiency on the cerebellar cortex morphology of rats from birth to weaning. In these works, the authors showed that zinc deficiency produces severe undernutrition in dams, and their pups display a reduced brain size and an underdeveloped cerebellum with changes in the thickness of external granule cells and molecular layers of this region. Consequently, this affection in cerebellum development resulted in reduced dendritic field area, total length, and the number of branches per neuron of the cerebellar basket and stellate neurons of the molecular layer, which suggested alterations in dendritic differentiation and growth of this brain region. (Dvergsten et al., 1983; Dvergsten et al., 1984a; Dvergsten et al., 1984b). Importantly, mossy fibers in the dentate gyrus of the hippocampal formation contain zinc which is necessary for cognitive processes. These structures are vulnerable to alterations by perinatal malnutrition with 6-8% casein diets (Cintra et al., 1990; Granados et al., 2002; Cintra et al., 1997). Therefore, combined deficiency intake of protein and zinc can be highly harmful to hippocampal-dependent processes, as it interferes with the plasticity of dendritic spines and mossy cells in the hippocampal formation.

Vitamin Deficiency

Vitamin B6 (pyridoxamine) is an essential co-factor for neurotransmitter synthesis during normal neurodevelopment (Kirksey et al., 1990; Gerster, 1996). Previous research demonstrated a disruption in dendritic arborization of molecular and granular pyramidal neurons in the cerebellum of pups caused by vitamin B6 deficiency during pregnancy and lactation (Chang et al., 1981). A later study found that preweaning pyridoxamine restriction in rats reduced the number of dendritic spines in the hippocampus in comparison to the non-restricted group (Krishna & Ramakrishna, 2004).

Methyl donors are a group of nutrients including folate, vitamin B12, methionine, zinc, betaine, and choline, which are important for DNA methylation but also play a role in dendritic growth and synapse formation in Purkinje cells of the cerebellum modulated by estradiol during neurodevelopment (Bake et al., 2008). A study in rats fed with a methyl donor-deficient diet during gestation and lactation found a decrement in the expression of estradiol receptors involved in the dendritic growth and maturation of Purkinje cells of female pups (El Hajj Chehadeh et al., 2014), suggesting that reduced consumption of methyl donors can impact dendritic plasticity and cell migration in the cerebellum.

Reduced vitamin B9 or folate intake during pregnancy is widely known to be linked with neural tube defects in the fetus (Nelson et al., 1952). The effects of vitamin B9 deficiency on dendritic spines in animal models have been less studied. However, in hippocampal cell cultures, it has been shown that a folate-deficient environment alters the gene expression of cytoskeleton-related proteins (Akchiche et al., 2012), whereas SH-SY5Y cells in an environment with excessive folate display detrimental effects on dendritic spine densities that resemble alterations in autism (Al Sayed et al., 2019; Clark et al., 2021).

Hypercaloric Diets

HFD is a critical factor in neurodevelopment, as it modifies the *in utero* environment, thus affecting nutrient supply throughout the placenta. Importantly, such adaptations by this diet during gestation and lactation affect dendritic spine plasticity in newborns. For instance, in a study by Hatanaka et al. (2016), genetically manipulated mice whose layer V pyramidal neurons expressed yellow fluorescent protein were exposed pre-gestationally, during pregnancy, and until lactation to an HFD. After weaning, all groups of pups were fed with chow until early adulthood, when the effect of the maternal HFD on dendritic spine morphology in the cerebral cortex was evaluated by *in vivo* transcranial two-photon imaging. Through this approach, Hatanaka and colleagues measured the formation and elimination of dendritic spines and filopodia within one hour, finding that HFD increased the elimination rate of dendritic spines and both formation and elimination rates of filopodia in layer I dendrites of the primary somatosensory cortex of the offspring. This finding demonstrates that maternal HFD intake induces lasting modifications in the dendritic spine plasticity of the offspring despite following a balanced diet.

In another study, rats were subjected to an HFD from gestational day 1 to postpartum day 21. The effect of this diet on the dendritic morphology of the hippocampus and amygdala of the offspring was evaluated on postnatal day 15 and after seven months (Janthakhin et al., 2017). The authors found that adult offspring from mothers exposed to HFD displayed morphological alterations (length and number of branching points) in the basal dendrites of pyramidal neurons in the CA1 region of the hippocampus and basolateral amygdala. Furthermore, spine density tended to increase in CA1 pyramidal neurons.

A recent study compared the offspring's brain responses to HFD, a high-carbohydrate diet, and a combined diet rich in fats and carbohydrates during pregnancy and lactation (Gawlińska et al., 2021). The authors did not perform histological techniques; however, they found that genetic expression of TAOK2, which is partially involved in dendrite and dendritic spine development, is downregulated in the prefrontal cortex and hippocampus of female rats at postnatal day 28. On the other hand, male juvenile offspring showed increased levels of SHANK1, which is related to the structural and functional organization of the dendritic spine.

Childhood

After weaning, children transition from breast milk to other sources of nourishment, which can be solid or liquid. During this transition, adults present different types of food to the child. Unfortunately, such feeding practices in children from poor sectors of populations in countries like Mexico are typically characterized by the introduction of nonhuman milk and nonnutritive liquids before the age of 6 months. Moreover, late introduction of nutritive liquids and solid foods has also been reported (Gonzalez de Cossio et al., 2006). These variations in feeding practices can contribute to malnutrition and further alterations in normal dendritic spine processes during early life. In agreement, a Mexican study revealed that post-mortem brains of infants who died from severe malnutrition had shorter apical dendrites, reduced number of dendritic spines per dendrite, and other morphological abnormalities in dendritic spines (Benítez-Bribiesca et al., 1999). Importantly, an MRI study has shown that abnormalities in brain structure caused by severe malnutrition can be corrected after nutritional interventions in early life to prevent further neurological dysfunctions (Bhargava & Singh, 2020).

Starvation

Animal studies have proven that postnatal undernutrition has a greater impact than maternal undernutrition on brain development and dendritic spine morphology. Leuba and Rabinowicz have performed histological studies in the cerebral cortex of albino mice offspring. In these studies, the authors induced postnatal undernutrition by restricted feeding during lactation, then submitted pups to post-weaning rehabilitation with a normal chow diet to evaluate cellular alteration at different time points. After 10 days of postnatal starvation, pups had an increased neuronal density in all cortex layers compared to controls, which was considerably decreased at postnatal day 30. A continuous decrease was detected until postnatal day 60 and even 180 with the peculiarity that neuronal density in starved animals and further rehabilitated with standard diet was not lower for different layers of the cerebral cortex than that of fully lactated animals (Leuba & Rabinowicz, 1979b). Under the same experimental scheme, it was shown that undernutrition by postnatal restriction to mother's milk heavily interfered with normal dendritic spine formation in thinner and thicker dendrites of apical neurons in the cerebral cortex, leading to a reduction of these structures in layers III and V. This damage persisted during the childhood, adolescence, and adulthood of mice (Leuba & Rabinowicz, 1979a). Consistent with previous findings, male rats with restrictions to mother's milk during lactation displayed lower spine density in layer V pyramidal cells receiving projections in the frontal, parietal, and occipital cortices which was related to impairment of integrative brain functions (Salas, 1980).

Protein Deficiency

The rat model of undernourishment by a low-protein diet was extensively used by West and Kemper in the eighties to demonstrate how deficient protein intake in pups during lactation and post-weaning alters dendritic spines (West & Kemper, 1976).

In this work, the authors found differences in size proportion compared with total brain weight, particularly in the neocortex of protein-deficient newborns. Moreover, through the morphological analysis of drawings from neurons stained by application of Rapid Golgi, West and Kemper showed an increased number of spines and dendritic formations in cortical neurons between postnatal days 10 and 30, and a slight reduction in dendrite thickness at 1 month of age. In the visual cortex, dendrite thickness and the number of apical dendrites were proportionally reduced at postnatal day 30. The latter changes consequently reduced the spine density in the apical terminal, oblique and basal branches, and apical main shaft, as well as the dendritic length in the oblique and basal branches in the visual cortex. In their findings, West and Kemper described that a low-protein diet also led to decreased thickness of the cerebellar cortical plate primarily affecting the molecular layer. Such structural abnormality in this region produced proportional morphological alterations like reduced dendritic length for all types of cells, lower dendritic arbors in Purkinje cells, and decreased dendritic length of basket cell oblique branches and those from granule cell dendrites. Interestingly, all these differences were undetectable at postnatal day 90, indicating that, at least for protein-deficient malnutrition, brain growth and maturation processes seem to contemplate mechanisms for recuperation of neural arrangement and brain architecture.

In addition, 30-day-old rats fed with a high-protein diet displayed elevated dendritic spines in the frontal, parietal and entorhinal cortices, striatum, and medial septum, whereas those fed with a low-protein diet had increased density of these structures in the entorhinal cortex and striatum (Brock & Prasad, 1992).

Mineral Deficiency

Zinc is one of the mineral deficiencies with a higher incidence among the pediatric population (Ahsan et al., 2021). In 2004, zinc deficiency was the cause of death for almost 0.5 million children between 6 months and 5 years of age in Africa, Asia, and Latin America (Fischer Walker et al., 2009). Moreover, it has been found that healthy 1-3-year-old children from Western European countries commonly exhibit zinc deficiency (Vreugdenhil et al.,

2021). A study evaluating the effect of low doses of zinc (15 ppm) in the drinking water of mice from weaning to adulthood found a reduction in dendritic spines together with a downregulation of synaptic proteins in the hippocampus (Yang et al., 2013). Therefore, zinc deficiency during childhood affects dendritic spine plasticity as it does during pregnancy. However, more studies are required to associate zinc deficiency in childhood with future health outcomes.

Vitamin Deficiency

According to a study by Nishida and colleagues, the post-weaning restriction of folate in ddY mice from weeks 3 to 9 reduced the number of spines in pyramidal neurons of the dentate gyrus but not in the apical or basal dendrites of neurons in the CA1 and CA3 regions. Moreover, thin and mature spines, like filopodia, were increased, whereas stubby and mushroom spines were decreased in these same hippocampal subregions (Nishida et al., 2020). Therefore, reduced intake of this vitamin at postnatal age is critical for dendritic spines in the infant brain.

Hypercaloric Diets

The ability of making healthy food choices is underdeveloped in infants and children, which makes them vulnerable to consuming foods rich in fat and sugars, the most important factor in the prevalence of obesity among the pediatric population. A study has examined the effect of maternal HFD and reduced access to mother's milk during the suckling period on dendrite morphology within the mPFC (Rincel et al., 2018). The authors found that both maternal HFD and restricted lactation produced a significant reduction of dendritic spines in basal dendrites of layer II/III pyramidal neurons of the mPFC at postnatal day 21. Furthermore, restricted lactation increased thin spines in both basal and apical dendrites independently of HFD. On the other hand, when the offspring were undernourished by restricted lactation in concomitance with maternal HFD, the dendritic spine density was reestablished similarly to that of controls, and no differences in immature spines were found. Importantly, such restoration in dendrite morphology and spine density of the mPFC in early life persisted until adulthood, and additionally, the changes observed with maternal separation (restricted lactation) and HFD alone were also normalized but gave place to morphological changes like reduced dendritic length of apical dendrites at this age. Therefore, there is a complex interaction between the nutritional status of the mother and the offspring on dendritic spine alterations, where dietary styles like an HFD could even prevent undernutrition effects in these structures under certain circumstances like interrupted lactation.

Health Outcomes and Malnutrition-Induced Dendritic Spine Alterations

Visual Dysfunction

Visual development is highly sensitive to malnutrition, which is associated with eye dysfunction (Lien & Hammong, 2011), astigmatism, myopia, and other optical alterations in the optic nerve and retina (Dantas et al., 2005). Anatomic-functional changes in the brain, including dendritic spine plasticity and synaptic connectivity of the visual cortex, are actively observed from the first months of life following a gradual decrease until an age close to adolescence (Klaver et al., 2011). Dendritic spines in the visual cortex exhibit motility at postnatal day 21, which decreases until postnatal day 28; such activity is then stabilized until postnatal day 42 (Majewska & Sur, 2003).

Morphological alterations of dendritic spines in the visual cortex have been described as a result of protein deficiency during pregnancy (Leuba & Rabinowicz, 1979b; Díaz-Cintra et al., 1990) and severe malnutrition in childhood (Benítez-Bribiesca et al., 1999). Studies covering the periods before mating, gestation, and lactation in rats fed with diets of 6, 8 and 25% (control group) casein have been performed to analyze dendritic spines in the visual cortex (Díaz-Cintra et al., unpublished data). In these works, the 6 and 8% diets were low-protein and isocaloric diets, and the density of dendritic spines in the visual cortex was analyzed at the ages of 30, 90, and 220 days in the three groups. An increase was found in secondary dendrites in animals fed with 8% casein between postnatal days 30 and 90. By contrast, the 6% casein diet group showed a decrease in primary dendrites at postnatal day 90 and in secondary dendrites at postnatal days 90 and 220. This indicated that chronic malnutrition pre- and postnatally installed in the rat reveals adaptive plasticity phenomena caused by the diet in pyramidal neurons of the visual cortex, which can translate into important functional alterations in the integration of behavioral and motor visual responses.

Recently, Almeida and colleagues assessed the cognitive status of children with and without protein-deficient intake during early life through an eye-tracking task (Almeida et al., 2022). Malnourished children exhibited deficits in this task, which was associated with failures in eye movement. Therefore, alterations in dendritic spines of the visual cortex by some types of undernutrition, like a low-protein diet, could be associated with underdevelopment in this region, which could induce visual dysfunction throughout life. Importantly, magnesium has been shown to enhance visual cortex plasticity and recovery of visual function in adult mice (Liu et al., 2015). Hence, dietary interventions with minerals like magnesium could help to combat dendritic plasticity alterations in the visual cortex caused by protein malnourishment during neurodevelopment.

Auditory Dysfunctions

Previous evidence has evaluated the auditory function in healthy and malnourished infants revealing that malnourishment affects auditory skills modulated by the cortex (Flinn et al., 1993). Moreover, a current study found that early stunting, underweight, and wasting were associated with hearing loss in young adults from a rural zone of Nepal (Emmett et al., 2018). In addition, a clinical study from Brazil showed that malnourished children displayed altered auditory memory (Caldas et al., 2014).

The effects of nutritional deficiencies on dendritic spines in the primary auditory cortex and their functionality have been less studied. In the auditory cortex of rats, the appearance of dendritic spine protrusions of both mature and immature structures follows a specific pattern initiated at a slow rate on postnatal day 4 and at a high rate during postnatal days 9-19. Mature protrusions are fully achieved on postnatal day 21, which correlates with the evolution of auditory function (Schachtele et al., 2011). Moreover, a study in mice showed that dendritic spine plasticity and distribution in the visual and auditory cortices exhibit sexual dimorphism during brain maturation from adolescence to adulthood. According to this study, the density of mushroom spines and short and long stubby spines was lower in females than in males (Parker et al., 2020). Neurotransmission involved in sound processing can be modulated by zinc (Anderson et al., 2017) through inhibition or enhancement of excitatory synaptic plasticity (Vogler et al., 2020). Therefore, some types of malnutrition, like zinc deficiency, could severely impact auditory functions by altering dendritic spine plasticity and impairing neurotransmission signaling during neurodevelopment or maturation of the auditory cortices.

Motor Dysfunctions

In largely undeveloped countries, like Papua New Guinea, it has been reported that chronic malnutrition in children is associated with delayed development of motor skills such as sitting, crawling, and independent walking (Groos, 1991; Heywood et al., 1991). Moreover, a recent cross-sectional study from Burkina Faso showed that some features, including higher fat mass index or anemia in children exposed to moderate acute malnutrition, are associated with underdeveloped motor and language skills (Olsen et al., 2020). These findings reveal the underlying role of overnutrition with hypercaloric diets.

In the first days after birth, the primary motor cortex of rats suffers different adaptive changes and responses to establish frameworks later required for motor control functions (Dooley & Blumberg, 2018). Dendritic spine formation and stability in this region play a crucial role in motor learning (Albarran et al., 2021). In the offspring, different alterations in cell morphology or dendritic spines have been observed in the motor cortex after postnatal food restriction (Torrero et al., 2005) and in the cerebellum as a consequence of maternal low-protein diet (West & Kemper, 1976; Hillman & Chen, 1981), deficient intake of vitamin B6 (Chang et al., 1981) and methyl donors during pregnancy (Bake et al., 2008; El Hajj Chehadeh et al., 2014), and dietary zinc deficiency in early life (Dvergsten et al., 1983; Dvergsten et al., 1984a; Dvergsten et al., 1984b). Therefore, motor dysfunctions caused by malnutrition could rely on the alterations of dendritic spine plasticity of brain areas involved in motricity.

Neuropsychiatric Illnesses

Undernutrition *in utero* is one of the causes of low birth weight, which is a risk factor for several diseases in the offspring (Barker, 2007; Kondo et al., 2023). Long-term consequences of low birth weight due to perinatal malnutrition include elevated glucose concentrations, blood pressure, and lipid profiles in adults. Low birth weight also contributes to the appearance of mental disorders (Victora et al., 2008). In addition, maternal malnutrition predisposes the offspring to develop obesity, type 2 diabetes, and hypertension in adulthood (Barker et al., 1990; Hales et al., 2001), all of which are associated with an increased risk for Alzheimer's disease (Chatterjee et al., 2018; Lennon et al., 2019; Ma et al., 2020). For these reasons, we can propose that malnutrition may produce neuropsychiatric disorders.

Schizophrenia

Most of the knowledge about the physiopathology of schizophrenia has been obtained from the maternal immune activation (MIA) model, which is based on the fact that microorganisms play an important role in the development of psychiatric phenotypes by activating the immune response after an infection in pregnancy phenotypes (Estes & McAllister, 2016). Therefore, in this model, pregnant animals are subjected to molecules that induce the immune response to further evaluate psychological and physiological features in the offspring in order to identify correlations between infection and the development of psychiatric illnesses.

A common feature of schizophrenia is a decrease in dendritic spines which disrupts synaptic connectivity, leading to the behavioral changes commonly seen in these patients (Moyer et al., 2015). Such reduction in dendritic spines has been observed in layer III of the primary auditory cortex (MacDonald et al., 2017), dorsolateral prefrontal cortex (Glantz & Lewis, 2000), primary visual cortex (Konopaske et al., 2017), and subiculum of the hippocampal formation (Rosoklija et al., 2000). However, it is in layer III of the neocortex where spines are extensively decreased (Glausier & Lewis, 2013). Using the MIA mouse model and *in vivo* two-photon laser scanning time-lapse microscopy, researchers found diminished dendritic spine density in layer V of the primary somatosensory cortex at postnatal days 17-19 for apical dendrites and at postnatal day 30 for basal dendrites (Coiro et al., 2015). Furthermore, such alterations correlated with deficits in the marble burying test, a non-conditional task used to study repetitive and compulsive-like behaviors, which are positive symptoms of schizophrenia.

The prenatal low-protein diet (Leuba & Rabinowicz, 1979b Díaz-Cintra et al., 1990) and iodine deficiency (Chen et al., 1986) are types of malnutrition that share an issue with schizophrenia: the reduction of dendritic spine density in the brain cortices. Moreover, important data have associated severe prenatal malnutrition during the Dutch Hunger Winter (1944-1945) and Chinese (1959-1961) famines with increased cases of schizophrenia (Susser et al., 1996; He et al., 2018). Therefore, severe malnutrition during pregnancy seems to promote schizophrenia development in the offspring partly by inducing alterations in the dendritic spine plasticity of relevant brain regions for the illness.

Alzheimer's Disease and Synaptopathy

Patients with AD lose their memory and cognitive abilities, and their personalities can change dramatically due to the dysfunction and progressive death of nerve cells that are responsible for storing and processing information. These alterations are identified as a synaptopathy and encompass dystrophic neurites and dendritic spine modifications. The latter presumably occurs after a neurotoxic form of β-peptide elevation and further aggregation in limbic and cortical brain areas, whose deposits produce cognitive decline and memory dysfunction in AD (Stroo et al., 2017).

On the other hand, postsynaptic dendritic spines are specialized structures that contribute to synaptic efficiency in neuronal processes where excitatory synaptic contacts take place; thus, alterations are directly correlated to the loss synaptic function. Although neurotransmitter/receptor interactions are essential to determine where spine formation or elimination occurs on dendritic stretchers, other external molecules have recently been shown to co-regulate spine dynamics; for example, in the hippocampus, dopamine and serotonin are neuromodulators in spine elongation (Yagishita et al., 2014).

Cognitive impairment in AD is hypothesized to be linked to loss of dendritic spine density (Walker and Herskowitz, 2020). It has been observed that simultaneous expression of mutant amyloid precursor protein of the β-amyloid and tau results in a 35% reduction in dendritic spine density. However, the expression of this protein and tau alone do not result in dendritic spine loss (Chabrier et al., 2014). Thus, it was concluded that β-amyloid has a degenerative effect on dendritic spines, and tau serves to silence the synapses, preventing them from their functioning properly.

Conclusion

Throughout this chapter, we have shown how nutrient deficiencies or excesses cause alterations in dendritic spine morphology and plasticity. Most of the existing information is based on rodent studies and focuses on malnutrition during pregnancy and lactation. Furthermore, it is important to consider that nutrient limitations and overconsumption in humans is related to a wide variety of these substances and occur in all periods of life. This is because our diet is affected by socio-economic, psychological, and technological factors.

We have also described how malnutrition can impact brain neurodevelopment through dendritic spine alterations which impair synaptic connectivity. The presence of such modifications in specific brain regions can disrupt essential functions like audition, vision and motricity. Importantly, neuropsychiatric illnesses like schizophrenia could be triggered by malnutrition, possibly by causing dendritic spine alterations. Moreover, brain areas involved in feeding regulation are also altered by malnutrition (Rudebeck & Izquierdo, 2022), supporting unbalanced feeding behavior.

Malnutrition interferes with learning and memory processes (Rubio et al., 2018). Overnutrition, such as maternal-HFD, also induces lasting effects in the hippocampus, which regulates episodic-like memory in the male offspring (Cordner et al., 2019). Particularly, early protein malnutrition during pregnancy and lactation induces deficits in encoding and consolidation of long-term memory in the offspring at adult age (Pérez-García et al., 2016).

Furthermore, low-protein diet installed during gestation propitiates effects later in adult life, like performance impairment of attentional tasks (McGaughy et al., 2014). Chronic malnutrition with a low casein diet impairs mossy fibers of the dentate gyrus (Cintra et al., 1990; Granados et al., 2002) which can persist until adulthood (Cintra et al., 1997). Therefore, postnatal hypo-protein malnutrition can cause loss of flexibility in behavioral response strategies in the face of a harmful experience, also causing alterations in learning and memory processes, possibly due to limited inter- and extra-hippocampal communication, as well as integrative brain functions.

Acknowledgments

We thank Dr. Jessica González Norris for proofreading the manuscript. Humberto Martínez-Orozco received postdoctoral fellowship from CONACYT, Mexico for the completion of this chapter. Some fundings were provided by PAPIIT-DGAPA-UNAM.

References

Ahsan, A. K., Tebha, S. S., Sangi, R., Kamran, A., Zaidi, Z. A., Haque, T., Ali Hamza, M. S., 2021. Zinc micronutrient deficiency and its prevalence in malnourished pediatric children as compared to well-nourished children: A nutritional emergency. *Global Ped. Health* 8, 2333794X211050316.

Akchiche, N., Bossenmeyer-Pourié, C., Kerek, R., Martin, N., Pourié, G., Koziel, V., Helle, D., Alberto, J. M., Ortiou S., Camadro, J. M., Léger, T., Guéant, J. L., Daval, J. L.,

2012. Homocysteinylation of neuronal proteins contributes to folate deficiency-associated alterations of differentiation, vesicular transport, and plasticity in hippocampal neuronal cells. *FASEB J.* 26, 3980–3992.

Albarran, E., Raissi, A., Jáidar, O., Shatz, C. J., Ding, J. B., 2021. Enhancing motor learning by increasing the stability of newly formed dendritic spines in the motor cortex. *Neuron* 109, 3298–3311.e4.

Almeida, N. L., Silva, J. B. S., Oliveira, M. E. C., Fernandes, T. P., Santos, N. A., 2022. Eye movement impairments in children with malnutrition. *Int. J. Psychol.* 57, 644–651.

Alonso-Alonso, M., Pascual-Leone, A., 2007. The right brain hypothesis for obesity. *JAMA* 297, 1819–1822.

Al Sayed, R., Smith, W., Rogers, N., Smith, N., Clark, D., Castillo, G., McLeod, H., Glenister, S., Shorter, K. R., 2019. A 2x folic acid treatment affects epigenetics and dendritic spine densities in SHSY5Y cells. *Biochem. Biophys. Rep.* 20, 100681.

Anderson, C. T., Kumar, M., Xiong, S., Tzounopoulos, T., 2017. Cell-specific gain modulation by synaptically released zinc in cortical circuits of audition. *eLife* 6, e29893.

Bake, S., Ma, L., Sohrabji, F., 2008. Estrogen receptor-alpha overexpression suppresses 17beta-estradiol-mediated vascular endothelial growth factor expression and activation of survival kinases. *Endocrinology* 149, 3881–3889.

Barker, D. J., Bull, A. R., Osmond, C., Simmonds, S. J., 1990. Fetal and placental size and risk of hypertension in adult life. *BMJ* 301, 259–262.

Barker D. J., 1998. In utero programming of chronic disease. *Clin. Sci.* 95, 115–128.

Barker D. J., 2007. The origins of the developmental origin's theory. *J. Int. Med.* 261, 412–417.

Benítez-Bribiesca, L., De la Rosa-Alvarez, I., Mansilla-Olivares, A., 1999. Dendritic spine pathology in infants with severe protein-calorie malnutrition. *Pediatrics* 104, e21.

Bhargava, H., Singh, J., 2020. Structural changes in the brain on cranial magnetic resonance imaging in severely malnourished children before and after treatment. *Nutrition* 74, 110753.

Bleker, L. S., de Rooij, S. R., Roseboom, T. J., 2019. Malnutrition and depression in pregnancy and associations with child behaviour and cognitive function: a review of recent evidence on unique and joint effects 1. *Can. J. Physiol. Pharmacol.* 97, 158–173.

Brock, J. W., Prasad, C., 1992. Alterations in dendritic spine density in the rat brain associated with protein malnutrition. *Brain Res. Dev. Brain Res.* 66, 266–269.

Brunette, K. E., Tran, P. V., Wobken, J. D., Carlson, E. S., Georgieff, M. K., 2010. Gestational and neonatal iron deficiency alters apical dendrite structure of CA1 pyramidal neurons in adult rat hippocampus. *Dev. Neurosci.* 32, 238–248.

Caldas, P. A., Giacheti, C. M., Capellini, S. A., 2014. Habilidade auditiva em criança desnutrida [Hearing ability in unnourished children]. *Audiol. Commun. Res.* 19 19, 3.

Cakir, M., Senyuva, S., Kul, S., Sag, E., Cansu, A., Yucesan, F. B., Yaman, S. O., Orem, A., 2019. Neurocognitive functions in infants with malnutrition; relation with long-chain polyunsaturated fatty acids, micronutrients levels and magnetic resonance spectroscopy. Pediatric gastroenterology, *Hepatol. Nutr.* 22, 171–180.

Chabrier, M. A., Cheng, D., Castello, N. A., Green, K. N., LaFerla, F. M., 2014. Synergistic effects of amyloid-beta and wild-type human tau on dendritic spine loss in a floxed double transgenic model of Alzheimer's disease. *Neurobiol. Dis.* 64, 107–117.

Chang, S. J., Kirksey, A., Morré, D. M., 1981. Effects of vitamin B-6 deficiency on morphological changes in dendritic trees of Purkinje cells in developing cerebellum of rats. *J. Nutr.* 111, 848–857.

Chatterjee, S., Mudher, A., 2018. Alzheimer's disease and Type 2 diabetes: A critical assessment of the shared pathological traits. *Front. Neurosci.* 12, 383.

Chen, Z. P., Chen, X. X., Dong, L., Hu, X., Zhao, W. D., Wang, D., Ma, T., 1986. The iodine deficient rat, in: Medeiros-Neto, G., Maciel, R. M. B., Halpern, A. (Eds.), *Iodine Deficiency Diseases and Congenital Hypothyroidism.* Ache Press, Sao Paulo, pp. 46–51.

Cintra, L., Aguilar, A., Granados, L., Galván, A., Kemper, T., DeBassio, W., Galler, J., Morgane, P., Durán, P., Díaz-Cintra, S., 1997. Effects of prenatal protein malnutrition on hippocampal CA1 pyramidal cells in rats of four age groups. *Hippocampus* 7, 192–203.

Cintra, L., Díaz-Cintra, S., Galván, A., Kemper, T., Morgane, P. J., 1990. Effects of protein undernutrition on the dentate gyrus in rats of three age groups. *Brain Res.* 532, 271–277.

Cintra, L., Durán, P., Guevara. M. A., Aguilar, A., Castañón-Cervantes, O., 2002. Pre- and post-natal protein malnutrition alters the effect of rapid eye movements sleep-deprivation by the platform-technique upon the electrocorticogram of the circadian sleep-wake cycle and its frequency bands in the rat. *Nutr. Neurosci.* 5, 91–101.

Clark, D. F., Schmelz, R., Rogers, N., Smith, N. E., Shorter, K. R., 2021. Acute high folic acid treatment in SH-SY5Y cells with and without MTHFR function leads to gene expression changes in epigenetic modifying enzymes, changes in epigenetic marks, and changes in dendritic spine densities. *PloS One* 16, e0245005.

Coiro, P., Padmashri, R., Suresh, A., Spartz, E., Pendyala, G., Chou, S., Jung, Y., Meays, B., Roy, S., Gautam, N., Alnouti, Y., Li, M., Dunaevsky, A., 2015. Impaired synaptic development in a maternal immune activation mouse model of neurodevelopmental disorders. *Brain Behavior Immun.* 50, 249–258.

Cordner, Z. A., Khambadkone, S. G., Boersma, G. J., Song, L., Summers, T. N., Moran, T. H., Tamashiro, K. L. K., 2019. Maternal high-fat diet results in cognitive impairment and hippocampal gene expression changes in rat offspring. *Exp. Neurol.* 318, 92–100.

Cortés-Albornoz, M. C., García-Guáqueta, D. P., Velez-van-Meerbeke, A., Talero-Gutiérrez, C., 2021. Maternal nutrition and neurodevelopment: A scoping review. *Nutrients* 13, 3530.

Cusick, S. E., Georgieff, M. K., 2016. The role of nutrition in brain development: The golden opportunity of the "First 1000 Days." *J. Pediatrics* 175, 16–21

Dantas, A. P., Brandt, C. T., Leal, D. N., 2005. Manifestações oculares em pacientes que tiveram desnutrição nos primeiros seis meses de vida [Ocular manifestations in patients who had malnutrition in the first six months of life]. *Arq. Brasil. Oftalm.* 68, 753–756.

D'Amelio, M., Cavallucci, V., Cecconi, F., 2010. Neuronal caspase-3 signaling: not only cell death. *Cell Death Differ.* 17, 1104–1114.

Díaz-Cintra, S., Cintra, L., Galván, A., Aguilar, A., Kemper, T., Morgane, P. J., 1991. Effects of prenatal protein deprivation on postnatal development of granule cells in the fascia dentata. *J. Comp. Neurol.* 310, 356–364.

Díaz-Cintra, S., Cintra, L., Kemper, T., Resnick, O., Morgane, P. J., 1981. Nucleus raphe dorsalis: a morphometric Golgi study in rats of three age groups. *Brain Res.* 207, 1–16.

Díaz-Cintra, S., Cintra, L., Kemper, T., Resnick, O., Morgane, P. J., 1984. The effects of protein deprivation on the nucleus locus coeruleus: a morphometric Golgi study in rats of three age groups. *Brain Res.* 304, 243–253.

Díaz-Cintra, S., Cintra, L., Ortega, A., Kemper, T., Morgane, P. J., 1990. Effects of protein deprivation on pyramidal cells of the visual cortex in rats of three age groups. *J. Comp. Neurol.* 292, 117–126.

Doi, M., Usui, N., Shimada, S., 2022. Prenatal environment and neurodevelopmental disorders. *Front. Endocrinol.* 13, 860110.

Dooley, J. C., Blumberg, M. S., 2018. Developmental 'awakening' of primary motor cortex to the sensory consequences of movement. *eLife* 7, e41841.

Dvergsten, C. L., Fosmire, G. J., Ollerich, D. A., Sandstead, H. H., 1983. Alterations in the postnatal development of the cerebellar cortex due to zinc deficiency. I. Impaired acquisition of granule cells. *Brain Res.* 271, 217–226.

Dvergsten, C. L., Fosmire, G. J., Ollerich, D. A., Sandstead, H. H., 1984a. Alterations in the postnatal development of the cerebellar cortex due to zinc deficiency. II. Impaired maturation of Purkinje cells. *Brain Res.* 318, 11–20.

Dvergsten, C. L., Johnson, L. A., Sandstead, H. H., 1984b. Alterations in the postnatal development of the cerebellar cortex due to zinc deficiency. III. Impaired dendritic differentiation of basket and stellate cells. *Brain Res.* 318, 21–26.

El Hajj Chehadeh, S., Dreumont, N., Willekens, J., Canabady-Rochelle, L., Jeannesson, E., Alberto, J. M., Daval, J. L., Guéant, J. L., Leininger-Muller, B., 2014. Early methyl donor deficiency alters cAMP signaling pathway and neurosteroidogenesis in the cerebellum of female rat pups. American journal of physiology. *Endocrinol. Metab.* 307, E1009–E1019.

Emmett, S. D., Schmitz, J., Karna, S. L., Khatry, S. K., Wu, L., LeClerq, S. C., Pillion, J., West, K. P. Jr., 2018. Early childhood undernutrition increases risk of hearing loss in young adulthood in rural Nepal. *Am. J. Clin. Nutr.* 107, 268–277.

Estes, M. L., McAllister, A. K., 2016. Maternal immune activation: Implications for neuropsychiatric disorders. *Science* 353, 772–777.

Fischer Walker, C. L., Ezzati, M., Black, R. E., 2009. Global and regional child mortality and burden of disease attributable to zinc deficiency. *European journal of clinical nutrition,* 63, 591–597.

Flinn, J. M., Barnet, A. B., Lydick, S., Lackner, J., 1993. Infant malnutrition affects cortical auditory evoked potentials. *Perc. Motor Skills* 76(3 Pt 2), 1359–1362.

Funahashi S., 2017. Prefrontal contribution to decision-making under free-choice conditions. *Front. Neurosci.* 11, 431.

Galler, J. R., Bringas-Vega, M. L., Tang, Q., Rabinowitz, A. G., Musa, K. I., Chai, W. J., Omar, H., Abdul Rahman, M. R., Abd Hamid, A. I., Abdullah, J. M., Valdés-Sosa, P.

A., 2021. Neurodevelopmental effects of childhood malnutrition: A neuroimaging perspective. *NeuroImage* 231, 117828.

Galler, J. R., Koethe, J. R., Yolken, R. H., 2017. Neurodevelopment: The impact of nutrition and inflammation during adolescence in low-resource settings. *Pediatrics* 139(Suppl 1), S72–S84.

Gawlińska, K., Gawliński, D., Borczyk, M., Korostyński, M., Przegaliński, E., Filip, M., 2021. A maternal high-fat diet during early development provokes molecular changes related to autism spectrum disorder in the rat offspring brain. *Nutrients* 13, 3212.

Georgieff M. K., 2020. Iron deficiency in pregnancy. *Am. J. Obstet. Gynecol.* 223, 516–524.

Gerster H., 1996. Die bedeutung von vitamin b6 für die entwicklung des säuglings. Humanmedizinische und tierexperimentelle studien [The importance of vitamin B 6 for development of the infant. Human medical and animal experiment studies]. *Z. Ernahrungswiss.* 35, 309–317.

Glantz, L. A., Lewis, D. A., 2000. Decreased dendritic spine density on prefrontal cortical pyramidal neurons in schizophrenia. *Arch. Gen. Psychiatry* 57, 65–73.

Glausier, J. R., Lewis, D. A., 2013. Dendritic spine pathology in schizophrenia. *Neuroscience* 251 90–107.

Goetsch, C., 1962. An evaluation of aminopterin as an abortifacient. *Am. J. Obstet. Gynecol.* 83, 1474–1477.

Gonzalez de Cosio, T., Escobar-Zaragoza, L., González-Castell, L. D., Rivera-Dommarco, J., Á., 2013. Infant feeding practices and deterioration of breastfeeding in Mexico. *Salud Publica de Mexico* 55, S170–S179.

Granados-Rojas, L., Serrano, N., Gutiérrez-Ospina, G., Díaz-Cintra, S., 2002. Prenatal protein malnutrition differentially affects the volume of the granule layer and mossy fibers in young male and female rats. *Proc. West. Pharmacol. Soc.* 45, 53–54.

Groos A. D., 1991. Delayed motor development in relation to nutritional status among children under two years of age in two districts of Simbu Province. *Papua and New Guinea Med. J.* 34, 238–245.

Gustafson, C. R., 2022. Active consideration of future health can be prompted by simple health messages and improves nutritional quality of food choices. *Front. Nutr.* 9, 926643.

Hales, C. N., Barker, D. J., 2001. The thrifty phenotype hypothesis. *Brit. Med. Bull.* 60, 5–20.

Hatanaka, Y., Wada, K., Kabuta, T., 2016. Maternal high-fat diet leads to persistent synaptic instability in mouse offspring via oxidative stress during lactation. *Neurochem. Int.* 97, 99–108.

He, P., Chen, G., Guo, C., Wen, X., Song, X., Zheng, X., 2018. Long-term effect of prenatal exposure to malnutrition on risk of schizophrenia in adulthood: Evidence from the Chinese famine of 1959-1961. *Eur. Psychiatry* 51, 42–47.

Herms, J., Dorostkar, M. M., 2016. Dendritic spine pathology in neurodegenerative diseases. *Ann. Rev. Pathol* 11, 221–250.

Heywood, A. H., Marshall, T., Heywood, P. F., 1991. Motor development and nutritional status of young children in Madang, Papua New Guinea. *Papua and New Guinea Med. J.* 34, 109–116.

Hillman, D. E., Chen, S., 1981. Vulnerability of cerebellar development in malnutrition-II. Intrinsic determination of total synaptic area on Purkinje cell spines. *Neuroscience* 6, 1263–1275.

Hortensius, L. M., van Elburg, R. M., Nijboer, C. H., Benders, M. J. N. L., de Theije, C. G. M., 2019. Postnatal nutrition to improve brain development in the preterm infant: A systematic review from bench to bedside. *Front. Physiol.* 10, 961.

Janthakhin, Y., Rincel, M., Costa, A. M., Darnaudéry, M., Ferreira, G., 2017. Maternal high-fat diet leads to hippocampal and amygdala dendritic remodeling in adult male offspring. *Psychoneuroendocrinology* 83, 49–57.

Jernigan, T. L., Baaré, W. F., Stiles, J., Madsen, K. S., 2011. Postnatal brain development: structural imaging of dynamic neurodevelopmental processes. *Prog. Brain Res.* 189, 77–92.

Jernigan, T. L., Tallal, P., 1990. Late childhood changes in brain morphology observable with MRI. *Dev. Med. Child Neurol.* 32, 379–385.

Jernigan, T. L., Trauner, D. A., Hesselink, J. R., Tallal, P. A., 1991. Maturation of human cerebrum observed in vivo during adolescence. *Brain* 114 (Pt 5), 2037–2049.

Jorgenson, L. A., Wobken, J. D., Georgieff, M. K., 2003. Perinatal iron deficiency alters apical dendritic growth in hippocampal CA1 pyramidal neurons. *Dev. Neurosci.* 25, 412–420.

Kemper, T. L., Pasquier, D. A., Drazen, S., 1978. Effect of a low protein diet on the anatomical development of subcortical formations. *Brain Res. Bull.* 3, 443–450.

Kirksey, A., Morré, D. M., Wasynczuk, A. Z., 1990. Neuronal development in vitamin B6 deficiency. *Ann. NY Acad. Sci.* 585, 202–218.

Klaver, P., Marcar, V., Martin, E., 2011. Neurodevelopment of the visual system in typically developing children. *Prog. Brain Res.* 189, 113–136.

Kondo, R., Ozawa, R., Satomi, T., Funabayashi, K., Iwata, H., Kuwayama, T., Shirasuna, K., 2023. Severe maternal stress alters placental function, resulting in adipose tissue and liver dysfunction in offspring of mice. *Mol. Cell. Endocrinol.* 560, 111814.

Konopaske, G. T., Lange, N., Coyle, J. T., Benes, F. M., 2014. Prefrontal cortical dendritic spine pathology in schizophrenia and bipolar disorder. *JAMA Psychiatry* 71, 1323–1331.

Krakowiak, P., Walker, C. K., Bremer, A. A., Baker, A. S., Ozonoff, S., Hansen, R. L., Hertz-Picciotto, I., 2012. Maternal metabolic conditions and risk for autism and other neurodevelopmental disorders. *Pediatrics* 129, e1121–e1128.

Krishna, A. P., Ramakrishna, T., 2004. Effect of pyridoxine deficiency on the structural and functional development of hippocampus. *Indian J. Physiol. Pharmacol.* 48, 304–310.

Lennon, M. J., Makkar, S. R., Crawford, J. D., Sachdev, P. S., 2019. Midlife hypertension and Alzheimer's disease: A systematic review and meta-analysis. *J. Alz. Dis.* 71, 307–316.

Leuba, G., Rabinowicz, T., 1979a. Long-term effects of postnatal undernutrition and maternal malnutrition on mouse cerebral cortex. II. Evolution of dendritic branchings and spines in the visual region. *Exp. Brain Res.* 37, 299–308.

Leuba, G., Rabinowicz, T., 1979b. Long-term effects of postnatal undernutrition and maternal malnutrition on mouse cerebral cortex. I. Cellular densities, cortical volume and total numbers of cells. *Exp. Brain Res.* 37(2), 283–298.

Lien, E. L., Hammond, B. R., 2011. Nutritional influences on visual development and function. *Prog. Ret. Eye Res.* 30, 188–203.

Liu, H., Li, Y., Wang, Y., Wang, X., An, X., Wang, S., Chen, L., Liu, G., Yang, Y., 2015. The distinct role of NR2B subunit in the enhancement of visual plasticity in adulthood. *Mol. Brain* 8, 49.

Ma, Y., Ajnakina, O., Steptoe, A., Cadar, D., 2020. Higher risk of dementia in English older individuals who are overweight or obese. *Int. J. Epidemiol.* 49, 1353–1365.

MacDonald, M. L., Alhassan, J., Newman, J. T., Richard, M., Gu, H., Kelly, R. M., Sampson, A. R., Fish, K. N., Penzes, P., Wills, Z. P., Lewis, D. A., Sweet, R. A., 2017. Selective loss of smaller spines in Schizophrenia. *Am. J. Psychiatry* 174, 586–594.

McGaughy, J. A., Amaral, A. C., Rushmore, R. J., Mokler, D. J., Morgane, P. J., Rosene, D. L., Galler, J. R., 2014. Prenatal malnutrition leads to deficits in attentional set shifting and decreases metabolic activity in prefrontal subregions that control executive function. *Dev. Neurosci.* 36, 532–541.

Majewska, A., Sur, M., 2003. Motility of dendritic spines in visual cortex in vivo: changes during the critical period and effects of visual deprivation. *Proc. Natl. Acad. Sci. USA* 100, 16024–16029.

Mathur, P., Pillai, R., 2019. Overnutrition: Current scenario & combat strategies. *Indian J. Med. Res.* 149, 695–705.

Min, H., Dong, J., Wang, Y., Wang, Y., Yu, Y., Shan, Z., Xi, Q., Teng, W., Chen, J., 2017. Marginal iodine deficiency affects dendritic spine development by disturbing the function of Rac1 signaling pathway on cytoskeleton. *Mol. Neurobiol.* 54, 437–449.

Monk, C., Georgieff, M. K., Xu, D., Hao, X., Bansal, R., Gustafsson, H., Spicer, J., Peterson, B. S., 2016. Maternal prenatal iron status and tissue organization in the neonatal brain. *Ped. Res.* 79, 482–488.

Moyer, C. E., Shelton, M. A., Sweet, R. A., 2015. Dendritic spine alterations in schizophrenia. *Neurosci. Lett.* 601, 46–53.

MRC Vitamin Study Research Group., 1991. Prevention of neural tube defects: results of the Medical Research Council Vitamin Study. *Lancet* 338, 131–137.

Nelson, M. M., Asling, C. W. Evans, H. M., 1952. Production of multiple congenital abnormalities in young by maternal pteroylglutamic acid deficiency during gestation. *J. Nutr.* 48, 61–79.

Nguyen, L. D., Nguyen, L. H., Ninh, L. T., Nguyen, H. T. T., Nguyen, A. D., Vu, L. G., Nguyen, H. S. A., Nguyen, S. H., Doan, L. P., Vu, T. M. T., Tran, B. X., Latkin, C. A., Ho, C. S. H., Ho, R. C. M., 2022. Women's holistic self-care behaviors during pregnancy and associations with psychological well-being: implications for maternal care facilities. *BMC Pregn. Childbirth* 22, 631.

Nishida, S., Araki, R., Baba, A., Asari, S., Tachibana, S., Nakajima, Y., Iwakumo, A., Yabe, T., 2020. Post-weaning folate deficiency induces a depression-like state via neuronal immaturity of the dentate gyrus in mice. *J. Pharmacol. Sci.* 143, 97–105.

Nishiyama J., 2019. Plasticity of dendritic spines: Molecular function and dysfunction in neurodevelopmental disorders. Psychiat. *Clin. Neurosci.* 73, 541–550.

Olsen, M. F., Iuel-Brockdorff, A. S., Yaméogo, C. W., Cichon, B., Fabiansen, C., Filteau, S., Phelan, K., Ouédraogo, A., Wells, J. C., Briend, A., Michaelsen, K. F., Lauritzen, L., Ritz, C., Ashorn, P., Christensen, V. B., Gladstone, M., Friis, H., 2020. Early development in children with moderate acute malnutrition: A cross-sectional study in Burkina Faso. *Mat. Child Nutr.* 16, e12928.

Parker, E. M., Kindja, N. L., Cheetham, C. E. J., Sweet, R. A., 2020. Sex differences in dendritic spine density and morphology in auditory and visual cortices in adolescence and adulthood. *Sci. Rep.* 10, 9442.

Pérez-García, G., Guzmán-Quevedo, O., Da Silva Aragão, R., Bolaños-Jiménez, F., 2016. Early malnutrition results in long-lasting impairments in pattern-separation for overlapping novel object and novel location memories and reduced hippocampal neurogenesis. *Sci. Rep.* 6, 21275.

Rincel, M., Lépinay, A. L., Janthakhin, Y., Soudain, G., Yvon, S., Da Silva, S., Joffre, C., Aubert, A., Séré, A., Layé, S., Theodorou, V., Ferreira, G., Darnaudéry, M., 2018. Maternal high-fat diet and early life stress differentially modulate spine density and dendritic morphology in the medial prefrontal cortex of juvenile and adult rats. *Brain Struct. Funct.* 223, 883–895.

Rodriguez, A., Miettunen, J., Henriksen, T. B., Olsen, J., Obel, C., Taanila, A., Ebeling, H., Linnet, K. M., Moilanen, I., Järvelin, M. R., 2008. Maternal adiposity prior to pregnancy is associated with ADHD symptoms in offspring: evidence from three prospective pregnancy cohorts. *Int. J. Obesity* 32, 550–557.

Rosoklija, G., Toomayan, G., Ellis, S. P., Keilp, J., Mann, J. J., Latov, N., Hays, A. P., Dwork, A. J., 2000. Structural abnormalities of subicular dendrites in subjects with schizophrenia and mood disorders: preliminary findings. *Arch. Gen. Psychiatry* 57, 349–356.

Rubio, L., Téllez, L., Regalado, M., Torrero, C., Salas, M., 2018. Effects of perinatal undernutrition on social transmission of food preference in adult male Wistar rats. *Int. J. Dev. Neurosci.* 71, 105–110.

Rubio, L., Torrero, C., Regalado, M., Salas, M., 2004. Alterations in the solitary tract nucleus of the rat following perinatal food restriction and subsequent nutritional rehabilitation. *Nutr. Neurosci.* 7, 291–300.

Rudebeck, P. H., Izquierdo, A., 2022. Foraging with the frontal cortex: A cross-species evaluation of reward-guided behavior. *Neuropsychopharmacology* 47, 134–146.

Runge, K., Cardoso, C., de Chevigny, A., 2020. Dendritic spine plasticity: Function and mechanisms. *Front. Synaptic Neurosci.* 12, 36.

Salas M., 1980. Effects of early undernutrition on dendritic spines of cortical pyramidal cells in the rat. *Dev. Neurosci.* 3, 109–117.

Salas, M., Díaz, S., Nieto, A., 1974. Effects of neonatal food deprivation on cortical spines and dendritic development of the rat. *Brain Res.* 73, 139–144.

Salas, M., Torrero, C., Regalado, M., Loranca, A., 1998. The hindlimb hyperextension associated to the urogenital response in newborn underfed rats. *Arch. Med. Res.* 29, 125–132.

Salas, M., Torrero, C., Regalado, M., Rubio, L., 2015. Effects of perinatal undernutrition on the basilar dendritic arbor of the anterior cingulate pyramidal neurons in lactating dams. *Acta Neurobiol. Exp.* 75, 372–380.

Saunders, J., Smith, T., 2010. Malnutrition: causes and consequences. *Clin. Med.* 10, 624–627.

Schachtele, S. J., Losh, J., Dailey, M. E., Green, S. H., 2011. Spine formation and maturation in the developing rat auditory cortex. *J. Comp. Neurol.* 519, 3327–3345.

Schönheit, B., Haensel, P., 1984. Der Einfluss einer unspezifischen mangelernährung auf die spinemorphologie von lamina V-pyramidenzellen der regio cingularis juveniler und erwachsener ratten [Effect of nonspecific malnutrition on spine morphology of lamina V pyramidal cells of the cingulate area of juvenile and adult rats]. *J. Hirnfors.* 25, 617–631.

Stroo, E., Koopman, M., Nollen, E. A., Mata-Cabana, A., 2017. Cellular regulation of amyloid formation in aging and disease. *Front. Neurosci.* 11, 64.

Sullivan, E. L., Nousen, E. K., Chamlou, K. A., Grove, K. L., 2012. The impact of maternal high-fat diet consumption on neural development and behavior of offspring. *Int. J. Obesity Suppl.* 2(Suppl 2), S7–S13.

Susser, E., Neugebauer, R., Hoek, H. W., Brown, A. S., Lin, S., Labovitz, D., Gorman, J. M., 1996. Schizophrenia after prenatal famine. *Further evidence. Arch. Gen. Psychiatry* 53, 25–31.

Torrero, C., Regalado, M., Perez, E., Rubio, L., Salas, M., 2005. Neonatal food restriction and binaural ear occlusion interfere with the maturation of cortical motor pyramids in the rat. Nutr. Neurosci. 8, 63–66.

Torrero, C., Regalado, M., Rubio, L., Ruiz-Marcos, A., Salas, M., 2003. Neonatal food restriction induces hypoplasia in developing facial motor neurons of rats. *Nutr. Neurosci.* 6, 317–324.

Urbonaite, G., Knyzeliene, A., Bunn, F. S., Smalskys, A., Neniskyte, U., 2022. The impact of maternal high-fat diet on offspring neurodevelopment. *Front. Neurosci.* 16, 909762.

Valoppi, F., Agustin, M., Abik, F., Morais de Carvalho, D., Sithole, J., Bhattarai, M., Varis, J. J., Arzami, A. N. A. B., Pulkkinen, E., Mikkonen, K. S., 2021. Insight on current advances in food science and technology for feeding the world population. *Front. Sust. Food Syst.* 5, 385.

van Meer, F., van der Laan, L. N., Viergever, M. A., Adan, R. A. H., Smeets, P. A. M., Family Consortium, 2017. Considering healthiness promotes healthier choices but modulates medial prefrontal cortex differently in children compared with adults. *NeuroImage* 159, 325–333.

Victora, C. G., Adair, L., Fall, C., Hallal, P. C., Martorell, R., Richter, L., Sachdev, H. S., Maternal and Child Undernutrition Study Group, 2008. Maternal and child undernutrition: consequences for adult health and human capital. *Lancet* 371, 340–357.

Vogler, N. W., Betti, V. M., Goldberg, J. M., Tzounopoulos, T., 2020. Mechanisms underlying long-term synaptic zinc plasticity at mouse dorsal cochlear nucleus glutamatergic synapses. *J. Neurosci.* 40, 4981–4996.

Vreugdenhil, M., Akkermans, M. D., van der Merwe, L. F., van Elburg, R. M., van Goudoever, J. B., Brus, F., 2021. Prevalence of zinc deficiency in healthy 1-3-year-old children from three western european countries. *Nutrients* 13, 3713.

Walker, C. K., Herskowitz, J. H., 2021. Dendritic spines: Mediators of cognitive resilience in aging and Alzheimer's disease. *Neuroscientist* 27, 487–505.

West, C. D., Kemper, T. L., 1976. The effect of a low protein diet on the anatomical development of the rat brain. *Brain Res.* 107, 221–237.

WHO. World Health Organization. What is malnutrition?. 2020. Available from: https://www.who.int/features/qa/malnutrition/en/. [updated Apr, 15th 2020; cited 2022 Dec, 28nd].

Yagishita, S., Hayashi-Takagi, A., Ellis-Davies, G. C., Urakubo, H., Ishii, S., Kasai, H., 2014. A critical time window for dopamine actions on the structural plasticity of dendritic spines. *Science* 345, 1616–1620.

Yang, Y., Jing, X. P., Zhang, S. P., Gu, R. X., Tang, F. X., Wang, X. L., Xiong, Y., Qiu, M., Sun, X. Y., Ke, D., Wang, J. Z., Liu, R., 2013. High dose zinc supplementation induces hippocampal zinc deficiency and memory impairment with inhibition of BDNF signaling. *PloS One* 8, e55384.

Yapici, N., Cohn, R., Schusterreiter, C., Ruta, V., Vosshall, L. B., 2016. A taste circuit that regulates ingestion by integrating food and hunger signals. *Cell* 165, 715–729.

Young, M. F., Ramakrishnan, U., 2021. Maternal Undernutrition before and during pregnancy and offspring health and development. *Ann. Nutr. Metab.* 1–13.

Zhang, Y., Wei, J., Yang, Z., 2013. Perinatal undernutrition attenuates field excitatory postsynaptic potentials and influences dendritic spine density and morphology in hippocampus of male rat offspring. *Neuroscience* 244, 31–41.

Index

D

E

F

G

H

N

R

S

T

U

V